MW00986931

Green Supply Chain Management

About the Authors

Hsiao-Fan Wang, Ph.D., is Distinguished Chair Professor in the Department of Industrial Engineering and Engineering Management (IEEM) at National Tsing Hua University (NTHU), Taiwan, Republic of China. At NTHU she previously was Vice Dean of the College of Engineering and Head of the IEEM Department. She has also been President of the Chinese Fuzzy Systems Association, Vice President of the International Fuzzy Systems Association, and an Erskine Fellow at Canterbury University in New Zealand. Dr. Wang has received a Distinguished Research Award from the National Science Council of Taiwan and a Distinguished Teaching Award from NTSU's College of Engineering. She is an area editor for several international journals, has published more than 120 journal papers, and is the author or coauthor of 12 books. Her research interests include multicriteria decision making, fuzzy set theory, and green value chain management.

Surendra M. Gupta, Ph.D., is Professor of Mechanical and Industrial Engineering and the Director of the Laboratory for Responsible Manufacturing at Northeastern University, Boston, Massachusetts. He has authored or coauthored well over 400 technical papers published in books, journals, and international conference proceedings. He has taught over 100 courses in such areas as operations research, inventory theory, queuing theory, engineering economy, supply chain management, and production planning and control. Dr. Gupta has received an Outstanding Research Award and an Outstanding Industrial Engineering Professor Award for Teaching Excellence from Northeastern University and a national outstanding doctoral dissertation advisor award, among many others.

Green Supply Chain Management

Product Life Cycle Approach

Hsiao-Fan Wang, Ph.D.
Surendra M. Gupta, Ph.D.

New York Chicago San Francisco
Lisbon London Madrid Mexico City
Milan New Delhi San Juan
Seoul Singapore Sydney Toronto

The McGraw·Hill Companies

Green Supply Chain Management: Product Life Cycle Approach

1 2 3 4 5 6 7 8 9 0 QFR/QFR 1 7 6 5 4 3 2 1

ISBN 978-0-07-162283-7
MHID 0-07-162283-7

The pages within this book were printed on acid-free paper containing 100% postconsumer fiber.

Sponsoring Editor
Michael Penn

Editing Supervisor
Stephen M. Smith

Production Supervisor
Pamela A. Pelton

Acquisitions Coordinator
Michael Mulcahy

Project Manager
Sandhya Joshi, Aptara, Inc.

Copy Editor
Steven Pisano

Proofreader
Laura Lawrie

Indexer
Kathrin Unger

Art Director, Cover
Jeff Weeks

Composition
Aptara, Inc.

Dedicated to our families:

I-Fan and Tao-Fan

Hsiao-Fan Wang

Sharda, Monica, and Neil

Surendra M. Gupta

Contents

Preface xv
Acknowledgments xvii

Part 1 Basic Concepts and Background

1 Introduction 3
1.1 Development of Green Supply Chain Management 3
1.2 Evolution of GSCM from SCM 4
1.3 Impact of GSCM on Industry 6
1.3.1 Impact of GSCM on Industry Tactics 7
1.3.2 Impact of the Green Supply Chain on Industrial Administration 8
1.3.3 Intensification of Competition by GSCM 10
1.4 Summary and Conclusion 10
References 12

2 Mathematical Background 13
2.1 Fuzzy Numbers and Arithmetic 13
2.2 Utility Theory 15
2.3 The Analytic Hierarchy Process (AHP) 17
2.3.1 Basic Concepts and Pairwise Comparison 18
2.3.2 The Procedure 23
2.3.3 An Example: Determining Consumer Preference 24
2.4 Optimization Programming 30
2.4.1 Multi-Objective Linear Program (MOLP) 31
2.4.2 Illustrative Example 32
References 33

Part 2 Green Engineering Technology

3 Green Engineering 37
3.1 Introduction 37

3.2 Green Design 38
3.2.1 Design for X 38
3.2.2 Life Cycle Analysis 39
3.2.3 Material Selection 39
3.3 Some Green Design Guidelines 40
3.3.1 Modular Product Structure 40
3.3.2 Design of Functional Units 40
3.3.3 Material Selection 41
3.3.4 Minimize Waste and Harmful Contaminating Materials 41
3.3.5 Ease of Separation 41
3.3.6 Steps for Designing Green Products 41
3.4 Product Recovery at the End-of-Life 43
3.5 The 12 Principles of Green Engineering 45
3.5.1 Inherent Rather Than Circumstantial 45
3.5.2 Prevention Instead of Treatment 45
3.5.3 Design for Separation 46
3.5.4 Maximize Efficiency 46
3.5.5 Output-Pulled Versus Input-Pushed 46
3.5.6 Conserve Complexity 46
3.5.7 Durability Rather Than Immortality 46
3.5.8 Meet Need, Minimize Excess 46
3.5.9 Minimize Material Diversity 46
3.5.10 Integrate Material and Energy Flows 46
3.5.11 Design for Commercial "Afterlife" 46
3.5.12 Renewable Rather Than Depleting 46
References 47

4 Green Materials 51
4.1 Introduction 51
4.2 WEEE and RoHS Directives 51
4.3 Selection of Materials 52
4.3.1 Metals 52
4.3.2 Ceramics 52
4.3.3 Polymer Thermoplastics 53
4.3.4 Polymer Thermosets 53
4.3.5 Elastomers 53
4.3.6 Natural Organic Materials 53
4.3.7 Composites 54
4.4 Conclusions 54
References 54

5 Environmental Design 55
5.1 Introduction 55

5.2 Design for Disassembly Index 56
5.2.1 Nomenclature 57
5.2.2 The Cost–Benefit Function 58
5.2.3 DfDI Calculation Procedure 60
5.2.4 An Application of the Procedure 60
5.2.5 The Optimization Model 67
5.2.6 An Application of the Optimization Procedure 69
5.3 Use of Sensor Embedded Products 71
5.3.1 Sensor-Embedded Products (SEPs) 71
5.3.2 Remote Monitoring Center (RMC) 72
5.3.3 Maintenance Center 73
5.3.4 Disassembly Center 73
5.3.5 Recycling Center 74
5.3.6 Remanufacturing Center 74
5.3.7 Disposal Center 74
5.3.8 Sensor Data Mining 74
5.4 Benefits from SEP and Product Monitoring Framework 74
References 75

Part 3 **Green Value Chain Management**

6 **Green Procurement: Vendor Selection with Risk Analysis 79**
6.1 Introduction 79
6.2 Risk Analysis of Green Vendor Selection 80
6.2.1 Criteria of Selection with Their Hierarchical Relations 81
6.2.2 Weighting of the Criteria 81
6.2.3 Measures of the Attributes 83
6.3 Vendor Evaluation and Selection 89
6.3.1 Risk Aggregation Method 89
6.3.2 Ranking Method 90
6.4 Sensitivity Analysis and Alliance Development 91
6.4.1 Issues of Sensitivity Analysis 91
6.4.2 Sensitivity Analysis on AHP 92
6.5 Summary of the Selection Procedure 92
6.6 Numerical Example 93
6.6.1 Estimation of Risks and Ranking of Two Suppliers 93

6.6.2 Sensitivity Analysis 99
6.6.3 Conclusion of the Example 101
6.7 Summary and Conclusion 101
References 102
Appendix: Pairwise Matrices Given for AHP of the Numerical Example 102

7 Green Production: Manufacture and Remanufacture in Certain and Uncertain Environments 105
7.1 Introduction 105
7.2 Current Development 107
7.2.1 Elements of the Lot-Sizing Model 107
7.2.2 Current Lot-Sizing Models 108
7.2.3 Conclusion 109
7.3 The Green Lot-Sizing Production Model 111
7.3.1 Framework of the Periodic Closed-Loop Production System 111
7.3.2 Modeling in a Certain Environment 111
7.3.3 Modeling in an Uncertain Environment . . 115
7.4 Numerical Illustration 120
7.5 Summary and Conclusion 123
References 124

8 Green Logistics: Recycling with Certain and Uncertain Situations 125
8.1 Introduction 125
8.2 Deterministic Modeling of Closed-Loop Logistics 127
8.2.1 The Deterministic Closed-Loop Logistics Model (DCLL) 128
8.2.2 The Transformed Integer Linear Programming Model 131
8.2.3 An Illustrative Example 134
8.3 Closed-Loop Modeling for Uncertain Logistic 137
8.3.1 The Fuzzy Programming Model 137
8.3.2 Resolution of Uncertainly 140
8.3.3 Numerical Illustration 155
8.4 Conclusions 164
References 165
Appendix: Comparison of the Expected Objective Value Between θ_1 and θ_2 167

9 Green Customers: Features and Identification 169
9.1 Introduction 169
9.2 Features of the Green Consumers 169
9.2.1 Demographic Characteristics 170
9.2.2 Psychographic Characteristics 171
9.3 Questionnaire Design 172
9.4 Analytical Methods 172
9.4.1 Sample Size Determination 172
9.4.2 Data Analysis 181
9.4.3 Cluster Analysis 182
9.5 Target Consumer Identification: Numerical Illustration 187
9.5.1 Sample Size Determination 187
9.5.2 Quantification of the Survey Data 189
9.5.3 Distributions of Sociodemographic Variables 189
9.5.4 Factor Analysis 189
9.5.5 Target Customer Identification 190
9.5.6 Conclusion of the Numerical Example . . . 192
9.6 Summary and Conclusion 194
References 196

10 End-of-Life Management: Disassembly and Reuse 197
10.1 Introduction 197
10.2 Current Development 198
10.2.1 Issues with a Closed-Loop Supply Chain 198
10.2.2 Life-Cycle Effects on the Quantity and Quality of Returned Products 199
10.2.3 EOL and End-of-Use Recovery Selection 201
10.3 Concept of Disassembly 202
10.3.1 Design for Disassembly Representation 204
10.3.2 Demand-Driven Disassembly Planning 208
10.4 Disassembly to Demand (D2D): Modeling and Analysis 208
10.4.1 Representation of Product Structure 209
10.4.2 Disassembly Configurations for Modules 210

10.4.3 Restrictions of Recovery Options 211
10.4.4 Mathematical Model 211
10.5 An Illustrative Example 220
10.6 Conclusion and Future Research Direction 221
References 223

Part 4 Green Information Management Systems

11 Database for Life Cycle Assessment: Procedure with Database 229
11.1 Introduction 229
11.1.1 Applicable International Standards on Product Carbon Footprint 230
11.1.2 Available Software for LCA 231
11.1.3 Inventory of Product Carbon Footprint . . 235
11.1.4 Summary and Conclusion 236
11.2 Procedure of LCA 237
11.2.1 Setting the Inventory Target for the Selected Product 237
11.2.2 Setting the Borderline 237
11.2.3 Identification of Emission Source 238
11.3 Data Collection 239
11.4 Quantification of Emission 239
11.4.1 Methods of Quantification 239
11.4.2 Selection of the Emission Coefficient 241
11.4.3 Summary of Quantification Results 241
11.4.4 Construction of Inventory Database 241
11.5 Impartial Third-Party Verification 241
11.5.1 General Requirements 242
11.5.2 Monitoring Mechanism of Impartiality . . 244
11.6 An Illustrative Example 244
11.6.1 Setting the Borderline 244
11.6.2 Identification of Emission Source 245
11.6.3 Application Simulation 245
11.6.4 Quantification of Emission from Storage and Transport 248
11.6.5 Summary and Discussion 250
11.7 Conclusion 251
References 251
Appendix: Emission Factors and Coefficient Charts of Different Industries and Nations 252

12 Web-Based Information Support Systems 265
12.1 Introduction 265
12.1.1 Infrastructure of Recommender Systems 267
12.1.2 Recommendation Methods 268
12.1.3 Roles and Their Goals in a Recommender System 270
12.1.4 Summary and Discussion 271
12.2 Operations in the Submodules of the System . . . 271
12.2.1 Offline Operations 271
12.2.2 Online Operations 276
12.2.3 Measures of Recommendation Performance 279
12.2.4 Summary of Offline and Online Operation Procedures 279
12.3 An Illustrative Case: Laptops RS of a 3C Retailer 280
12.3.1 Experiments of Strategy Implementation 280
12.3.2 Summary and Remarks of Experiments 284
12.4 Conclusion 288
References 289

Index 291

Preface

Energy and environmental concerns are intricately linked to the supply chains of various goods. Increased public awareness of such concerns is reflected in the contemporary business environment as well as in government legislation. The green supply chain is a new trend in industrial development. Because of various pieces of legislation such as restriction of hazardous substances (RoHS), waste electrical and electronic equipment (WEEE), and eco-design of energy-using products (EuP) announced by the European Union (EU), many industries today need to incorporate environmental factors into their supply chain management. This requires implementation of various techniques to quantify the environmental impact on supply chains, and to identify opportunities for making improvements. This leads to both green engineering and green management of a product.

While global warming has become an urgent issue, educating the general public how to achieve a green environment is essential. In addition, it is not only essential to produce a green product or just manage and control the whole supply chain under the required green guidelines, but it is also obligatory to treat the used materials and products properly. Coping with such complicated and correlated issues such as "4R" (reduce, redesign, remanufacture, and reuse) requires that practitioners utilize more analytical and scientific methodologies. The purpose of this comprehensive text is to address such issues. The book provides a stage-by-stage production methodology within the life cycle of a product to ensure environmental compliance and economical goals. The book also provides coverage for the development of green product information management and retrieval systems to facilitate the understanding and application of green product life cycle assessment and promotion.

This book is designed for senior undergraduate students and graduate students from both engineering and management disciplines. The book contains 12 chapters organized into four parts, viz., Basic Concepts and Background; Green Engineering Technologies; Green

Value Chain Management; and Green Information Management Systems.

This book will be useful to both researchers and practitioners who are interested in a comprehensive overview of a green supply chain as it relates to the life cycle of a green product. In particular, students with an engineering or management background will learn about:

1. the importance of green engineering and management with respect to enterprise competence, environmental protection, product sustainability, and legislation;
2. the resolution and assessment of issues related to a green product using a life cycle approach;
3. the basic concepts of green engineering with manageable solutions;
4. the basic principles of both supply chain and demand chain management; and
5. the elements of designing an information management system.

Finally, it is expected that by utilizing the philosophy, techniques and methodologies covered in this book, we can push forward towards a cleaner planet.

Hsiao-Fan Wang, Ph.D.
National Tsing Hua University
Hsinchu, Taiwan

Surendra M. Gupta, Ph.D.
Northeastern University
Boston, Massachusetts

Acknowledgments

We would like to acknowledge the help and encouragement of numerous people whose support was vital in making this book a reality. Many thanks to Mr. Wei-Liang Kao, Mr. Chun-Yuan Fu, Mr. Hsin-Wei Hsu, Dr. Miao-Lin Wang, Mr. Yen-Shan Huang, Mr. Cheng-Ting Wu, Dr. Elif Kongar, Dr. Mehmet Ali Ilgin, Dr. Kishore M. Pochampally, Dr. Seamus M. McGovern, and Mr. Onder Ondemir for their help, contributions, and creative ideas in developing the book.

We also thank McGraw-Hill for their commitment to encourage innovative ideas, and express our appreciation to its staff for providing seamless support in making it possible to complete this timely book on schedule.

Most importantly, we are indebted to our families, to whom this book is lovingly dedicated, for providing us with unconditional support during the development of this book.

PART 1

Basic Concepts and Background

Chapter 1
Introduction

Chapter 2
Mathematical Background

CHAPTER 1
Introduction

In this introductory chapter, we review the development and evolution of green supply chain management and its impact on industry.

1.1 Development of Green Supply Chain Management

As the effects of environmental problems on the living conditions of the world's population become more apparent, an emphasis on environmental awareness has become more prominent. The general public has started to pay more attention to the potential consequences of this global environmental problem.

Some of the most pressing environmental issues include ozone layer depletion, global warming, and hazardous wastes. In an effort to mitigate the negative impacts of such environmental problems, many nations have passed laws and regulations and have set environmental standards aimed at reducing industry carbon and greenhouse gas emissions to the atmosphere. Some of these standards include *end-of-life vehicle* (ELV); *restriction of hazardous substances* (RoHS); *waste electrical and electronic equipment* (WEEE); eco-design of *energy-using products* (EuP); and *registration, evaluation, authorization, and restriction of chemicals* (REACH).

In addition, industries have been increasingly creating environmentally friendly products to satisfy consumers' demand for green products. For instance, businesses have lessened the use of raw materials, have begun recycling old products, and have started using renewable energy.

Supply chain management (SCM) covers industry planning and control activity that relates to trade, exchange, and logistics management. More importantly, it includes collaboration among suppliers, agents, and new customers. With increasing customer awareness and more stringent regulations, industries have started to integrate environmental factors throughout their organizations. Furthermore, industries have been gradually shifting towards environmentally friendly supply chains by integrating green technologies into their product designs, production, and distribution processes. These efforts, together with the desire to incorporate *extended production responsibility* (EPR), have led to the evolution of *green supply chain management* (GSCM).

In 1969, the Coca-Cola Company commissioned a study on life-cycle analysis to assess the environmental impact of its product packages, laying the framework for the *life cycle assessment* (LCA) practiced today. During this period, LCA emphasized reducing the negative environmental impact from production, transportation, and disposal of goods. However, progression in recent years has gradually transformed and integrated this methodology into the entire SCM system.

1.2 Evolution of GSCM from SCM

SCM was first discussed by a group of professional consultants back in the 1980s. This discussion and later developments have made people more aware of SCM (Oliver and Webber, 1982). In his book titled *Competitive Advantage: Creating and Sustaining Superior Performance,* Porter (1985) introduces the concept of value chain. He emphasizes that industry can perform a series of primary and support activities to increase the added value of products, which could in turn add value for customers. The linkages of these activities to add value to the products and services that an organization produces is known as *the value chain.* According to Porter, the support activities include inbound logistics, operations, outbound logistics, marketing, sales, and services.

SCM is recognized as a network of interconnected businesses that form a tight linkage among raw material resources, production, transportation, and distribution of material resources, information, and financial flows for the ultimate provision of goods and services. Consequently, the Global Supply Chain Forum of 1999 defined SCM as, "An integration of procedures from suppliers to consumers to provide products, services and information in order to add the values of the customers and the related roles." Therefore, the SCM complexity revolves around three factors: (1) products, (2) suppliers, and (3) raw materials.

In terms of sustainability and the international green initiatives, focus is given to the three major issues—dematerialization, detoxification, and decarbonization—which lead to the 4R's (reduction, redesign, reuse, and remanufacture) in practice. Most industries strive for dematerialization to reduce the amount of materials or time needed to produce and deliver products and services required by the customers. The production of poisonous materials can be the best example of people who suffer from their own actions because they ignore the potentially global effects of such systems such as oil spill or some other poisonous material spill.

One of the most pressing environmental needs is the detoxification of industrial pollutants. Much of the natural resources around the world have been contaminated by industrial waste and pollutants due to the use of hazardous materials in industrial products. The profound effect of environmental degradation threatens the living conditions of all organisms. In this context, detoxification becomes a more important and serious challenge to individuals, governments, and industries. In addition, carbon emissions from the production of hydrogen to produce power from fossil fuels release greenhouse gases into the atmosphere, resulting in the climate change problem. Hence, decarbonization, also known as de-energization, is necessary to mitigate this problem. To achieve these goals, redesign of a product for reducing energy consumption and ease of disassembly for reuse and remanufacture becomes a necessity.

Given the global green initiatives, new methods have emerged for adopting the practices and standards for analyzing sustainable development in leading enterprises operating in developed countries. These methods generally focus on the three most important aspects of the enterprise, viz., (1) product design, (2) production process, and (3) the organization itself.

In 1994, the Asian Productivity Organization (APO) stated that if Asia was to fully develop, sustainability should be part of a new paradigm to integrate the environmental factors with production. These recommendations have led *green productivity* (GP) to prevail in Asia. The fundamental goal of GP is to develop higher-level production processes to protect society while simultaneously increasing industry product qualities and maintaining profit targets for industry.

Thus, green productivity focuses on enhancing productivity and environmental performance for sustainable developments in industry to achieve a competitive advantage. Wu and Dunn (1995) use environment and energy problems together with Porter's statement on value chain integration to emphasize that industry not only has to increase product value, but it also needs to use the value chain. The main idea is to decrease environmental impact throughout the value chain, from raw materials to the final product. In this case, environmental impact includes reduction in energy use, consumption of natural resources, and pollution-related problems. Although Wu and Dunn do not fully describe the framework of the green supply chain, they tackle pollution produced by industry at different levels in the value chain, as well as postconsumer consumption. The authors point out that there should be an increase in product recycling to increase the further use of raw material and supply. To that end, GSCM should augment the activities of reverse logistic management (which includes decreasing production source, renewing natural resources, recycling materials, cleaning waste material, and managing hazardous substances) to integrate all aspects of environmental management into its domain (see, for example, Pochampally et al., 2009).

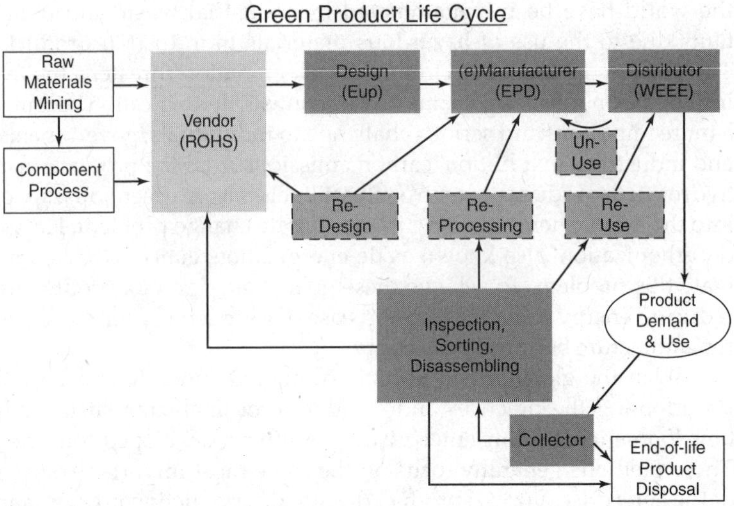

Figure 1.1 Framework of a green supply chain with the involved forward and reverse activities and their corresponding directives.

This concept is adopted in the book by Wang (2009). In particular, Wang points out that market expectations have been an essential motivation to encourage industries to focus on creating and adding value at all levels of the value chain to enhance their competitive advantages. This has led to the establishment of communication channels between customers and suppliers. With the rapid development of Internet technology, such channels can be effectively built through online platforms. Thus, organizations must establish reliable databases and effective retrieval schemes to make this information available via search engines.

In this book, we shall discuss the green supply chain in the framework shown in Fig. 1.1, addressing both the forward activities of procurement, design, manufacture, and distribution to the consumer, as well as the reverse activities of inspection, sorting, and disassembly for the purposes of reuse, reprocessing, and redesign for the demand of primary and secondary markets.

1.3 Impact of GSCM on Industry

Life-cycle analysis evaluates various aspects of the product system, including technology and its potential impact to the environment, throughout its life cycle. It involves examining all stages of the development of a product or service, from the selection of raw materials, through product outputs and usage, to product processing after use. The most common tool used in life-cycle analysis is called *life cycle*

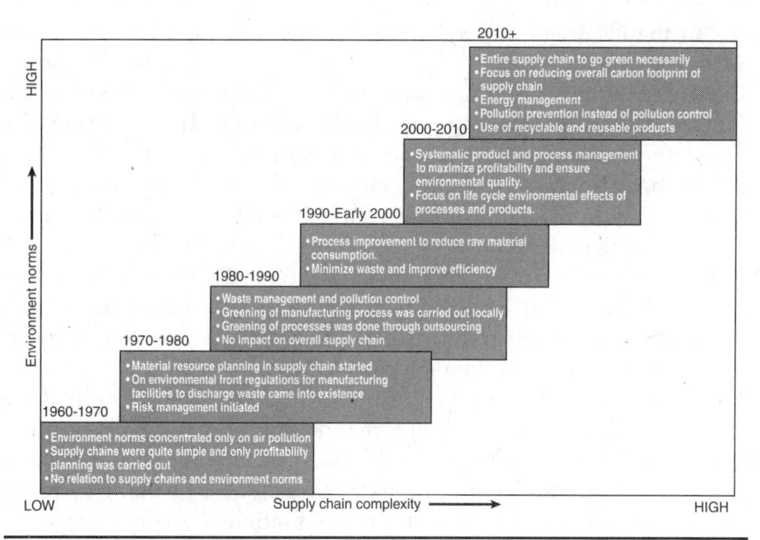

FIGURE 1.2 Evolution of the green supply chain.

assessment (LCA), which is used to assist designers at the beginning of product design to identify techniques that ensure that the product components and designs are attuned to environmental standards.

Standards must be set when conducting an LCA of the amount of energy and natural resources utilized, the amount of emissions caused by air and water pollution, and the waste products resulting from consumption of the product output.

Although the theory of GSCM was developed in the 1990s, due to numerous challenges, most industries did not begin to adopt GP until 2000 or even later. Figure 1.2 shows the evolution of green supply chain technologies. One can see in the course of development that while consciousness concerning the environment increased, the complexity and scale of GSC technologies also increased. Each country has varying environmental protection standards integrated into its own regulatory system. Furthermore, these laws need to be acceptable to industries for them to be adopted in *original design manufacturing* (ODM) and *original equipment manufacturing* (OEM). The impact of GSCM on industries will be discussed in the next section.

1.3.1 Impact of GSCM on Industry Tactics

The effect of GSCM on industry is significant, in that the green procurement system is built to control suppliers, especially for ODM, so that they satisfy environmental protection standards for their products. For example, in order to satisfy the ELV standards, automobile companies collaborated and created the International Material Data System (IMDS), which ensures that the materials used by OEMs and their suppliers meet international standards.

In the electrical and electronics industries, suppliers use green procurement procedures and documentation to specifically control the use of various hazardous materials. However, GSCM aims to satisfy the "green requirement" for ODMs throughout the entire SCM. One famous example that can be referred to is the event of SONY, in which, because of the forbidden metal Cd found in the connectors of their play stations, their cargo was not allowed to enter the harbor of Holland, which caused enormous damage to both the products as well as the reputation. Thus, SONY developed SS-00259 technical standard (SONY, 2004) to restrict all of their component suppliers to follow a standard flow in partnership, in order to ensure that no toxic material will ever appear along their supply chain.

Also using SCM, Dell can estimate its Supply Chain Carbon Output (InfoWorld website). Dell demands that all first-level suppliers sign the Electronic Industry Code of Conduct (EICC), and encourages these suppliers to conform to these standards by being certified. The EICC establishes standards and regulations for corporate social responsibility in the electronic industry's global supply chain. These regulations are developed from reviews of basic standards, which involve laborers and employers, safety, environmental responsibility, system management, and ethics (Dell website).

1.3.2 Impact of the Green Supply Chain on Industrial Administration

In order to achieve the goal of GSCM, industries must gather all existing material systems, information management systems, and process operation systems, and integrate all the green requirements into tasks and procedures to achieve greening. For this purpose, industries strive through the GSCM system to evaluate the green performance of suppliers, assess the quality of green products, and direct overall management of suppliers. In addition, industry leaders use *information technology* (IT) to assist other industries to solve the problems and challenges of applying GSCM. Efforts have been made to ensure that the overall supply chain conforms to the standards of GSCM through management systems, production processes, technique standards, verification of work, and analysis of product components. Some of the benefits of introducing GSCM into industry are discussed in the following subsections.

Increased Interaction Between Client and Supplier

With increasing customer awareness and regulatory norms, organizations with greener supply chain management practices increase client interaction to gain understanding of their clients' needs and wants, providing opportunity for future product development. Clients also encourage suppliers to strengthen their design capability and production control process in order to meet requirements based on standards.

Collaboration Between the Company and the Supplier to Create Innovative Product Design

In the development of green products, innovation is needed, such as identifying substitute materials (e.g., non-lead material fabrication and non-halogenic plastic material and fabrication), creating energy-saving designs, and upgrading energy-transforming efficiency. All of these tasks rely on collaborative and innovative design efforts between the industry and its suppliers.

Requirement for Suppliers to Provide Environmental Information

To ensure that products conform to environmental standards, information on products and their suppliers should be based on the content of a survey that the collaborative suppliers are required to provide.

- In order to satisfy the recycling rate and the recovery rate that the WEEE standard has set, the supplier must accomplish a survey regarding information on the location of the hazardous component, the amount and component *bill of manufacturing* (BOM) form, the means of transportation, and the weight of the product, in order to analyze the disassembling rate and the possibility of recycling. In addition, this information helps recycling companies to increase the recycling rate of waste products of the electronic and electrical machinery components.
- In order to understand the potential environmental impact of the production of goods, the brand company can use SCM to ensure that suppliers cooperate in providing information on the material component and content of the product.

Collaboration Between the Company and the Supplier in Creating Green Product Design

In order to meet standards for green product design, companies must collaborate with suppliers in developing product design. For example, the company should inform its suppliers of the product requirements to meet certain environmental standards, such as developing non-lead products and non-halogenic products, or to provide postconsumer plastic materials.

Auditing the Suppliers' Green Performance

Companies should also set up a green audit management or environmental audit management system to ensure that suppliers satisfy the product requirement. This allows the supplier to gain understanding of the industry's environmental strategy or the green standards for the product to which the company is adhering, so that the supplier is able to assist the company in achieving its target *corporate social responsibility* (CSR).

Challenge of Increased Prime Costs

To satisfy the standards of green purchasing for the company, suppliers need not only send products to one laboratory for testing and supply test results but should also provide and install their own testing equipment (e.g., XRF) and controlling systems (e.g., QC 080000, ISO 9000, ISO 14000, OSAS 18000). Moreover, some of the tools for integrating information systems (e.g., ERP, PLM, CAD, CRM) are costly, which can lead to an increase in the cost of production.

1.3.3 Intensification of Competition by GSCM

As organizations face the challenges of adhering to environmental protection regulations, many domestic and international industries are finding opportunities in green product development. These include searching for new raw material supplies, developing new processes, and expanding new markets. This industrial strategy is similar to the shift from the so-called "red ocean strategy" (i.e., low-price competition in existing markets) to a "blue ocean strategy" (i.e., developing new market opportunities). These results prove Porter's analysis of competition:

- Attempt to get out of present factory competition;
- Increase the barrier facing the new competitor when trying to become involved in the burgeoning market;
- Pose competition to rivals.

Furthermore, Porter's analysis can be extended. Rather than considering the situation from the point of view of competition, GSCM can be viewed from the point of view of collaboration, thereby creating more opportunities in the global industry. A strategic alliance both upstream and downstream can lead to the expansion of a much bigger energy industry. This allows an enterprise to gain understanding of its customer needs and to actively devote its efforts to the development of green products. However, the enterprise needs to expand its supplier base in order to strictly adhere to standards set to ensure quality control and to allow price negotiation.

1.4 Summary and Conclusion

Supply chain management facilitates interaction among the supplier, the shipping industry, the customer, the retail merchant, the quality controller, and the manager of the final product. This interaction may change from the upstream or the downstream side depending on the influence of the enterprise on the supply chain. The enterprise can enhance the relationship within the supply chain through effective communication to avoid misunderstanding. Thus, the enterprise serves as the linkage that facilitates interaction in the supply chain.

With increasing interest and development in the areas of green procurement, technology, and product development, local enterprises have already adopted some initial measures. The challenge, however, is how to effectively implement and manage this initiative in order to achieve significant results. Figure 1.3 shows an implementation procedure for extending the supply chain that includes activities focusing on operations analysis, measurement, and continuous improvement. All of these major issues will be discussed in this book, and their value will be measured against corresponding performance.

In summary, as environmental protection continues to grow as a global concern, organizations are paying more attention to ensuring that green standards are adhered to at all levels of the value chain, from manufacturing, through social welfare, to sustainability of the product development system. The green value chain framework is built based on future trends and environmental requirements for a sustainable design. For related issues of development along the life cycle of a product, the reader is referred to the handbook published by the Global Green Supply Chain Group (2007).

This book is organized into four parts: "Basic Concepts and Background" (which includes this chapter and Chapter 2, which introduces

IDENTIFY PROCESS	DEVELOP PERFORMANCE MEASUREMENT SYSTEM	MEASURE THE SUPPLY CHAIN SYSTEM	PRIORITIZE AND DEVELOP ALTERNATIVES	ESTABLISH IMPROVEMENT PROCEDURES
Procurement and Supply	- Planning - Training - Implementation	- Mixed contracts with more profit in less consumption - Involving suppliers in decision-making process	- Prioritize the process steps in order of increasing composite performance - Develop alternatives for performance improvement	- Organizational alignment towards continuous improvement - Involvement of suppliers/vendors in key decision-making process - Regular audit and performance review - Clear communication between suppliers and logistics providers - Aligned incentives
Production	- Environmental cost accounting - Life cycle costing - Process design - Life cycle engineering	- Categorizing and identifying environmental costs. - Developing most economical and environment-friendly process	- Achieve process improvement through conducting interviews with process personnel and suppliers	
Packaging	- Resource minimization - Reduced hazards - Recycling and energy recovery	- Financial savings - Local and global environmental benefits	- Usage of reusable and recyclable packaging - Customers collaboration - Waste reduction	
Product sales and marketing	- Creating awareness among consumers for green products	- Consumers buying green products	- Making eco-friendly products as prime business objective	
Logistics	- Mode of travel used - 3PL performance evaluation	- Use of rail transport - 3PL collaboration	- Logistics optimization - Use of clean fuel - Optimized truck load	
Product End-of-life Management	- Product design - Supply loop effectiveness	- Reprocessing end-of-life products	- Collecting waste material for economic value recovery	

FIGURE 1.3 Development of a GSCM system with the related issues.

the basic mathematical background needed to follow this book), "Green Engineering Technology" (which is mainly concerned with the engineering issues of developing a GSC), "Green Value Chain Management" (which considers GSC from both sides of demand and supply as a closed-loop system including the roles indicated in Fig. 1.3) and "Green Policy and Information Management System" (which discusses information support for such system to be found on the Internet).

References

Global Green Supply Chain Group, *Handbook of Green Supply Chain*. Australia, 2007.

Oliver, R. K., and Webber, M. D. (1982). Supply-chain management: Logistics catches up with strategy. Outlook, Booz, Allen, and Hamilton; Reprinted in Christopher, M. G., (Ed.), (1992). *Logistics: The Strategic Issues*. London: Chapman and Hall, pp. 63-75.

Pochampally, K. K., Nukala, S. and Gupta, S. M. (2009). *Strategic Planning Models for Reverse and Closed-loop Supply Chains*. Boca Raton, FL: CRC Press.

Porter, M. E. (1985). *Competitive Advantage: Creating and Sustaining Superior Performance*. New York: Free Press.

SONY Corporation. (2004). Management Regulations for the Environment-related Substances to be Controlled which are included in Parts and Materials (SS-00259), version 3, January 2004.

Wang, H. F. (2009). *Web-Based Green Products Life Cycle Management Systems: Reverse Supply Chain Utilization*, Hershey, NY: Information Science Reference.

Wu, H. J., and Dunn, S. (1995). Environmentally responsible logistics systems, *International Journal of Physical Distribution & Logistics Management*, 25(2), 20-38.

Websites

Dell website: http://www.businessinsider.com/is-dell-the-worlds-most-social-company-scorecard-2011-6

EUP website: EUP, 25 April (2010): http://www.eup-network.de

InfoWorld website: http://weblog.infoworld.com/sustainableit/archives/green business/green supply chain/index.html

Korea National Cleaner Production Center website: http://www.kncpc.re.kr/eng/topics/Lci.asp

ROHS website: ROHS, 25 April (2010): http://www.rohs.gov.uk

CHAPTER 2
Mathematical Background

This chapter reviews basic mathematical concepts that are relied on throughout the text.

2.1 Fuzzy Numbers and Arithmetic

Fuzzy sets were introduced by L. A. Zadeh (1965) both as a new way to represent vagueness in everyday life and also as an attempt to model human reasoning and thinking processes. Fuzzy sets have the flexibility to capture the incompleteness or imprecision in information.

Let X be a universal set. A *fuzzy set* $\tilde{A}$ in X is a set of ordered pairs:

$$\tilde{A} = \{(x, \mu_{\tilde{A}}(x)) \,|\, x \in X, \mu_{\tilde{A}}(x) \in [0, 1]\} \tag{2.1}$$

Therefore, fuzzy sets are the generalization of crisp sets by a mapping from the universe of discourse X to the unit interval [0, 1]. Each element belongs to a fuzzy set with a certain degree, known as *membership degree,* between 0 and 1. The *membership function* of a fuzzy set $\tilde{A}$ is denoted by $\mu_{\tilde{A}}$, $\mu_{\tilde{A}} : X \to [0, 1]$. Fuzzy sets characterize imprecise properties so that they can be effectively used to model vagueness associated with real-life systems.

Definition 2.1: Level Set A crisp set of elements that belong to a fuzzy set $\tilde{A}$, at least to a degree γ, is called an *γ-level set* of $\tilde{A}$ defined by

$$A_\gamma = \{x \in X | \mu_{\tilde{A}}(x) \geq \gamma\}, 0 \leq \gamma \leq 1 \tag{2.2}$$

Definition 2.2: Fuzzy Number If $\tilde{I}$ is a normal fuzzy set on R and I_γ is a closed interval for each $0 \leq \gamma \leq 1$, then $\tilde{I}$ is a fuzzy number.

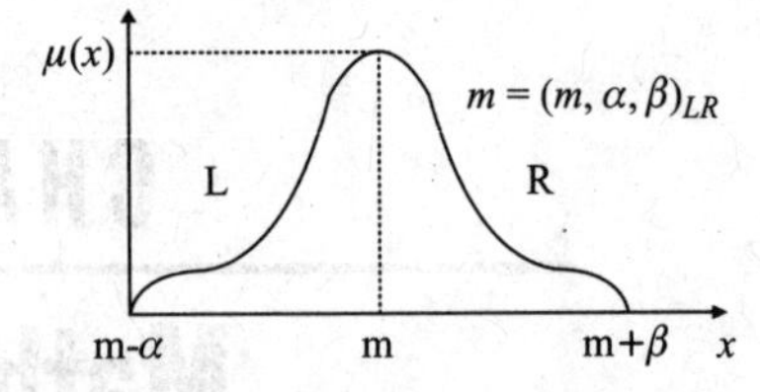

Figure 2.1 L-R fuzzy number.

There are some commonly used fuzzy numbers as below:

1. *L-R Fuzzy Number* An *L-R fuzzy number* $\tilde{I}$ denoted by (m, α, β) is a fuzzy set that has a membership function defined for all $x \in R$ by:

$$\mu_{\tilde{I}}(x) = \begin{cases} L\left(\dfrac{m-x}{\alpha}\right), & \text{if } x \leq m; \\ R\left(\dfrac{x-m}{\beta}\right), & \text{if } x \geq m. \end{cases} \quad (2.3)$$

   where $\alpha > 0$, $\beta > 0$ with respect to left and right spreads. Thus, L is monotonically increasing toward 1; R is monotonically decreasing from 1 with $L(0) = R(0) = 1$ and $L(1) = R(1) = 0$; and the highest membership value 1 is at $x = m$, as shown in Fig. 2.1. If $m < 0$, we have a left translation, and for $m > 0$, we have a right translation.

2. *Triangular Fuzzy Number* If both L and R are linear, we have a *triangular fuzzy number* denoted by $\tilde{I} = (x, \alpha, \beta)$ where α is a left spread from x and β is a right spread from x.

3. *Trapezoidal Fuzzy Numbers* If a linear *L-R* fuzzy number has $m_1 \leq m \leq m_2$, then it becomes a *trapezoidal fuzzy number* denoted by $(m_1, m_2, \alpha, \beta)$.

Figure 2.2 shows the example of triangular and trapezoidal fuzzy numbers.

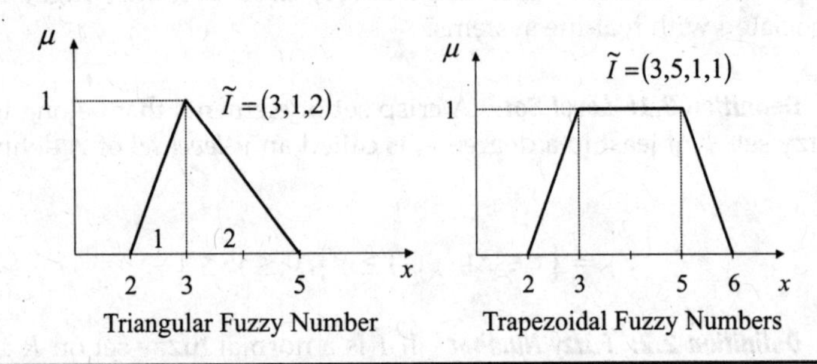

Figure 2.2 Examples of linear fuzzy numbers.

Definition 2.3: Fuzzy Arithmetic The arithmetic of two linear fuzzy numbers $\widetilde{X}$ and $\widetilde{Y}$ is defined by their level set as below:

1. *Fuzzy addition* defined by $\widetilde{X}(+)\widetilde{Y}\triangleq\widetilde{Z}$,

$$Z_\gamma = X_\gamma + Y_\gamma = [X_\gamma^L + Y_\gamma^L, X_\gamma^U + Y_\gamma^U], \forall\gamma \in [0, 1]. \quad (2.4)$$

2. *Fuzzy subtraction* defined by $\widetilde{X}(-)\widetilde{Y}\triangleq\widetilde{Z}$, where

$$Z_\gamma = X_\gamma - Y_\gamma = [X_\gamma^L - Y_\gamma^U, X_\gamma^U - Y_\gamma^L], \forall\gamma \in [0, 1] \quad (2.5)$$

3. *Fuzzy multiplication* defined by $\widetilde{Z}\triangleq\widetilde{X}(\bullet)\widetilde{Y}$ if and only if

$$Z_\gamma = X_\gamma \cdot Y_\gamma = \begin{bmatrix} \text{Min}\left((X_\gamma^U Y_\gamma^L, X_\gamma^L Y_\gamma^U, X_\gamma^U Y_\gamma^U, X_\gamma^L Y_\gamma^L)\right), \\ \text{Max}\left(X_\gamma^U Y_\gamma^L, X_\gamma^L Y_\gamma^U, X_\gamma^U Y_\gamma^U, X_\gamma^L Y_\gamma^L\right) \end{bmatrix}, \forall\gamma \in [0, 1] \quad (2.6)$$

4. *Fuzzy division* defined by $\widetilde{Z}\triangleq\widetilde{X}(:)\widetilde{Y}$ if and only if

$$Z_\gamma = X_\gamma : Y_\gamma = \begin{bmatrix} \text{Min}\left(X_\gamma^U/Y_\gamma^L, X_\gamma^L/Y_\gamma^U, X_\gamma^U/Y_\gamma^U, X_\gamma^L/Y_\gamma^L\right), \\ \text{Max}\left(X_\gamma^U/Y_\gamma^L, X_\gamma^L/Y_\gamma^U, X_\gamma^U Y_\gamma^U, X_\gamma^L/Y_\gamma^L\right) \end{bmatrix}, \forall\gamma \in [0, 1] \quad (2.7)$$

2.2 Utility Theory

A utility function is a measure of desirability or satisfaction of a DM (decision maker), which provides a uniform scale to compare and/or combine criteria. A utility function is a device that quantifies the preferences of a DM by assigning a numerical index, which is comprised of varying levels of satisfaction as criteria. In other words, a utility function facilitates the transformation of contractor performance into measured levels of DM satisfaction.

Generally, decisions are comprised of three types—namely, risk aversion, neutral risk, and risk-prone—and the DM's risk attitude is reflected in the shape of the utility curve. When establishing different risk utility functions for each criterion, the DM will be asked several questions to obtain information regarding their risk attitude. First, this system will establish the best and worst scores of each criterion; then the best score's utility value is $U = 1$, and the worst score's utility value is $U = 0$. After deciding the best and worst scores, the mid-score C between these two scores is computed. Depending on the mid-score value, the DM is asked the following question (Zedan and Skitmore, 1998):

> You are offered two routes, as shown in Fig. 2.3. The first route is R1 and will provide you the outcome score of C, with a probability of $P = 1$. The

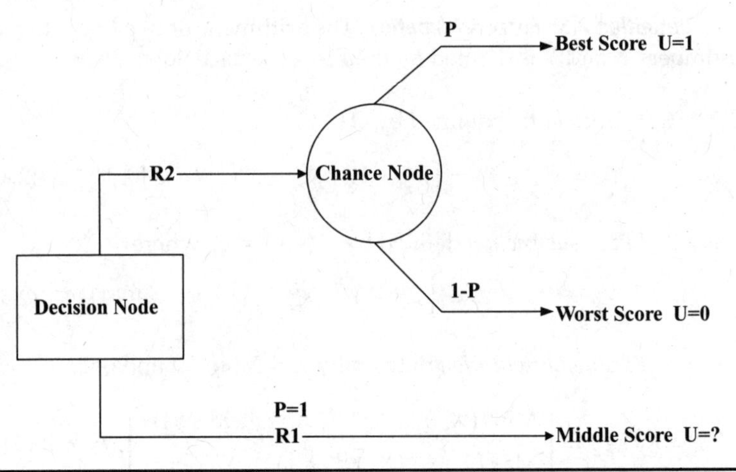

FIGURE 2.3 The routes of a decision process (Zedan and Skitmore, 1998).

second route is R2. Here you either receive the best outcome of best score, which has a utility $U = 1$ and a probability of P, or you will receive the worst score, which has a utility $U = 0$ and a probability of $1 - P$. Which probability score will you chose between the two routes?

According to the probability, we can compute the mid-score C utility value, U_c, the expected value of R2 can be computed as shown in Eq. (2.8):

$$U_c = P \times 1 + (1 - P) \times 0 = P \tag{2.8}$$

When three points are obtained for the utility function, the utility function can be computed. Yasutaka and Tawara (2006) proposed the multiple-attribute utility theory method for assessing a green supply chain.

Let the utility function be $U(x) = x^a$, where x is normalized into the unit interval of [0,1]; then, from the articulation above, we may obtain three types of utility function that represent different risk attitude as follows:

1. $0 < a < 1$: If the curve of the utility function is convex, it means that the DM is conservative regarding the criterion; in other words, the DM is risk-averse, and has low tolerance in the criterion.
2. $a > 1$: If the curve of the utility function is concave, it means that the DM is radical regarding the criterion; in other words, the DM is risk-prone, with high tolerance in the criterion.
3. $a = 1$: Between these two is the neutral risk, which will result in a straight line for the utility function. These three kinds of utility functions are shown in Fig. 2.4.

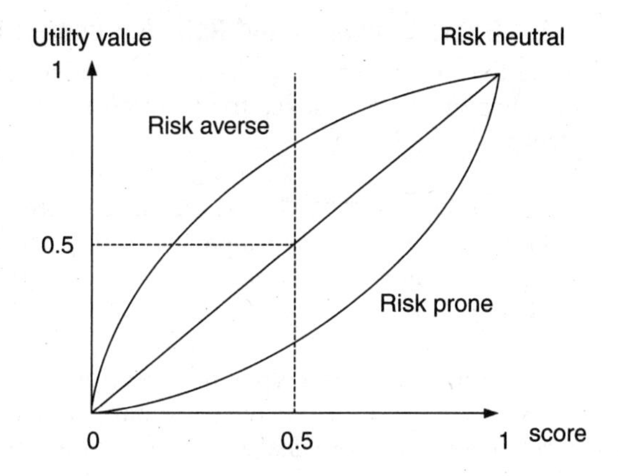

FIGURE 2.4 The utility functions of different risk attitudes.

2.3 The Analytic Hierarchy Process (AHP)

Since Saaty (1980) proposed the analytic hierarchy process (AHP) to decompose a complex multiple-attribute decision making (MADM) problem into a simpler system in hierarchy, it has drawn much attention both in research and in practice. The importance of AHP and its variants in decision making can be seen by the more than 1,000 references cited in Saaty (1994); a number of special issues in the refereed journals such as the recent publication at the *European Journal of Operational Research* (48(1), 1990) and *Mathematical and Computer Modeling* (17(4/5), 1993); and a biannual International Symposia called *ISAHP*.

Based on pairwise comparison, three basic properties of the analytic hierarchy process can be characterized:

1. *Hierarchic representation and decomposition* This is a *hierarchic structuring* of breaking down the problem into separate elements.
2. *Priority discrimination and synthesis* This is called the *priority setting* for ranking the elements by relative importance.
3. *Logical consistency* This is to ensure that elements are grouped logically and are ranked consistently according to logical criteria.

Therefore, to begin with the AHP, we first lay out the elements of a problem as a hierarchy. We then make paired comparisons among the elements of a level as required by the attributes of the next higher level. These comparisons give rise to priorities and, finally, through synthesis, to overall priorities. Then consistent and interdependent tests are performed. The basic steps of the process are outlined in Sec. 2.3.2 after the main concepts are introduced and discussed in Sec. 2.3.1.

2.3.1 Basic Concepts and Pairwise Comparison

Denote that, in general, a pairwise comparison matrix represents a preference matrix that takes the form of a multiplicative preference matrix as defined below:

Definition 2.4: A Multiplicative Preference Matrix (MPM) (Saaty, 1980) $A = [a_{ij}], i, j = 1, 2, \ldots, n$, is a pairwise comparison matrix whose element a_{ij} estimates the dominance of alternative i over j, and meets $a_{ij} > 0, a_{ji} = 1/a_{ij}$.

Therefore, this matrix is positive reciprocal and is denoted as PRM. In order to proceed the pairwise comparison effectively, Saaty (1980) has based on psychological theory proposed by Weber (1846) and Miller (1956) to suggest a scale listed in Table 2.1, in which 1 is the lower limit and 9 is the upper limit. Because *a change in sensation is noticed*

Intensity of Importance	Definition	Explanation
1	Equal importance of both elements.	Two elements contribute equally to the property.
3	Weak importance of one element over another.	Experience and judgment slightly favor one element over another.
5	Essential or strong importance of one element over another.	Experience and judgment strongly favor one element over another.
7	Demonstrated importance of one element over another.	An element is strongly favored and its dominance is demonstrated in practice.
9	Absolute importance of one element over another.	The evidence favoring one element over another is of the highest possible order of affirmation.
2, 4, 6, 8	Intermediate values between two adjacent judgments.	Compromise is needed between two judgments.
Reciprocals	If activity a_i has one of the preceding numbers assigned to it when compared with activity a_j, then a_j has the reciprocal value when compared with a_i.	

TABLE 2.1 The Pairwise Comparison Scale

if the stimulus is increased by a constant percentage of the stimulus itself, a unit difference between successive scale values is used. Besides, the experiment has shown that most individuals cannot simultaneously compare more than seven objects (plus or minus two); thus the total number of scales is defined.

Based on the instruction of Table 2.1, the role of a decision maker is to perform pairwise comparisons that are coded in a matrix form by the following rules:

Let $A = \{a_1, a_2, \ldots, a_n\}$ be the set of attributes. The quantified judgments on pairs of attributes (a_i, a_j) are represented by an $n \times n$ matrix,

$$R = [r_{ij}], i,\ j = 1, 2, \ldots, n.$$

The entries r_{ij} take the values from Table 2.1 and are defined by the following entry rules to indicate the strength of a_i when compared with a_j.

Rule 1. If $r_{ij} = \alpha$, then $r_{ji} = 1/\alpha, \alpha \neq 0$.

Rule 2. If a_i is judged to be of equal relative importance a_j, then $r_{ij} = r_{ji} = 1$; in particular, $r_{ij} = 1$ for all i.

Thus, the matrix R is a *multiplicative preference matrix* and has the form below in which the available values are members of the set: $\{9, 8, 7, 6, 5, 4, 3, 2, 1, 1/2, 1/3, 1/4, 1/5, 1/6, 1/7, 1/8, 1/9\}$:

$$R = \begin{bmatrix} 1 & r_{12} & \cdots & r_{1n} \\ 1/r_{12} & 1 & \cdots & r_{2n} \\ \vdots & \vdots & \vdots & \vdots \\ 1/r_{1n} & 1/r_{2n} & \ldots & 1 \end{bmatrix} \tag{2.9}$$

Definition 2.5: If R is a multiplicative preference matrix, it is *consistent.*

$$\text{If } r_{ik} = r_{ij}r_{jk}, \forall i, j, k = 1, 2, \ldots, n. \tag{2.10}$$

Therefore, having recorded the quantified judgments on pairs (a_i, a_j) as numerical entries r_{ij} in the matrix R, according to the rules and Definition 2.5, a multiplicative preference matrix is obtained. Now the task of the analyst is twofold: one is to examine whether R is consistent, and the other is to assign n contingencies $a_1, a_2, \ldots, a_n$ to a set of numerical weights $w_1, w_2, \ldots, w_n$ that would reflect the decision maker's preference

Derivation of Weight of Importance

Since the matrix R is reciprocal, if our judgment is consistent in all comparisons, $r_{ik} = r_{ij}r_{jk}$ should be held for all i, j, k. An obvious case of a consistent matrix is the one in which the comparisons are based on exact measurements, that is, the weights $w_1, \ldots, w_n$ are already known.

Then

$$r_{ij} = \frac{w_i}{w_j} \quad i, j = 1, \ldots, n \tag{2.11}$$

And thus

$$r_{ij}r_{jk} = \frac{w_i}{w_j} \cdot \frac{w_j}{w_k} = \frac{w_i}{w_k} = r_{ik}$$

Also, of course,

$$r_{ji} = \frac{w_j}{w_i} = \frac{1}{w_i/w_j} = \frac{1}{r_{ij}}$$

Let us consider this paradigm case further with the matrix equation

$$\mathbf{A}\mathbf{x} = \mathbf{y} \tag{2.12}$$

where $\mathbf{x} = (x_1, \ldots, x_n)$ and $\mathbf{y} = (y_1, \ldots, y_n)$. This is a shorthand notation for the set of equations

$$\sum_{j=1}^{n} a_{ij}x_i = y_i \quad i = 1, \ldots, n \tag{2.13}$$

Now, we observe that from Eq. (2.11), we obtain

$$r_{ij} \cdot \frac{w_j}{w_i} = 1, i, j = 1, \ldots, n$$

and consequently

$$\sum_{j=1}^{n} r_{ij}w_j \frac{1}{w_i} = n, i = 1, \ldots, n$$

or

$$\sum_{j=1}^{n} r_{ij}w_j = nw_i, i = 1, \ldots, n$$

which is equivalent to

$$R\mathbf{w} = n\mathbf{w} \tag{2.14}$$

	a_1	a_2	...	a_n	
a_1	$\frac{w_1}{w_1}$	$\frac{w_1}{w_2}$	...	$\frac{w_1}{w_n}$	
a_2	$\frac{w_2}{w_1}$	$\frac{w_2}{w_2}$	...	$\frac{w_2}{w_n}$	$\begin{bmatrix} w_1 \\ w_2 \\ \vdots \\ w_n \end{bmatrix} =$
$\vdots$	$\vdots$	$\vdots$	...	$\vdots$	
a_n	$\frac{w_n}{w_1}$	$\frac{w_n}{w_2}$	...	$\frac{w_n}{w_n}$	$n \begin{bmatrix} w_1 \\ w_2 \\ \vdots \\ w_n \end{bmatrix}$

TABLE 2.2 Eigensystem Tableau

In matrix theory, this formula expresses the fact that **w** is an eigenvector of R with respect to the eigenvalue n. When fully writing out this equation, it has the form shown in Table 2.2.

Therefore, by solving this eigensystem, we can obtain the weight of importance as the priorities in Saaty's original term. Proceeding to each level of hierarchy, in the end of the process, an $m \times n$ matrix $R = [r_{ij}]$ for n alternatives with m attributes and their weight of importance, $w_i, i = 1, \ldots, m$ will be obtained. Thus the best alternative is the one with

$$a^*_{AHP} \equiv \arg \operatorname{Max}_j \sum_{i=1}^{n} w_i r_{ij} \tag{2.15}$$

Measure of inconsistency Let us turn to the practical case, in which the r_{ij} are not based on exact measurements, but on subjective judgments. Thus, r_{ij} will deviate from the "ideal" ratio w_i/w_j, and therefore Definition 2.5 will no longer hold. Two facts of matrix theory come to our rescue.

Lemma 2.1 Let $R = [r_{ij}]_{nxn}$ be a positive reciprocal matrix. If $\lambda_1, \ldots, \lambda_n$ are the numbers satisfying the equation $R\mathbf{x} = \lambda x$, that is, are the eigenvalues of R, and if $r_{ii} = 1$ for all i, then $\sum_{i=1}^{n} \lambda_i = n$.

Therefore, if Eq. (2.12) holds, then all eigenvalues are zero, except one, which is n. In the consistent case, n is the largest eigenvalue of R as $\lambda_{\max} = n$.

Lemma 2.2: (Saaty, 1980) If one changes the entries r_{ij} of a positive reciprocal matrix R by small amounts, then the eigenvalues change by small amounts.

Combining these results we find that if the diagonal of a matrix R consists of ones ($r_{ij} = 1$), and if R is consistent, then small variations of the r_{ij} keep the largest eigenvalue, λ_{max}, *close* to n, and the remaining eigenvalues *close* to zero. Therefore, our problem is this:

If R is a multiplicative preference matrix, in order to find the priority vector, how do we find the vector $\mathbf{w}$ *that satisfies* $R\mathbf{w} = \lambda_{max}\mathbf{w}$?

Since it is desirable to have a normalized solution, if w is altered slightly by setting $\alpha = \sum_{i=1}^{n} w_i$ and w is replaced by $(1/\alpha)w$. This ensures uniqueness, and also that $\sum_{i=1}^{n} w_i = 1$.

Observe that since small changes in r_{ij} imply a small change in λ_{max}, the deviation of the latter from n is a measure of consistency. Saaty thus expresses the inconsistency of a pairwise comparison matrix in terms of the *consistency index* defined by $\mathrm{CI} = \frac{\lambda_{max}-n}{n-1}$.

Therefore the inconsistency may appear regardless of the scale used to quantify the pairwise comparisons (Triantaphyllou, 2000).

Saaty used a *random index* (RI) as a reference of the consistency index of a randomly generated reciprocal matrix on a scale of 1 to 9, with reciprocals forced. Table 2.3 gives the order of the matrix and the average RI:

n	1	2	3	4	5	6	7	8	9	10	11	12	13	14	15
RI	0.00	0.00	0.58	0.90	1.12	1.24	1.32	1.41	1.45	1.49	1.51	1.48	1.56	1.57	1.59

Table 2.3 The Mean Consistency Index of Randomly Generated Matrices

The ratio of CI to average RI for the same-order matrix is called the *consistency ratio* (CR). A consistency ratio $\mathrm{CR} \leq 0.10$ is considered acceptable.

Improvement of consistency If $\mathrm{CR} > 0.1$, the multiplicative preference matrix R is consistently unacceptable. Xu (2000) and Xu and Wei (1999) have studied this issue and have proposed a calibration method to obtain a weighted geometric mean matrix or weighted arithmetic mean matrix, which has been shown to be consistently acceptable.

Theorem 2.1 Let $R = [r_{ij}]_{n\times n}$ be a positive reciprocal matrix, λ_{max} be the maximal eigenvalue of R, and $\mathbf{w} = (w_1, w_2, \ldots, w_n)^T$ be the eigenvector of R. Let $R^* = [r_{ij}^*]_{n\times n}$, with values either taking the weighted geometric mean of

$$r_{ij}^* = r_{ij}^{\delta}\left(\frac{w_i}{w_j}\right)^{1-\delta}, \quad i, j \in N, \tag{2.16}$$

or the weighted arithmetic mean of

$$r_{ij}^{*} = \begin{cases} \delta r_{ij} + (1-\delta)\dfrac{w_i}{w_j}, & i \in N; j = i, i+1, \ldots, n. \\ \dfrac{1}{\delta r_{ji} + (1-\delta)\dfrac{w_j}{w_i}}, & i = 2, 3, \ldots, n; j = 1, 2, \ldots, i-1. \end{cases} \tag{2.17}$$

where $0 < \delta < 1$ and $N = \{1, 2, \ldots, n\}$.

Let $\mu_{\max}$ be the maximal eigenvalue of R^*; then $\mu_{\max} \leq \lambda_{\max}$, with equality if and only if R is consistent.

Discussion In practice, there is one shortcut to ensure consistency. That is, the pairwise matrix is derived based only on the data of one row or one column only. Then, based on Rule 1 and the consistency requirement of Definition 2.5, one may derive the rest of the data to obtain a consistent matrix.

2.3.2 The Procedure

Based on the above concept, let us outline the procedure to illustrate how it works:

1. Define the problem and specify the solution desired.
2. Structure the hierarchy from the overall managerial viewpoint (i.e., from the top levels to the level at which intervention to solve the problem is possible).
3. Use Table 2.1 and the two rules given earlier to construct a multiplicative preference matrix $R^{(0)}$, $k = 0$, of the relevant contribution or impact of each element on each governing criterion at the next highest level. In this matrix, pairs of elements are compared with respect to a criterion in the superior level. If there are many people participating, the multiple judgements can be synthesized by using their geometric means.
4. Obtain the priorities, $\mathbf{w}$, by solving Eq. (2.14).
5. Calculate CI and CR, if CR > 0.1, continue the next step; otherwise go to Step 8.
6. Let $R^{(k+1)} = [r_{ij}]_{n\times n}^{(k+1)}$ be a matrix obtained by either Eqs. (2.16) or (2.17).
7. Let $k = k + 1$, and return to Step 4.
8. Perform Steps 3, 4, and 5 for all levels and clusters in the hierarchy.

9. Calculate $u_j = \sum_i w_i r_{ij}$ to obtain an overall priority vector for the lowest level of the hierarchy. If there are several outcomes, their arithmetic average may be taken.
10. Output the final choice with $a^* \equiv \arg \mathrm{Max}_j u_j$.

Let us illustrate the above procedure step-by-step with the following example.

2.3.3 An Example: Determining Consumer Preference

The AHP is basically a simple, efficient technique for problem solving. The following step-by-step example demonstrates this simplicity; it can also serve as a model for using the process to solve other problems.

A firm wants to determine consumer preferences for three different kinds of mobile phones. The attributes considered most relevant from the consumer's perspective are (1) ease of use, (2) function, (3) price, (4) weight, (5) design, and (6) safety. The three kinds of mobile phones, X, Y, and Z, possess all these attributes, but at different levels of intensity: high (H), medium (M), and low (L). Given the consumer's "bounded rationality"—that is, the fact that consumers do not act on perfect or complete information and are satisfied with less than the economically most rational choice—we can best distinguish among the attributes by dividing them into this small number of intensity categories. The resulting hierarchy is shown in Fig. 2.5. The selection proceeds in the following manner:

Step 1. Determine consumer preference among the attributes: Develop a matrix that compares attributes in pairs with respect to product desirability (Table 2.4).

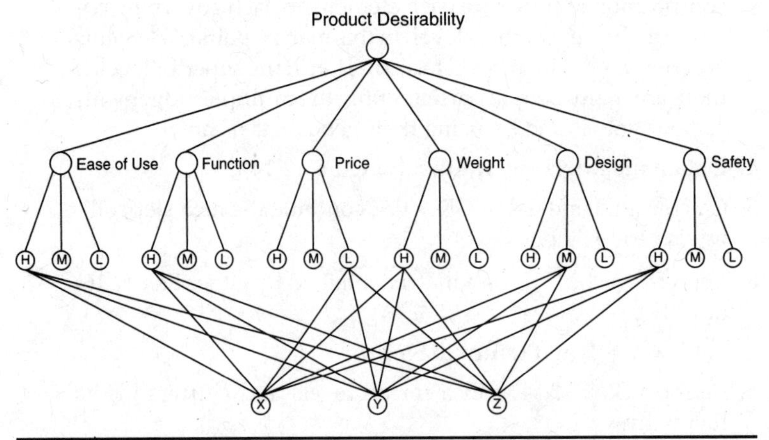

FIGURE 2.5 Hierarchical structure of the attributes.

Product Desirability	The Given Values						Priority	1st Calibration						Priority
	E	F	P	W	D	S	w	E	F	P	W	D	S	w
E	1	1/4	1/5	1/4	5	1/6	0.057	1	0.28	0.18	0.35	3.55	0.18	0.057
F	4	1	1/3	3	6	1/2	0.168	3.54	1	0.37	2.38	6.10	0.55	0.168
P	5	3	1	4	7	3	0.384	5.63	2.69	1	3.93	9.31	2.24	0.384
W	4	1/3	1/4	1	5	1/5	0.100	2.88	0.42	0.25	1	1.45	0.26	0.100
D	1/5	1/6	1/7	1/5	1	1/7	0.027	0.28	0.16	0.11	0.22	1	0.12	0.027
S	6	2	1/3	5	7	1	0.264	5.41	1.82	0.45	3.87	8.02	1	0.264
$\lambda_{max} = 6.66$; CI = 0.13, CR = 0.106								$\lambda_{max} = 6.23$; CI = 0.05, CR = 0.037						

TABLE 2.4 Matrix of Attribute Comparison

Step 2. Determine consumer preference among the intensities of the attribute: Develop six matrices that compare intensity levels in pairs with respect to each attribute (Table 2.5).

Now we want to synthesize these judgments to obtain the set of overall priorities that will indicate which product consumers prefer. The remaining steps take us through this process:

Ease of Use	Given Data			Priority	1st Calibration			Priority
	H	M	L	w	H	M	L	w
H	1	5	8	0.726	1	4.295	9.313	0.726
M	1/5	1	5	0.212	0.233	1	4.295	0.212
L	1/8	1/5	1	0.062	0.107	0.233	1	0.062
$\lambda_{max} = 3.146$; CI = 0.073; CR = 0.126					$\lambda_{max} = 3.052$; CI = 0.026; CR = 0.045			

Function	Given Data			Priority	1st Calibration			Priority
	H	M	L	w	H	M	L	w
H	1	7	9	0.761	1	5.584	11.281	0.761
M	1/7	1	7	0.191	0.179	1	5.584	0.191
L	1/9	1/7	1	0.048	0.089	0.179	1	0.048
$\lambda_{max} = 3.328$; CI = 0.164; CR = 0.283					$\lambda_{max} = 3.116$; CI = 0.058; CR = 0.100			

Price	Given Data			Priority	1st Calibration			Priority
	H	M	L	w	H	M	L	w
H	1	1/7	1/9	0.048	1	0.179	0.89	0.048
M	7	1	1/7	0.191	5.584	1	0.179	0.191
L	9	7	1	0.761	11.281	5.584	1	0.761
$\lambda_{max} = 3.328$; CI = 0.164; CR = 0.283					$\lambda_{max} = 3.116$; CI = 0.058; CR = 0.100			

Weight	Given Data			Priority
	H	M	L	w
H	1	3	5	0.627
M	1/3	1	4	0.280
L	1/5	1/4	1	0.094
$\lambda_{max} = 3.09$; CI = 0.04; CR = 0.07				

TABLE 2.5 Comparison Matrices of Intensity Level with Consistency Improvement

Design	Given Data H	M	L	Priority w
H	1	1/5	2	0.179
M	5	1	5	0.709
L	1/2	1/5	1	0.113
$\lambda_{max} = 3.05$; CI = 027; CR = 046				

Safety	Given Data H	M	L	Priority w	1st Calibration H	M	L	Priority w
H	1	7	9	0.761	1	5.584	11.281	0.761
M	1/7	1	7	0.191	0.179	1	5.584	0.191
L	1/9	1/7	1	0.048	0.089	0.179	1	0.048
$\lambda_{max} = 3.328$; CI = 0.164; CR = 0.283					$\lambda_{max} = 3.116$; CI = 0.058; CR = 0.100			

TABLE 2.5 Comparison Matrices of Intensity Level with Consistency Improvement (*Cont.*)

Step 3. Determine the first-level priority intensities: By grouping the priorities of the intensities (H, M, L) for each of the six attributes from the last columns of Table 2.5 and entering the priorities of the attributes taken from Table 2.4 above the rows of Table 2.6. Then multiply each column by the priority for the intensities to get Table 2.7 for the first level of intensities.

Step 4. Determine the vector of desired attribute intensities: By selecting from each column the element with the highest intensity attribute intensities as follows:

H Ease of Use	H Function	L Price	H Weight	M Design	H Safety	Total
0.041	0.128	0.292	0.063	0.019	0.201	0.744

	(0.057) Ease of Use	(0.168) Function	(0.384) Price	(0.100) Weight	(0.027) Design	(0.264) Safety
H	0.726	0.761	0.048	0.627	0.179	0.761
M	0.212	0.191	0.191	0.280	0.709	0.191
L	0.062	0.048	0.761	0.094	0.113	0.048

TABLE 2.6 Information Collected from Table 2.4 and Table 2.5

	Ease of Use	Function	Price	Weight	Design	Safety
H	0.041	0.128	0.018	0.063	0.005	0.201
M	0.012	0.032	0.073	0.028	0.019	0.051
L	0.004	0.008	0.292	0.009	0.003	0.013

TABLE 2.7 Vectors of Priority Intensities for the Attributes

Then add this row and divide each entry by the total to get the normalized vector of desired attribute intensities:

H Ease Use	H Function	L Price	H Weight	M Design	H Safety	Total
0.056	0.172	0.392	0.084	0.026	0.270	1.0

Step 5. Determine the perceived product standers by developing matrices that compare the three mobile phones (X, Y, and Z) in pairs with respect to the most desired attribute intensities, as shown in Table 2.8.

Step 6. Determine the second-level priority intensities by grouping the priorities of the mobile phones with respect to each

H Easy Use	Given Data			Priority	1st Calibration			Priority
	X	Y	Z	w	X	Y	Z	w
X	1	5	7	0.715	1	4.22	8.29	0.715
Y	1/5	1	5	0.219	0.24	1	4.22	0.219
Z	1/7	1/5	1	0.067	0.12	0.24	1	0.067
$\lambda_{max} = 3.18$; CI = 0.09; CR = 0.16					$\lambda_{max} = 3.07$; CI = 0.032; CR = 0.056			

H Function	Given Data			Priority
	X	Y	Z	w
X	1	2	7	0.566
Y	1/2	1	8	0.373
Z	1/7	1/8	1	0.061
$\lambda_{max} = 3.08$; CI = 0.04; CR = 0.07				

TABLE 2.8 Matrices of Comparing Three Papers Towels for Desired Attribute Intensities

L Price	Giving Data X	Y	Z	Priority w	1st Calibration X	Y	Z	Priority w
X	1	1/4	1/7	0.073	1	0.29	0.12	0.073
Y	4	1	1/5	0.205	3.48	1	0.23	0.205
Z	7	5	1	0.722	8.05	4.35	1	0.722
$\lambda_{max} = 3.12$; CI = 0.062; CR = 0.107					$\lambda_{max} = 3.04$; CI = 0.022; CR = 0.038			

H Weight	Given Data X	Y	Z	Priority w
X	1	2	1	0.413
Y	1/2	1	1	0.260
Z	1	1	1	0.328
$\lambda_{max} = 3.05$; CI = 0.027; CR = 0.046				

M Design	Given Data X	Y	Z	Priority w	1st Calibration X	Y	Z	Priority w
X	1	2	1	0.407	1	.58	.27	0.406
Y	1/2	1	3	0.370	.63	1	.36	0.369
Z	1	1/3	1	0.224	.78	.42	1	0.224
$\lambda_{max} = 3.37$; CI = 0.183; CR = 0.317					$\lambda_{max} = 3.13$; CI = 0.065; CR = 0.112			

	2nd Calibration X	Y	Z	Priority w
X	1	1.36	1.47	0.407
Y	0.73	1	2.05	0.370
Z	0.68	0.49	1	0.224
$\lambda_{max} = 3.01$; CI = 0.023; CR = 0.04				

H Safety	Given Data X	Y	Z	Priority w
X	1	4	6	0.682
Y	1/4	1	4	0.236
Z	1/6	1/4	1	0.082
$\lambda_{max} = 3.12$; CI = 0.054; CR = 0.09				

TABLE 2.8 Matrices of Comparing Three Papers Towels for Desired Attribute Intensities (*Cont.*)

	(0.056) H Ease of Use	(0.172) H Function	(0.392) L Price	(0.084) H Weight	(0.025) M Design	(0.270) H Safety
X	0.715	0.566	0.073	0.413	0.407	0.682
Y	0.219	0.373	0.205	0.260	0.370	0.236
Z	0.067	0.061	0.722	0.328	0.224	0.082

Table 2.9 Information Collected from Step 4 and Table 2.8

desired attribute intensity from Table 2.8 in columns and entering the normalized priorities from Step 4 above the columns as shown in Table 2.9. Then multiply each column by the normalized priority of the corresponding attribute intensity to obtain the weighted vectors of priority for the desired attribute intensities for each mobile phone as shown in Table 2.10.

	H Ease of Use	H Function	L Price	H Weight	M Design	H Safety	u_i
X	<u>0.040</u>	<u>0.097</u>	0.029	<u>0.035</u>	<u>0.010</u>	<u>0.194</u>	0.405
Y	0.012	0.064	0.080	0.022	<u>0.010</u>	0.064	0.252
Z	0.004	0.011	<u>0.283</u>	0.028	0.006	0.022	0.354

Table 2.10 Weighted Overall Product Attribute Perception

Step 7. Add each of the three rows to obtain the overall priorities of the three mobile phones. This synthesis yields the following priorities:

$$X = 0.395, \quad Y = 0.252, \quad Z = 0.353$$

From these results we would select product X as most desirable from the customer's perspective.

Even though low price was the desired attribute intensity with the highest priority, product X, whose priority was very low with respect to low price, was the final choice. The reason for this choice is clear: X dominated Y and Z on all other desired attribute intensities, and thus the overall utility is the largest. This leads the firm to market a superior but high-priced product.

2.4 Optimization Programming

There are many optimization models developed in the literature. For decision analysis, various forms based on multi-objective mathematical programming have been applied extensively (Wang, 2004). Among these, the multi-objective linear programming model is the most commonly adopted form. In Sec. 2.4.1, we shall briefly introduce this model with a solution method, namely *the weighting method*. An illustrated example is given in Sec. 2.4.2.

2.4.1 Multi-Objective Linear Program (MOLP)

A multiple-objective linear program (MOLP) is written as

$$\text{Max-dominance}\{Cx = z | x \in X\}$$

or

$$\begin{aligned} &\max \left\{c^1 x = z_1\right\} \\ &\max \left\{c^2 x = z_2\right\} \\ &\vdots \\ &\max \left\{c^K x = z_K\right\} \\ &s.t.\ x \in X = \{x | Ax = b, x \geq 0\} \end{aligned} \tag{2.18}$$

where:

C is the $K \times n$ criterion matrix (i.e., matrix of objective function coefficients) whose rows are the gradients c^k of the k objective functions, $k = 1, \ldots, K$;
z is the criterion vector;
A is the $m \times n$ constraint matrix;
c^k is the gradient (i.e., vector of objective function coefficients) of the kth objective function;
z_k is the criterion value of the kth objective.

"Max-dominance" indicates that the purpose is to maximize all objectives simultaneously. When $k = 1$, it indicates single-objective linear programming.

Definition 2.6: An Efficient Set A feasible point is efficient if its criterion vector is not dominated by the criterion vector of some other point in the feasible region. The set of all efficient points is the efficient set.

A frequently discussed method in MOLP is the weighting method with a point estimate weighted-sums approach. The method is as follows. Each objective $c^k x$ is multiplied by a strictly positive scalar weight w_k. Then, the K-weighted objectives are summed to form a composite (or weighted-sums) objective function. Let C be the $K \times n$ criterion matrix whose rows are the c^k, the composite objective function is $w^T Cx$. Without loss of generality, we shall assume that each weighting vector $w \in R^K$ is normalized so that its elements are summed to one.

Definition 2.7: The Weighting Method Let Λ denote the set of all such weighting vectors where

$$\Lambda = \left\{ w \in R^K \,|\, w_k > 0, \sum_{k=1}^{K} w_k = 1 \right\}$$

Then, by doing one's best to estimate a $w \in \Lambda$, it is hoped that the composite (or weighted-sums) LP presented by $\mathrm{Max}\{w^T Cx | x \in X\}$ will produce a solution that is optimal; or if not, one that is close enough to being optimal to be useful. Thus, the point-estimated weighted-sums technique can be viewed as a method that experiments with strictly convex combinations of the objectives. That is, if x_1 and x_2 are adjacent efficient extreme points, then a set of efficient points can be generated by taking

$$\theta x_1 + (1 - \theta)x_2, 0 < \theta < 1$$

The above analysis can be supported by the following facts.

Theorem 2.2 Let $x^* \in X$ maximize the weighted-sums LP as $\mathrm{Max}\{\overline{w}^T Cx | x \in X\}$ where $\overline{w} \in \Lambda$. Then, x^* is efficient.

Theorem 2.3 $x^* \in X$ is efficient if and only if there exists a $\overline{w} \in \Lambda$ such that x^* is a maximal solution of $\mathrm{Max}\{\overline{w}^T Cx | x \in X\}$.

Therefore, if one can articulate the weights of importance for different objectives from a decision maker, then from Theorems 2.2 and 2.3 one can obtain the best solution of the MOLP model via single-objective linear programming, which may use commercial software such as Lingo (LINGO 12.0) or CPLEX (ILOG, CPLEX, 9.0).

2.4.2 Illustrative Example

A production plan for resource allocation was considered with the following weighted model:

$$\begin{aligned} \mathrm{Max}\ z(x) &= w_1 f_1(x) + w_2 f_2(x) \\ &= w_1(x_1 + 3x_2) + w_2(-3x_1 - 2x_2) \\ \text{s.t. } X &= \{x | 0.5x_1 + 0.25x_2 \le 8; \\ &\quad 0.2x_1 + 0.2x_2 \le 4; \\ &\quad x_1 + 5x_2 \le 72; x_1, x_2 \ge 0\}. \end{aligned}$$

By selecting different pairs of weightings, we have the results shown in Table 2.11 and Fig. 2.6.

Weights (w_1, w_2)	Efficient Extreme Point $x^* = (x_1, x_2)$	$f_1(x^*)$	$f_2(x^*)$	$z(x^*)$
(1, 0)	(7, 13)	46	−47	46
(1, 1)	(0,72/5)	216/5	−144/5	72/5
(1, 2)	(0, 0)	0.0	0.0	0.0
(1, 3)	(0, 0)	0.0	0.0	0.0
(0, 1)	(0, 0)	0.0	0.0	0.0

TABLE 2.11 Solutions Resulting from Different Weights

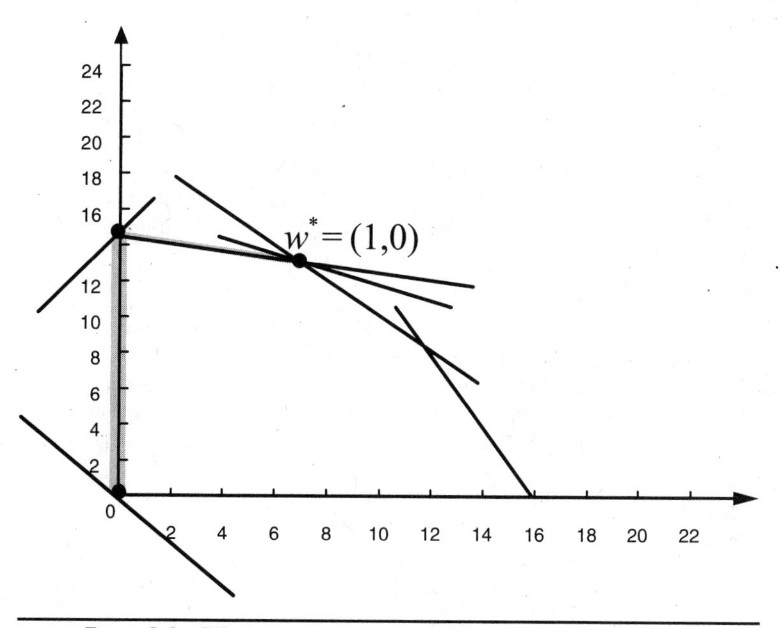

FIGURE 2.6 Efficient extreme points of the example.

Since all of the efficient extreme points are easily identified, an exact representation of the efficient set is obtained as the line segments between (0, 0) and (0, 72/5); and between (0, 72/5) and (7, 13). The decision maker can use this information to select a preferred solution.

References

Belton, V., and T. Gear. (1983). On a shortcoming of Saaty's method of analytic hierarchies. *Omega* 11, 228-230.

Ma, D., and X. Zheng. (1979). 9/9–9/1 scale method of the AHP. International Symposium on the AHP, Pittsburgh, PA, 197-202.

Miller, G. (1956). The magical number seven plus or minus two: some limits on our capacity for processing information. *Psych Rev* 3(2), 81-97.

Saaty, T. L. (1980). *The Analytic Hierarchical Process*, McGraw-Hill, New York.

Saaty, T. L. (1994). *Fundamentals of Decision Making and Priority Theory with the AHP*, RWS Publications, Pittsburgh, PA.

Steuer, R. E. (1986). *Multiple Criteria Optimization: Theory, Computation, and Application*, John Wiley, New York.

Triantaphyllou, E. (2000). *Multi-Criteria Decision Making Methods: A Comparative Study*, Kluwer, Dordrecht.

Wang, H. F. (2004), *Multi-Criteria Decision Analysis: From Certainty to Uncertainty*, Ting Lung Book, Taiwan.

Xu, Z. S. (2000). On consistency of the weighted geometric mean complex judgement matrix in AHP, *J Oper Res Soc* 126, 683-687.

Xu, Z. S., and Wei, C. P. (1999). A consistency improving method in the analytic hierarchy process, *Eur J Oper Res* 116, 443-449.

Yasutaka, K., and N. Tawara. (2006). A multiple-attribute utility theory approach to lean and green supply chain management, *Int J Prod Econ* 101, 99-108.

Zadeh, L. A. (1965). Fuzzy sets, *Information and Control* 8, 338-353.

Zedan, H., and M. Skitmore. (1998). Contractor selection using multicriteria utility theory: an additive model, *Building and Environment* 33, 105-115.

Software

ILOG, CPLEX 9.0, http://briian.com/?p=7356&utm_source=feedburner&utm_medium=feed&utm_campaign=Feed:+briian+(briian.com)

LINGO 12.0, Optimization Modeling Software for Linear, Nonlinear, and Integer Programming, http://www.lindo.com/index.php?option=com_content&view=article&id=2&Itemid=10

PART 2

Green Engineering Technology

CHAPTER 3
Green Engineering

CHAPTER 4
Green Materials

CHAPTER 5
Environmental Design

CHAPTER 3

Green Engineering

This chapter gives an overview of the basic elements of green engineering.

3.1 Introduction

Green engineering is the design and recovery of products that conserve natural resources and minimize their impact on the environment. Since the days of the Industrial Revolution in the late eighteenth and nineteenth centuries, green engineering has not been a general practice of the world's businesses, and has only become more common recently as new products have been designed and manufactured. This lack of concern for conserving natural resources was partly due to ignorance and partly due to the belief that this planet had an abundance of resources that would never run out.

While this controversial question is still debated amongst a small minority, about 20 years ago, many people started noticing a flaw in the assumption of limitless resources. As a result, many governments started to realize that green engineering—the elements of which include *environmentally conscious manufacturing* (ECM) and *product recovery* (ECMPRO)—must become an obligation if our planet is to continue enjoying a clean and sustainable environment (Gungor and Gupta, 1999; Ilgin and Gupta, 2010).

This way of thinking was fueled by the changing perspective of customers on environmental issues. Environmentally conscious manufacturing is concerned with developing methods for manufacturing new products from conceptual design to final delivery, and ultimately to *end-of-life* (EOL) disposal, such that environmental standards and requirements are satisfied. *Product recovery*, on the other hand, aims to minimize the amount of waste sent to landfills by recovering materials and parts from old or outdated products by means of recycling and remanufacturing (including reuse of parts and products). Figure 3.1 depicts the interactions among the activities that take place in a product life cycle (Gungor and Gupta, 1999).

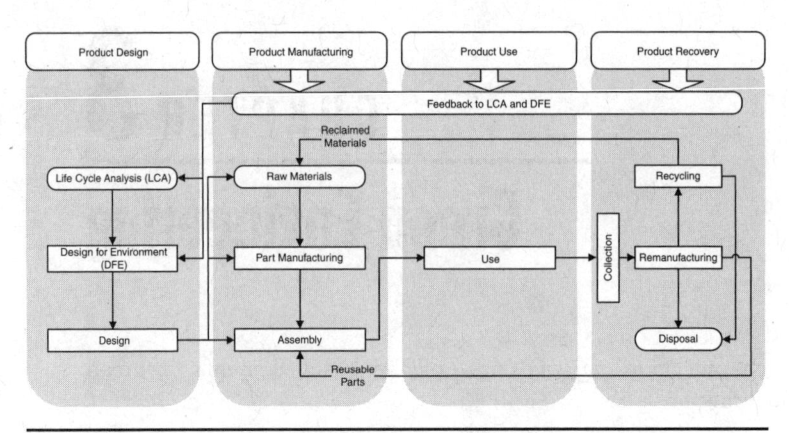

Figure 3.1 Interactions among the activities in a product life cycle.

3.2 Green Design

The increasing importance of green engineering requires product designers to consider environmental criteria in the design process. In order to help product designers make environmentally friendly design choices, a number of methodologies have been developed. This section presents an overview of some of these methodologies. For a thorough literature review, see Ilgin and Gupta (2010).

3.2.1 Design for X

Design for X (DFX) involves different design specialties such as design for manufacture, design for quality, design for environment, design for disassembly, and design for recycling. Veerakamolmal and Gupta (2000) and Kuo et al. (2001) give good overviews of the concepts, applications, and perspectives of various DFX methods. For detailed information on DFX, we refer the reader to the book by Huang (1996). Since this overview emphasizes green engineering, we only consider its related DFX tools here, viz., design for environment, design for disassembly and design for recycling.

Design for Environment

Design for environment (DFE) emphasizes designing a product in a way that the potential environmental impact throughout its life cycle is minimized (Giudice et al., 2006). Some researchers develop quality function deployment-based DFE methodologies for the simultaneous consideration of environmental criteria and customer requirements (e.g., see Cristofari et al., 1996; Zhang et al., 1999; Mehta and Wang, 2001; Bovea and Wang, 2007). The most commonly used technique in DFE methodologies to assess the environmental impact is life cycle analysis (e.g., see Veerakamolmal and Gupta, 2000; Bevilacqua et al., 2007).

Design for Disassembly

Design for disassembly (DFD) can be defined as the consideration of the ease of disassembly in the design process (Veerakamolmal and Gupta, 2000). Kroll and Hanft (1998) present a method for the evaluation of the ease of disassembly by using a spreadsheet-like chart and a catalog of task difficulty scores. The scores are determined based on the work-measurement analyses of standard disassembly tasks. Veerakamolmal and Gupta (1999) introduce the *design for disassembly index* (DfDI) to measure design efficiency. Kwak et al. (2009) develop a novel concept, called *eco-architecture analysis,* in which a product is represented as an assembly of EOL modules.

Design for Recycling

Design for recycling (DFR) focuses on the design attributes that support the cost-effective recycling and disaggregation of the materials embodied in the product (Masanet and Horvath, 2007). Boon et al. (2002) explore the electronic goods recycling infrastructure to identify the conditions required for profitable recycling of PCs. Ardente et al. (2003) present a computer-based tool called ENDLESS to calculate a *global recycling index* based on energy, environmental, technical, and economic indicators. Masanet and Horvath (2007) develop an analytical framework to quantify the economic and environmental benefits of DFR practices for plastic computer enclosures during the design process.

3.2.2 Life Cycle Analysis

Life cycle analysis (LCA) is a method used to evaluate the environmental impact of a product throughout its life cycle, encompassing extraction and processing of the raw materials, manufacturing, distribution, use, recycling, and final disposal. Most researchers use LCA within a DFE methodology as a tool to measure the environmental impact of a product's design (Zhang et al., 1999; Mehta and Wang, 2001; Bovea and Wang, 2003; Bovea and Wang, 2007; Bevilacqua et al., 2007; Boks and Stevels, 2007; Sakao, 2007; Grote et al., 2007).

3.2.3 Material Selection

Selection of materials for a particular application is affected mainly by factors such as weight, process ability, and cost. However, in recent years, environmental factors have also been considered in materials selection at an increasing rate. In order to deal with the environmental criteria in the design process, a number of tools and methodologies have been presented by researchers. For example, Holloway (1998) extends a conventional material selection technique—material selection charts—by integrating environmental concerns into it.

Giudice et al. (2005) develop a systematic method that minimizes the environmental impact of the selected materials while satisfying the functional and performance requirements. Recyclability of selected materials is an important factor considered by various researchers. Sodhi et al. (1999) and Knight and Sodhi (2000) develop mathematical models that can be used for the recycling-based evaluation of material content of product designs in the early design phase. Some studies investigate the recycling infrastructure to help designers select the materials suitable for recycling. Isaacs and Gupta (1997) use *goal programming* (GP) to investigate the effect of lighter materials on the profitability of the recycling infrastructure. In a follow-up study, Boon et al. (2003) expand Isaacs and Gupta's (1997) mathematical formulation for the recycling infrastructure to assess the materials streams and process profitabilities of several different clean vehicles with different material content. Boon et al. (2001) present a similar analysis for aluminum-intensive vehicles. Kumar and Sutherland (2008) provide a good overview of studies on automotive recovery infrastructure and identify future challenges.

3.3 Some Green Design Guidelines

In the following subsections, we present some guidelines for green design to facilitate disassembly, reuse, and recycling (Brennan et al., 1994; Veerakamolmal and Gupta, 2000).

3.3.1 Modular Product Structure

Consolidating components into modules can greatly simplify the disassembly tasks. This allows easy identification of the various components, which helps reduce unnecessary disassembly steps. Additionally, modular structure can help reduce the costs of assembly, maintenance, and end-of-life recovery. Ishii et al. (1994), for example, studied the design for product retirement and material life cycle, and successfully applied a technique called *clumping* to the design of a coffee maker.

3.3.2 Design of Functional Units

Design of electronic components should incorporate a companywide universal data interface standard [e.g., the *universal serial bus* (USB) standard used in the computer industry] and consolidate functional units in order to enable the simplicity in reuse.

3.3.3 Material Selection

The number of material varieties in a product influences the reuse and recycling efficiency. A product built with fewer components and materials greatly reduces the required logistics of sorting and processing. In addition, for ease of identification, labeling of parts by type of materials helps in the separation process, and thus preserves the materials' purity. Plastic parts should be marked according to U.S. recycling standards.

3.3.4 Minimize Waste and Harmful Contaminating Materials

Designers should refrain from using toxic substances whenever possible. Fragile or hazardous materials should be easily identified to prevent any harm to disassembly operators. Alternative materials that are environmentally benign and easy to recycle should be considered. Warning labels attached to the product, or to the component, can be very helpful in identifying harmful substances during disassembly and recycling. Similarly, materials that are corrosive and may contaminate other materials during the recycling process should be documented.

3.3.5 Ease of Separation

Products designed for ease of disassembly, reuse, and recycling should require minimal effort to take them apart. Important factors that may influence the ease of separation are the types and variety of fixtures used, the need of specific separation tools, and the orientation of the components' layout. Various types of fasteners (e.g., screws, glues, and welds) may undermine disassembly efforts. Fasteners made with a clip, snap-fit, or Velcro are easier to separate and require less energy to take apart. Fasteners that require special tools to remove can increase the setup time or impede the recovery process all together.

3.3.6 Steps for Designing Green Products

The above guidelines can help identify the basic requirements for green product design. In practice, however, each company may be motivated to individually create the resource for migrating to the new paradigm. There are always product-specific characteristics that each company must take into account. For example, while snap-fitted components used in the design of a refrigerator can be easily taken apart with a special separation tool, some computer companies have found it cumbersome to take apart snap-fitted components in their computer work stations. Furthermore, the migration process can be complex, and may require a different amount of developmental time for different product platforms. However, the experience gained from a particular disassembly operation may be helpful in the design of disassembly shops for other products. The steps required for the design of green products are given below (Steinhilper, 1994):

Step 1. *Thoroughly analyze the product platform and required end-of-life tasks.*
- Analyze and structure product platform.
- Analyze disassembly, reuse, and recycling tasks.
- Analyze material and information flow.
- Establish action plans.
- Point out potential improvements.

Step 2. *Determine the degree of disassembly, reusability, and recyclability of products and components.*
- Develop disassembly, reuse, and recycling criteria.
- Evaluate the materials for disassembly, reuse, and recycling potential.
- Evaluate various designs for disassembly, reuse, and recycling.
- Document the possible ways to reuse and recycle subassemblies and materials.
- Document potential improvements.
- Carry out the required improvements.

Step 3. *Plan return logistics and information flows between production and end-of-life processing.*
- Organize the return of products from consumers by selecting between the existing and alternative distribution/collection channels.
- Develop the technologies and equipment for the return logistics.
- Develop an information system for the return flows to provide feedback to designers.
- Generate a plan for internal and external material and information flows.
- Transfer knowledge from CIM (*computer-integrated manufacturing*) to CIR (computer-integrated recycling).

Step 4. *Develop and design reliable and visionary solutions for disassembly, reuse, and recycling.*
- Plan technical and time capabilities.
- Determine suitable disassembly, reuse, and recycling processes.

- Determine suitable disposal processes.
- Perform make or buy decisions for the recycling and disposal processes.
- Evaluate economic benefits.
- Establish future goals for technical, economical, and ecological recycling, and for a disposal-friendly environment.
- Integrate production, usage, and end-of-life processing.

Step 5. *Prepare comprehensive guidelines for future green product design.*

- Prepare product structure guidelines.
- Prepare material selection guidelines.
- Prepare fixture and joining technique guidelines.
- Recommend an appropriate configuration for the product platform.
- Prepare green manufacturing process guidelines.
- Prepare economic evaluation guidelines
- Establish clear guidelines for an environmentally friendly design.
- Document important design measures.
- Establish the curriculum to train product designers the techniques of green design and manufacturing.

3.4 Product Recovery at the End-of-Life

Determination of the best option for EOL products is an important problem faced by manufacturers. There are five commonly used options for a product at its EOL: (1) direct reuse, (2) repair, (3) remanufacturing, (4) recycling, and (5) disposal. *Direct reuse* involves the reuse of the whole product as is for its original task. In the *repair* option, damaged parts are changed in order to have a fully functional product. *Remanufacturing* consists of refurbishment of used products up to a quality level similar to a new product. The aim of *recycling* is to recover materials from the returned products. *Disposal* involves the landfill or incineration of the used products (Ilgin and Gupta, 2010).

Development of a decision model to select between these options requires the consideration of various qualitative and quantitative factors such as environmental impact, quality, legislative factors, cost, and so on. Mathematical programming models have been extensively used by researchers to develop EOL option-selection methodologies. Krikke et al. (1998) use stochastic *dynamic programming* (DP) to determine a product recovery and disposal strategy for one product type based on the maximization of net profit considering relevant technical, ecological, and commercial feasibility criteria at the product level. In follow-up papers, Krikke et al. (1999a,b) apply the methodology proposed in Krikke et al. (1998) to real-life cases on the recycling of copiers and monitors, respectively. Teunter (2006) extends Krikke et al. (1998) by considering multiple disassembly processes and partial disassembly. Lee et al. (2001b) determins the EOL option of each part by defining the objective function as the weighted sum of economic value and environmental impact. Das and Yedlarajiah (2002) propose a mixed-integer program to determine the optimal partial disposal strategy based on the maximization of the net profit. Jorjani et al. (2004) develop a piecewise linear concave program to determine the optimal allocation of disassembled parts to five disposal options (refurbish, resell, reuse, recycle, landfill) based on the maximization of the overall return. Ritchey et al. (2005) develop a mathematical model to evaluate the economic viability of remanufacturing option under a government-mandated take-back program. Willems et al. (2006) use Linear Programming to investigate the effect of reductions in the expected disassembly time and cost on the optimal EOL strategy. Tan and Kumar (2008) use an LP model to evaluate three EOL options for each part, namely, repair, repackage, or scrap.

In some studies, *multicriteria decision-making* (MCDM) methodologies are presented. Hula et al. (2003) present a multi-objective GA (*genetic algorithm*) to consider the trade-offs between environmental and economic variables in the selection of EOL alternatives. Bufardi et al. (2004) obtain a partial ranking of EOL options using the ELECTRE III MCDM methodology. Chan (2008) extends Bufardi et al. (2004) by developing an MCDM methodology based on *gray relational analysis* (GRA), which considers complete ranking of EOL options in an uncertain environment. Jun et al. (2007) develop a multi-objective evolutionary algorithm to select the best EOL options for maximizing the recovery value of an EOL product including recovery cost and quality. Fernandez et al. (2008) develop a fuzzy approach to evaluate five recovery options and one disposal option by considering four criteria: (1) product value, (2) recovery value, (3) useful life, and (4) level of sophistication. Wadhwa et al. (2009) propose a FL (*fuzzy logic*)-based MCDM methodology to consider the knowledge of experts (e.g., evaluators, or sortation specialists) in the selection of most appropriate alternative(s) for product reprocessing with respect to existing criteria. Iakovou et al. (2009) develop an MCDM methodology called the *multicriteria matrix*, which considers the residual value, environmental burden, weight, quantity, and ease of disassembly of each component in the evaluation of EOL alternatives for a product. Xanthopoulos and Iakovou (2009) use GP to determine the most attractive subassemblies and components to be disassembled for recovery from a set of different types of EOL products.

Gonzalez and Adenso-Diaz (2005) propose a BOM (*bill of material*)-based method for the joint determination of depth of disassembly, and an EOL option for disassembled parts based on the maximization of profits. Kleber (2006) tries to develop dynamic policies for three different new product development projects—(1) design for single use, (2) design for reuse, and (3) design for reuse with stock keeping—which differ with respect to different EOL recovery strategies. Shih et al. (2006) develop a CBR (*case-based reasoning*)-based methodology to determine the product EOL strategy. Staikos and Rahimifard (2007a,b) integrate AHP (*analytical hierarchy process*), LCA, and cost–benefit analysis to determine the most appropriate reuse, recovery, and recycling option for postconsumer shoes. Rahimifard et al. (2004) and Bakar and Rahimifard (2007) develop computer-aided decision support systems to support the EOL option selection process.

Some studies simultaneously investigate the EOL option selection problem and product design. Rose and Ishii (1999) and Gehin et al. (2008) develop tools to help product designers in the identification of appropriate EOL strategies in the early design phase. Mangun and Thurston (2002) develop a mathematical model in order to incorporate planning for component reuse, remanufacture, and recycle into product portfolio design. Kumar et al. (2007) present a value flow model considering different product life cycle stages to support the decision maker in selecting the best EOL option for a product. Zuidwijk and Krikke (2008) try to improve product recovery strategy by concerning the innovations in product design and recovery technologies.

3.5 The 12 Principles of Green Engineering

The following 12 principles of green engineering proposed by Anastas and Zimmerman (2003) incorporate life cycle thinking into engineering design. These excellent principles take into consideration all the stages in a product's life from resource reclamation to the product's disposal, so as to minimize its impact on the environment.

3.5.1 Inherent Rather Than Circumstantial

Designers should strive to ensure that all materials and energy inputs and outputs are as inherently nonhazardous as possible.

3.5.2 Prevention Instead of Treatment

It is better to prevent waste than to treat or clean up waste after it is formed.

3.5.3 Design for Separation

Separation and purification operations should be designed to minimize energy consumption and use of materials.

3.5.4 Maximize Efficiency

Products, processes, and systems should be designed to maximize mass, energy, space, and time efficiency.

3.5.5 Output-Pulled Versus Input-Pushed

Products, processes, and systems should be "output pulled" rather than "input pushed" through the use of energy and materials.

3.5.6 Conserve Complexity

Embedded entropy and complexity must be viewed as an investment when making design choices about recycle, reuse, or beneficial disposition.

3.5.7 Durability Rather Than Immortality

Targeted durability, not immortality, should be a design goal.

3.5.8 Meet Need, Minimize Excess

Design for unnecessary capacity or capability (e.g., "one size fits all") solutions should be considered a design flaw.

3.5.9 Minimize Material Diversity

Material diversity in multicomponent products should be minimized to promote disassembly and value retention.

3.5.10 Integrate Material and Energy Flows

Design of products, processes, and systems must include integration and interconnectivity with available energy and materials flows.

3.5.11 Design for Commercial "Afterlife"

Products, processes, and systems should be designed for performance in a commercial "afterlife."

3.5.12 Renewable Rather Than Depleting

Material and energy inputs should be renewable rather than depleting.

References

Anastas, P.T., Zimmerman, J.B., 2003. Design through the twelve principles of green engineering, *Environmental Science and Technology* 37, 95-101.

Ardente, F., Beccali, G., Cellura, M., 2003. Eco-sustainable energy and environmental strategies in design for recycling: the software "ENDLESS. *Ecological Modelling* 163, 101-118.

Bakar, M.S.A., Rahimifard, S., 2007. Computer-aided recycling process planning for end-of-life electrical and electronic equipment. *Proceedings of the Institution of Mechanical Engineers, Part B: Engineering Manufacture* 221, 1369-1374.

Bevilacqua, M., Ciarapica, F.E., Giacchetta, G., 2007. Development of a sustainable product lifecycle in manufacturing firms: A case study. *International Journal of Production Research* 45, 4073-4098.

Boks, C., Stevels, A., 2007. Essential perspectives for design for environment. Experiences from the electronics industry. *International Journal of Production Research* 45, 4021-4039.

Boon, J.E., Isaacs, J.A., Gupta, S.M., 2001. Economic impact of aluminum-intensive vehicles on the U.S. automotive recycling infrastructure. *Journal of Industrial Ecology* 4, 117-134.

Boon, J.E., Isaacs, J.A., Gupta, S.M., 2002. Economic sensitivity for end of life planning and processing of personal computers. *Journal of Electronics Manufacturing* 11, 81-93.

Boon, J.E., Isaacs, J.A., Gupta, S.M., 2003. End-of-life infrastructure economics for "clean vehicles" in the United States. *Journal of Industrial Ecology* 7, 25-45.

Bovea, M.D., Wang, B., 2003. Identifying environmental improvement options by combining life cycle assessment and fuzzy set theory. *International Journal of Production Research* 41, 593-609.

Bovea, M.D., Wang, B., 2007. Redesign methodology for developing environmentally conscious products. *International Journal of Production Research* 45, 4057-4072.

Brennan, L., Gupta, S. M., Taleb, K. N., 1994. Operations planning issues in an assembly/disassembly environment, *International Journal of Operations and Production Management* 14 (9), 57-67.

Bufardi, A., Gheorghe, R., Kiritsis, D., Xirouchakis, P., 2004. Multicriteria decision-aid approach for product end-of-life alternative selection. *International Journal of Production Research* 42, 3139-3157.

Chan, J.W.K., 2008. Product end-of-life options selection: Grey relational analysis approach. *International Journal of Production Research* 46, 2889-2912.

Cristofari, M., Deshmukh, A., Wang, B., 1996. Green quality function deployment. *In: Proceedings of the 4th International Conference on Environmentally Conscious Design and Manufacturing*. Cleveland, OH, July 23-25, 297-304.

Das, S., Yedlarajiah, D., 2002. An integer programming model for prescribing material recovery strategies. *In: Proceedings of the IEEE International Symposium on Electronics and the Environment*. San Francisco, CA, 118-122.

Fernandez, I., Puente, J., Garcia, N., Gomez, A., 2008. A decision-making support system on a products recovery management framework: A fuzzy approach. *Concurrent Engineering* 16, 129-138.

Gehin, A., Zwolinski, P., Brissaud, D., 2008. A tool to implement sustainable end-of-life strategies in the product development phase. *Journal of Cleaner Production* 16, 566-576.

Giudice, F., La Rosa, G., Risitano, A., 2005. Materials selection in the life-cycle design process: A method to integrate mechanical and environmental performances in optimal choice. *Materials & Design* 26, 9-20.

Giudice, F., La Rosa, G., Risitano, A., 2006. *Product Design for the Environment: A Life Cycle Approach*. Boca Raton, FL: CRC Press.

Gonzalez, B., Adenso-Diaz, B., 2005. A bill of materials–based approach for end-of-life decision making in design for the environment. *International Journal of Production Research* 43, 2071-2099.

Grote, C.A., Jones, R.M., Blount, G.N., Goodyer, J., Shayler, M., 2007. An approach to the EuP Directive and the application of the economic eco-design for complex products. *International Journal of Production Research* 45, 4099-4117.

Gungor, A., Gupta, S.M., 1999. Issues in environmentally conscious manufacturing and product recovery: A survey. *Computers & Industrial Engineering* 36, 811-853.

Holloway, L., 1998. Materials selection for optimal environmental impact in mechanical design. *Materials & Design* 19, 133-143.

Huang, G.Q., 1996. *Design for X: Concurrent Engineering Imperatives*. London: Chapman & Hall.

Hula, A., Jalali, K., Hamza, K., Skerlos, S.J., Saitou, K., 2003. Multi-criteria decision-making for optimization of product disassembly under multiple situations. *Environmental Science & Technology* 37, 5303-5313.

Iakovou, E., Moussiopoulos, N., Xanthopoulos, A., Achillas, C., Michailidis, N., Chatzipanagioti, M., Koroneos, C., Bouzakis, K.D., Kikis, V., 2009. A methodological framework for end-of-life management of electronic products. *Resources, Conservation and Recycling* 53, 329-339.

Ilgin, M. A. and Gupta, S. M., 2010. Environmentally conscious manufacturing and product recovery (ECMPRO): A review of the state of the art, *Journal of Environmental Management* 91(3), 563-591.

Isaacs, J.A., Gupta, S.M., 1997. Economic consequences of increasing polymer content for the U.S. automobile recycling infrastructure. *Journal of Industrial Ecology* 1, 19-33.

Ishii, K., Eubanks, C. F., Marco, P.D., 1994. Design for product retirement and material life-cycle, *Materials & Design* 15(4), 225-233.

Jorjani, S., Leu, J., Scott, C., 2004. Model for the allocation of electronics components to reuse options. *International Journal of Production Research* 42, 1131-1145.

Jun, H.B., Cusin, M., Kiritsis, D., Xirouchakis, P., 2007. A multi-objective evolutionary algorithm for EOL product recovery optimization: Turbocharger case study. *International Journal of Production Research* 45, 4573-4594.

Kleber, R., 2006. The integral decision on production/remanufacturing technology and investment time in product recovery. *OR Spectrum* 28, 21-51.

Knight, W.A., Sodhi, M., 2000. Design for bulk recycling: Analysis of materials separation. *CIRP Annals—Manufacturing Technology* 49, 83-86.

Krikke, H.R., van Harten, A., Schuur, P.C., 1998. On a medium-term product recovery and disposal strategy for durable assembly products. *International Journal of Production Research* 36, 111-140.

Krikke, H.R., van Harten, A., Schuur, P.C., 1999a. Business case Roteb: Recovery strategies for monitors. *Computers & Industrial Engineering* 36, 739-757.

Krikke, H.R., van Harten, A., Schuur, P.C., 1999b. Business case Océ: Reverse logistic network redesign for copiers. *OR Spectrum* 21, 381-409.

Kroll, E., Hanft, T.A., 1998. Quantitative evaluation of product disassembly for recycling. *Research in Engineering Design* 10, 1-14.

Kumar, V., Shirodkar, P.S., Camelio, J.A., Sutherland, J.W., 2007. Value flow characterization during product lifecycle to assist in recovery decisions. *International Journal of Production Research* 45, 4555-4572.

Kumar, V., Sutherland, J.W., 2008. Sustainability of the automotive recycling infrastructure: Review of current research and identification of future challenges. *International Journal of Sustainable Manufacturing* 1, 145-167.

Kuo, T.C., Huang, S.H., Zhang, H.C., 2001. Design for manufacture and design for "X": Concepts, applications, and perspectives. *Computers & Industrial Engineering* 41, 241-260.

Kwak, M.J., Hong, Y.S., Cho, N.W., 2009. Eco-architecture analysis for end-of-life decision making. *International Journal of Production Research* 47, 6233-6259.

Lee, S.G., Lye, S.W., Khoo, M.K., 2001b. A multi-objective methodology for evaluating product end-of-life options and disassembly. *The International Journal of Advanced Manufacturing Technology* 18, 148-156.

Mangun, D., Thurston, D.L., 2002. Incorporating component reuse, remanufacture, and recycle into product portfolio design. *IEEE Transactions on Engineering Management* 49, 479-490.

Masanet, E., Horvath, A., 2007. Assessing the benefits of design for recycling for plastics in electronics: A case study of computer enclosures. *Materials & Design* 28, 1801-1811.

Mehta, C., Wang, B., 2001. Green quality function deployment III: A methodology for developing environmentally conscious products. *Journal of Design and Manufacturing Automation* 1, 1-16.

Rahimifard, A., Newman, S.T., Rahimifard, S., 2004. A web-based information system to support end-of-life product recovery. *Proceedings of the Institution of Mechanical Engineers, Part B: Engineering Manufacture* 218, 1047-1057.

Ritchey, J.R., Mahmoodi, F., Frascatore, M.R., Zander, A.K., 2005. A framework to assess the economic viability of remanufacturing. *International Journal of Industrial Engineering* 12, 89-100.

Rose, C.M., Ishii, K., 1999. Product end-of-life strategy categorization design tool. *Journal of Electronics Manufacturing* 9, 41-51.

Sakao, T., 2007. A QFD-centred design methodology for environmentally conscious product design. *International Journal of Production Research* 45, 4143-4162.

Shih, L.-H., Chang, Y.-S., Lin, Y.-T., 2006. Intelligent evaluation approach for electronic product recycling via case-based reasoning. *Advanced Engineering Informatics* 20, 137-145.

Sodhi, M.S., Young, J., Knight, W.A., 1999. Modelling material separation processes in bulk recycling. *International Journal of Production Research* 37, 2239-2252.

Staikos, T., Rahimifard, S., 2007a. A decision-making model for waste management in the footwear industry. *International Journal of Production Research* 45, 4403-4422.

Staikos, T., Rahimifard, S., 2007b. An end-of-life decision support tool for product recovery considerations in the footwear industry. *International Journal of Computer Integrated Manufacturing* 20, 602-615.
Steinhilper, R., 1994. Design for Recycling and Remanufacturing of Mechatronic and Electronic Products: Challenges, Solutions and Practical Examples from the European Viewpoint, *Proceedings of the 1994 ASME National Design Engineering Conference* DE-Vol. 67, 65-67.
Tan, A., Kumar, A., 2008. A decision-making model to maximise the value of reverse logistics in the computer industry. *International Journal of Logistics Systems and Management* 4, 297-312.
Teunter, R.H., 2006. Determining optimal disassembly and recovery strategies. *Omega* 34, 533-537.
Veerakamolmal, P., Gupta, S.M., 1999. Analysis of design efficiency for the disassembly of modular electronic products. *Journal of Electronics Manufacturing* 9, 79-95.
Veerakamolmal, P., Gupta, S.M., 2000. Design for disassembly, reuse and recycling. In: Goldberg, L. (Ed.), *Green Electronics/Green Bottom Line: Environmentally Responsible Engineering*. Butterworth-Heinemann, pp. 69-82.
Wadhwa, S., Madaan, J., Chan, F.T.S., 2009. Flexible decision modeling of reverse logistics system: A value-adding MCDM approach for alternative selection. *Robotics and Computer-Integrated Manufacturing* 25, 460-469.
Willems, B., Dewulf, W., Duflou, J.R., 2006. Can large-scale disassembly be profitable? A linear programming approach to quantifying the turning point to make disassembly economically viable. *International Journal of Production Research* 44, 1125-1146.
Xanthopoulos, A., Iakovou, E., 2009. On the optimal design of the disassembly and recovery processes. *Waste Management* 29, 1702-1711.
Zhang, Y., Wang, H.P., Zhang, C., 1999. Green QFD-II: A life cycle approach for environmentally conscious manufacturing by integrating LCA and LCC into QFD matrices. *International Journal of Production Research* 37, 1075-1091.
Zuidwijk, R., Krikke, H., 2008. Strategic response to EEE returns: Product eco-design or new recovery processes? *European Journal of Operational Research* 191, 1206-1222.

CHAPTER 4
Green Materials

This chapter gives an overview of the basic materials in green engineering, and their roles.

4.1 Introduction

Green engineering stipulates that the products designed and produced use a minimum amount of resources, and that the process used to produce them have a minimum impact on the environment. In other words, the materials and energy consumption must be minimized. This includes minimizing the materials content and types, minimizing materials and energy consumption during usage, minimizing scraps during production, minimizing disposal at the products' ends-of-life, minimizing packaging materials, minimizing energy consumptions during product development and production stages, and so on. To address some of these issues, two important legislations, popularly known as WEEE and RoHS, were recently announced and adopted by the European Union and have since been given serious consideration by the rest of the world.

4.2 WEEE and RoHS Directives

The WEEE (Waste Electrical and Electronic Equipment) directive assigns the responsibility for the disposal of waste electrical and electronic equipment to the manufacturers of that equipment, who are required to establish an infrastructure for collecting WEEE in such a way that users can return WEEE free of charge. Also, the manufacturers are compelled to use the collected waste in an environmentally friendly manner, either by disposal or by reuse/refurbishment.

The RoHS (Restriction of Hazardous Substances) directive restricts the use of six hazardous materials in the manufacture of various types of electronic and electrical equipment. RoHS originated because of the use of lead in manufacturing and is often referred to as the lead-free directive. However, it also restricts the use of mercury, cadmium, hexavalent chromium (a carcinogen), polybrominated biphenyls, and polybrominated diphenyl (the latter two being flame retardants that may cause a variety of health issues). RoHS is closely linked with WEEE and is part of a legislative initiative to solve the problem of huge amounts of toxic e-waste.

4.3 Selection of Materials

While the selection of materials needed to produce products that satisfy the multitude of properties both desired by consumers and demanded by market competition is complicated enough, extending the requirements to incorporate environmental needs further complicates the decision process. There is a large variety of raw materials to choose from, every one of which has some advantages and some disadvantages and none of which can satisfy all the requirements perfectly. Nevertheless, it is important that designers select materials and processes that also minimize the impact on the environment.

Other considerations include reducing the amount of raw materials used, minimizing the number of components in the product, minimizing the energy consumption, increasing the useful life cycle, maximizing the use of renewable and recyclable materials, and minimizing the environmental footprint over the entire life cycle of the product.

The basic raw material types include metals, ceramics, synthetic polymers, elastomers, natural organic materials, and composites (Simoneau, 2008). In the following subsections, we will list some of the advantages, disadvantages, and sustainability properties of each.

4.3.1 Metals

Metals, which include such materials as steel, aluminum, and titanium, have an advantage in that they are strong, durable, inexpensive, easily molded/shaped/formed/machined, and have been thoroughly studied, with a vast amount of information readily available. However, the main disadvantages of metals are that they corrode easily and have a low strength-to-weight ratio. Still, metals are environmentally friendly in that they have a mature reclamation infrastructure (Isaacs and Gupta, 1997), they are easy to recycle, and they can be easily separated from one another.

4.3.2 Ceramics

Ceramics, which include such materials as porcelain, mineral glass, and metallic oxides, are nontoxic, hard, durable, noncorrosive, can tolerate high temperatures, and have been thoroughly studied with a vast amount of knowledge that is readily available. The disadvantages are that they tend to be brittle and difficult to machine. Environmentally, they tend to have low toxicity but have a limited reclamation infrastructure, and the separation technology is not mature.

4.3.3 Polymer Thermoplastics

Polymer thermoplastics, which include such materials as acrylic and polypropylene, are inexpensive, light, tough, easily molded/shaped/formed/machined, corrosion-resistant, and have been studied with a good understanding of design and performance issues. The disadvantages are that they have a low strength-to-weight ratio and a limited temperature range that is useful. The environmental advantages include their easy remelting property and maturing separation technology and reclamation infrastructure.

4.3.4 Polymer Thermosets

Polymer thermosets, which includes such materials as epoxy and polyurethane, are inexpensive, light, very tough, easily molded/shaped/formed/machined, corrosion- and high-temperature–resistant, and have been studied with a good understanding of design and performance issues. The disadvantages are that they have a low strength-to-weight ratio and they are toxic when burned. Environmentally, they are difficult to recycle, pulverize, or burn; additionally, they have a poor reclamation infrastructure and the separation technology is not yet mature.

4.3.5 Elastomers

Elastomers, which include isoprene, neoprene, and styrene butadiene rubber, are inexpensive, light, extremely tough, impact resistant, easily molded/shaped/formed, corrosion-resistant, and have been studied with a good understanding of design and performance issues. The disadvantages are that they have a low strength-to-weight ratio and a limited temperature range that is useful. Environmentally, they are difficult to recycle and have a limited reclamation infrastructure, and the separation technology is not yet mature.

4.3.6 Natural Organic Materials

Natural organic materials, which include such materials as wood, bamboo and cotton, are inexpensive, light, durable and strong, renewable, can be machined and woven, and have been studied with a mature understanding of design and performance issues. The disadvantage is that they have a low strength-to-weight ratio. Environmentally, they easily decompose and can be burned.

4.3.7 Composites

Composites, which include such materials as graphite epoxy, polyester fiberglass, have an extremely high strength-to-weight ratio, are highly corrosion-resistant, and have been studied with a good understanding of design and performance issues. The disadvantages include that it is difficult to predict their final properties during processing and it is expensive to produce them. Environmentally, they have no reclamation infrastructure and the separation technology is not yet mature.

4.4 Conclusions

As is clear from the discussion above that there is a large variety of materials to choose from, and new materials are constantly being innovated. Materials selection is a compromise of competing attributes. Yet, the environmental attribute is becoming more and more important if we are to preserve our planet.

References

Isaacs, J.A., Gupta, S.M., 1997. Economic consequences of increasing polymer content for the U.S. automobile recycling infrastructure. *Journal of Industrial Ecology* 1, 19–33.

Simoneau, R.W., 2008. *Material Selection for Sustainable Products*. PowerPoint presentation, Keene State College, Keene, NH.

CHAPTER 5
Environmental Design

5.1 Introduction

The basic conditions for achieving green engineering status suggest that governments, corporations, and societies must pledge to rethink, reduce, reuse, recycle, and redesign products in order to make a difference in reducing waste and pollution, and in sustaining quality of life. Some of the basic design concepts and guidelines for green engineering were discussed in Chap. 3. In this chapter, we discuss two novel techniques to analyze and improve the design of products for making them environmentally friendly.

The first technique is used to analyze the design efficiency of a product to study the effect on the environment of its EOL (*end-of-life*) disassembly and disposal. The design efficiency is measured using a *design for disassembly index* (DfDI). DfDI uses a *disassembly tree* (DT), which relies on the product's structural blueprint (Veerakamolmal and Gupta, 1999). The DT can be used to identify precedent relationships that define the structural constraints in terms of the order in which components can be retrieved.

The development of this index involves the analysis of a logic disassembly table to find the combination of components and materials together with their layout in the product so as to provide the optimum cost–benefit ratio for end-of-life retrieval. The cost considerations in this analysis include disposal and disassembly costs, while the benefit is derived from the sales of recovered components and materials in terms of reuse and recycling revenue. The result offers the best combination of components (with the highest net benefit) to recover from the product. If expensive components or materials were used to design products, then one would have to guarantee that it would be economically feasible to remove and reuse parts and materials from the products. Designers could incorporate and embed such information into the product.

The second technique discussed here involves sensors embedded into products during their production phase. These embedded sensors can improve the uncertainty of the disassembly yield by providing information on the condition and version of critical components before disassembling and processing products at the end of their lives. In addition, sensor data can be helpful in the prediction of component or product failures during the products' lives.

5.2 Design for Disassembly Index

In this section, we present a method to evaluate product design alternatives with respect to the costs and benefits of disassembly. A product design can make an enormous difference in the product's retirement strategy in an environmentally friendly way. A product designer must place equal importance on both designing products that accommodate disassembly, reuse, and recycling, and on the product's appeal and functionality. A designer is often faced with the dilemma of choosing among two or more design alternatives.

To judge if one design is better than the other, one must weigh the benefit from retrieving a set of components against the cost of disposing the residual product. For example, a feasible combination (set of components) for designing a computer for disassembly may be to retrieve the motherboard and hard disk drive, while recycling and disposing of the rest of the components.

To compare the merits of two (or more) designs, not only do we need to assess the feasible combinations, but we must also compare the combination with the highest cost–benefit ratio from one design against the others. To find the best combination, a general approach requires that all combinations be listed and enumerated with respect to the cost–benefit functions. For each combination, we assign X_{ij} with a value of "1" to indicate that the part is sold (reused) for its value, and a value of "0" to indicate that the part is recycled for its material content and/or the part is discarded.

We then present a technique to measure the design efficiency of products to study the effect on the environment of EOL disassembly and disposal. The design efficiency is measured using a DfDI, which analyzes the merits and drawbacks of different product designs (Veerakamolmal and Gupta, 1999). The index offers designers an important measure to help improve future products. We provide a comprehensive procedure for developing the index and demonstrate its application through an example.

Section 5.2.1 provides the nomenclature of variables used in the subsequent discussion. In Sec. 5.2.2, the cost–benefit function is developed. Section 5.2.3 provides the procedure to calculate the DfDI. In Sec. 5.2.4, an application of the DfDI procedure is presented. An optimization model is developed in Sec. 5.2.5, while Sec. 5.2.6 presents an application of the model.

5.2.1 Nomenclature

A_{ik}	subassembly node k in product i;
CF	recycling revenue factor (\$/unit of index scale);
CI_j	recycling revenue index of component P_j (index scale 0 = lowest, 10 = highest);
CRP_j	percentage (fraction) of recyclable contents by weight in component P_j;
D_j	vector representing the total demand for component P_j (unit);
DF	disposal cost factor (\$/unit of index scale);
DI_j	disposal cost index of component P_j (index scale 0 = lowest, 10 = highest);
DW_j	weight of component P_j (lb.);
$LS(A_{ik})$	leaf successor set of subassembly node k in product i;
$LS^s(A_{ik})$	set of selected leaf successors of subassembly node k in product i;
$LS(\text{Root}_i)$	leaf successor set of the root node in product i;
$LS^s(\text{Root}_i)$	set of selected leaf successors of the root node in product i;
m	total number of components in the problem space;
n	total number of products in the problem space;
P_j	component j;
PC	processing (e.g., disassembling, sorting, cleaning, identification and packaging) cost per unit time (\$/unit time);
Q_{ij}	multiplicity matrix representing the number of each type of component P_j obtained from each type of product i;
R_{ij}	matrix representing the proportion of units, with respect to multiplicity, of the selected component P_j (with $X_{ij} = 1$) retrieved from product i for reuse or recycle ($0 \leq R_{ij} \leq Q_{ij}$);
Root_i	root node of the product i;
RPN_{ij}	matrix representing the minimum mandated percentage (fraction) of units of component P_j retrieved from product i (the mandate can be either imposed by the government or may be set by the company);
RPX_{ij}	matrix representing the maximum mandated percentage (fraction) of units of component P_j retrieved from product i;
RV_j	resale value of component j (\$/unit);
s_i	total number of subassembly nodes in product i;

Sub_{ik}	subassembly node k of product i;
$T(A_{ik})$	time to disassemble subassembly k from product i (unit time);
$T_i(P_j)$	time to disassemble component P_j from product i (unit time);
$T(\text{Root}_i)$	time to disassemble root node of the product i (unit time);
TC_i	cost of acquisition and transportation for product i (\$/unit);
TD_i	total disassembly time for every component in product i (unit time);
TD_i^s	total disassembly time for a set of selected components in product i (unit time);
TCR	total recycling revenue (\$);
TDC	total disposal cost (\$);
TPC	total processing cost (\$);
TRR	total resale revenue (\$);
X_{ij}	matrix representing the (mutually exclusive combination) selection of component P_j retrieved from product i for reuse ($X_{ij} = 1$) or recycle and/or disposal ($X_{ij} = 0$);
Z	revenue of retrieval;
$\lceil \alpha \rceil$	gives the smallest integer that is larger than or equal to α;
$\lfloor \mu \rfloor$	gives the largest integer that is smaller than or equal to μ;
$\{\beta_{ij}\}$	element in row i and column j of matrix β_{ij};
$\{\gamma_i\}$	the ith element in vector γ_i.

5.2.2 The Cost-Benefit Function

In order to design products for environmental compatibility, we develop a cost–benefit function (representing the revenue of retrieval) that can be used to assess designs for disassembly. The cost–benefit function consists of four terms—total resale revenue (*TRR*), total recycling revenue (*TCR*), total processing cost (*TPC*), and total disposal cost (*TDC*)—as follows.

$$Z = TRR + TCR - TPC - TDC \tag{5.1}$$

Each term is described below.

Total Resale Revenue

TRR is directly influenced by RV_j and TC_i. RV_j is the resale value of component j, and TC_i is the cost per unit of acquiring and transporting product i from the distribution centers (or collection sources) to the disassembly facility. The revenue equation represents revenue less the cost of product acquisition, which can be formulated as

$$TRR = \sum_{j \exists P_j \in LS^S(\mathbf{Root}_i)} (RV_j \cdot \{Q_{ij}\} \cdot \{X_{ij}\}) - TC_i \tag{5.2}$$

Total Recycling Revenue

TCR is calculated by multiplying the component recycling revenue factors by the number of component units recycled for materials content as follows:

$$TCR = \sum_{j \exists P_j \in LS^S(\mathbf{Root}_i)} (CI_j \cdot DW_j \cdot CRP_j \cdot \{Q_{ij}\} \cdot (1 - \{X_{ij}\})) \cdot CF \tag{5.3}$$

Note that each component has a percentage of recyclable contents (CRP_j) (the portion not recycled must be properly discarded). CI_j is the recycling revenue index (varying in value from 1 to 10) representing the degree of benefit generated by the recycling of component P_j (the higher the value of index, the more profitable it is to recycle the component), DW_j is the weight of the component, and *CF* is the recycling revenue factor.

Total Processing Cost

TPC can be calculated from the process makespan ($TD_i^{\,s}$) and the processing cost per unit time (*PC*) as follows:

$$TPC = TD_i^{\,s} \cdot PC \tag{5.4}$$

and, in turn, $TD_i^{\,s}$ can be obtained using the following equation:

$$\begin{aligned} TD_i^{\,s} = {} & \left(\operatorname*{Max}_{\forall P_j \in LS^s(\mathbf{Root}_i)} \lceil \{X_{ij}\} \rceil \right) \left(T(\mathbf{Root}_i) \right) \\ & + \sum_{k=1}^{s_i} \left\{ \left(\operatorname*{Max}_{\forall P_j \in LS^s(A_{ik})} \lceil \{X_{ij}\} \rceil \right) \left(T(A_{ik}) \right) \right\} \end{aligned} \tag{5.5}$$

Total Disposal Cost

TDC is calculated by multiplying the component disposal cost by the number of component units disposed as follows:

$$TDC = \sum_{j \exists P_j \in LS^S(\mathbf{Root}_i)} (DI_j \cdot DW_j \cdot (1 - CRP_j) \cdot \{Q_{ij}\} \cdot (1 - \{X_{ij}\})) \cdot DF \tag{5.6}$$

Note that DI_j is the disposal cost index (varying in value from 1 to 10) representing the degree of nuisance created by the disposal of component P_j (the higher the value of index, the more nuisance the component creates, and thus the more it costs to dispose of it), DW_j is the weight of the component, and DF is the disposal cost factor.

5.2.3 DfDI Calculation Procedure

The DfDI can be calculated using the following steps:

Step 1. List each component by its ID, predecessor, resale value (RV_j), multiplicity ($\{Q_{ij}\}$), weight (DW_j), recyclable percentage (CRP_j), recycle index (CI_j), and disposal index (DI_j).

Step 2. Generate the mutually exclusive combination matrix (X_{ij}) for component(s) selection.

Step 3. Obtain the cost of acquiring each product (TC_i), processing cost/unit time (PC), disassembly times [$T(\text{Root}_i)$ and $T(A_{ik})$], recycling revenue factor (CF), and disposal cost factor (DF). Calculate TRR, TCR, TPC, and TDC with respect to the selection of reused components in each of the corresponding mutually exclusive combinations.

Step 4. Calculate total benefit, total cost, DfDI, and net benefit for each combination.

5.2.4 An Application of the Procedure

This example considers two environmentally friendly computer designs, $DX1$ and $DX2$ [Fig. 5.1 (a) and (b)], each consisting of six identical components—i.e., housing assembly (P_1), power supply (P_2), printed circuit boards (P_3), motherboard (P_4), floppy disk drive (P_5), and hard disk drive (P_6). Suppose $TC_1 = \$13$, $TC_2 = \$13$, $PC = 0.55$ \$/min, $CF = 0.14$ \$/lb., and $DF = 0.1$ \$/lb.

The following steps demonstrate the calculation of the DfDI for product design $DX1$.

Step 1. The component IDs, predecessor IDs, and other relevant information are listed in columns (A)–(H) of Table 5.1.

Step 2. Table 5.2 shows the mutually exclusive combinations in column (J). A value of "1" indicates that the part is sold (reused) for its value, and a value of "0" indicates that the part is recycled for its material content and/or the part is discarded.

Step 3. Calculate TRR, TCR, TPC, and TDC by assessing the selection of reused components in each of the corresponding mutually exclusive combinations. For instance, in combination number 29, the selection includes components P_2, P_3, and P_4. From Eq. (5.2),

$$TRR = (7.5)(1)(1) + (2.75)(4)(1) + (25.5)(1)(1) - 13 = \$31.00$$

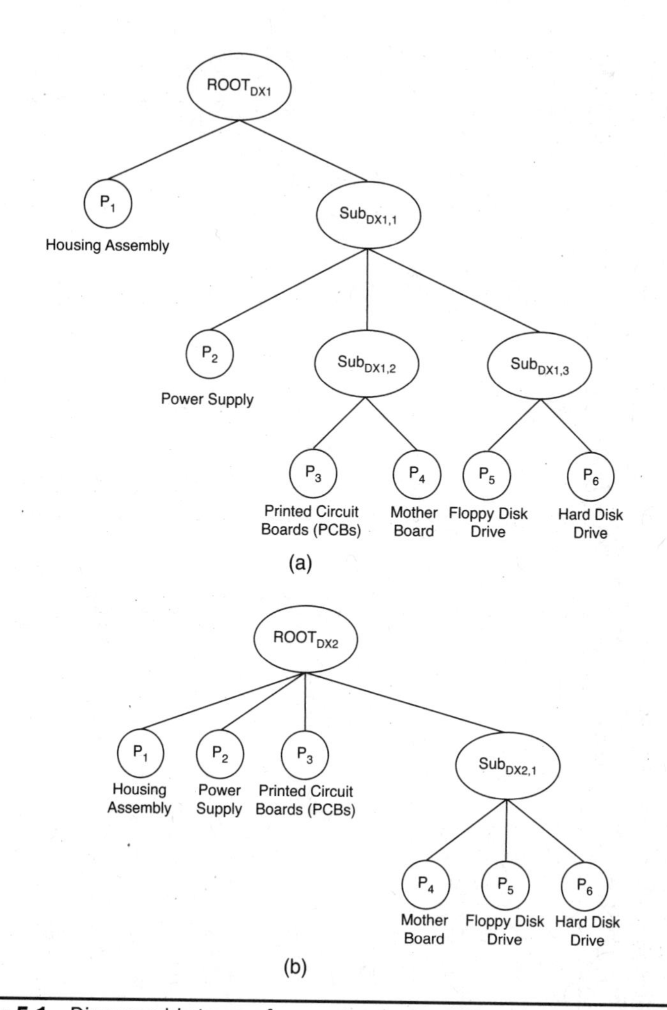

FIGURE 5.1 Disassembly trees of computer design *DX*1 and design *DX*2.

(A)	(B)	(C)	(D)	(E)	(F)	(G)	(H)
Component ID	**Predecessor ID**	**Resale Value (RV_j)**	**Multiplicity ($\{Q_{ij}\}$)**	**Weight (DW_j)**	**Recycle Percentage (CRP_j)**	**Recycle Index (CI_j)**	**Disposal Index (DI_j)**
P_1	$Root_{DX1}$	0.00	2	4.50	85%	7.5	3.0
P_2	$Sub_{DX1,1}$	7.50	1	10.50	60%	5.0	5.0
P_3	$Sub_{DX1,2}$	2.75	4	0.75	35%	6.5	5.5
P_4	$Sub_{DX1,2}$	25.50	1	1.50	25%	7.0	6.0
P_5	$Sub_{DX1,3}$	0.00	1	2.70	65%	7.5	1.2
P_6	$Sub_{DX1,3}$	3.75	1	3.75	45%	3.5	2.3

TABLE 5.1 Data of product design *DX1*

(I)	(J)						(K)	(L)	(M)	(N)	(O)	(P)	(Q)	(R)	(S)	(T)
	Parts Reused, Recycled and/or Disposed of (X_{ij})						Design DX_1								Design DX_2	
Combination	P_1	P_2	P_3	P_4	P_5	P_6	Reuse Revenue (*TRR*)	Recycling Revenue (*TCR*)	Processing Cost (*TPC*)	Disposal Cost (*TDC*)	Total Benefit (K) + (L)	Total Cost (M) + (N)	DFDI (O)/(P)	Net Benefit (O) − (P)	DFDI	Net Benefit
1	0	0	0	0	0	0	−13.00	16.44	0.00	4.84	3.44	4.84	0.71	−1.41	0.71	−1.41
2	0	0	0	0	0	1	−9.25	15.61	8.80	4.37	6.36	13.17	0.48	−6.81	0.87	−0.93
3	0	0	0	0	1	0	−13.00	14.59	8.80	4.73	1.59	13.53	0.12	−11.93	0.21	−6.06
4	0	0	0	0	1	1	−9.25	13.77	8.80	4.25	4.52	13.05	0.35	−8.54	0.63	−2.66
5	0	0	0	1	0	0	12.50	16.07	5.23	4.17	28.57	9.39	3.04	19.18	4.03	21.48
6	0	0	0	1	0	1	16.25	15.24	9.63	3.69	31.49	13.32	2.36	18.17	4.76	24.87
7	0	0	0	1	1	0	12.50	14.22	9.63	4.05	26.72	13.68	1.95	13.05	3.83	19.75
8	0	0	0	1	1	1	16.25	13.40	9.63	3.58	29.65	13.20	2.25	16.45	4.56	23.15
9	0	0	1	0	0	0	−2.00	15.48	5.23	3.77	13.48	8.99	1.50	4.49	2.52	8.14
10	0	0	1	0	0	1	1.75	14.65	9.63	3.29	16.40	12.92	1.27	3.48	2.10	8.61
11	0	0	1	0	1	0	−2.00	13.64	9.63	3.65	11.64	13.28	0.88	−1.64	1.43	3.48
12	0	0	1	0	1	1	1.75	12.81	9.63	3.18	14.56	12.81	1.14	1.76	1.90	6.88
13	0	0	1	1	0	0	23.50	15.11	5.23	3.09	38.61	8.32	4.64	30.29	5.09	31.02
14	0	0	1	1	0	1	27.25	14.29	9.63	2.62	41.54	12.24	3.39	29.29	5.83	34.42
15	0	0	1	1	1	0	23.50	13.27	9.63	2.98	36.77	12.60	2.92	24.17	4.92	29.29
16	0	0	1	1	1	1	27.25	12.44	9.63	2.51	39.69	12.13	3.27	27.56	5.67	32.69

17	0	1	0	0	0	0	−5.50	12.03	4.40	2.74	6.53	7.14	0.91	−0.62	1.51	2.21
18	0	1	0	0	0	1	−1.75	11.20	8.80	2.27	9.45	11.07	0.85	−1.62	1.40	2.68
19	0	1	0	0	1	0	−5.50	10.18	8.80	2.63	4.68	11.43	0.41	−6.74	0.66	−2.44
20	0	1	0	0	1	1	−1.75	9.36	8.80	2.15	7.61	10.95	0.69	−3.35	1.14	0.95
21	0	1	0	1	0	0	20.00	11.66	5.23	2.07	31.66	7.29	4.34	24.37	4.82	25.09
22	0	1	0	1	0	1	23.75	10.83	9.63	1.59	34.58	11.22	3.08	23.36	5.68	28.49
23	0	1	0	1	1	0	20.00	9.81	9.63	1.95	29.81	11.58	2.58	18.24	4.62	23.36
24	0	1	0	1	1	1	23.75	8.99	9.63	1.48	32.74	11.10	2.95	21.64	5.48	26.76
25	0	1	1	0	0	0	5.50	11.07	5.23	1.67	16.57	6.89	2.40	9.68	5.11	13.33
26	0	1	1	0	0	1	9.25	10.24	9.63	1.19	19.49	10.82	1.80	8.67	3.42	13.80
27	0	1	1	0	1	0	5.50	9.23	9.63	1.55	14.73	11.18	1.32	3.55	2.43	8.67
28	0	1	1	0	1	1	9.25	8.40	9.63	1.08	17.65	10.71	1.65	6.95	3.16	12.07
29	0	1	1	1	0	0	31.00	10.70	5.23	0.99	41.70	6.22	6.71*	35.48*	7.59	36.21
30	0	1	1	1	0	1	34.75	9.88	9.63	0.52	44.63	10.14	4.40	34.48	8.89*	39.61*
31	0	1	1	1	1	0	31.00	8.86	9.63	0.88	39.86	10.50	3.79	29.36	7.41	34.48
32	0	1	1	1	1	1	34.75	8.03	9.63	0.41	42.78	10.03	4.27	32.75	8.72	37.88
33	1	0	0	0	0	0	−13.00	8.40	1.65	4.44	−4.60	6.09	−0.76	−10.68	−0.76	−10.61
34	1	0	0	0	0	1	−9.25	7.58	8.80	3.96	−1.67	12.76	−0.13	−14.44	−0.20	−10.14
35	1	0	0	0	1	0	−13.00	6.56	8.80	4.32	−6.44	13.12	−0.49	−19.56	−0.73	−15.26

(continued)

(I)	(J)						(K)	(L)	(M)	(N)	(O)	(P)	(Q)	(R)	(S)	(T)
Combination	Parts Reused, Recycled and/or Disposed of (X_{ij})						Design DX_1								Design DX_2	
	P_1	P_2	P_3	P_4	P_5	P_6	Reuse Revenue (*TRR*)	Recycling Revenue (*TCR*)	Processing Cost (*TPC*)	Disposal Cost (*TDC*)	Total Benefit (K) + (L)	Total Cost (M) + (N)	DFDI (O)/(P)	Net Benefit (O) − (P)	DFDI	Net Benefit
36	1	0	0	0	1	1	−9.25	5.73	8.80	3.85	−3.52	12.65	−0.28	−16.16	−0.42	−11.86
37	1	0	0	1	0	0	12.50	8.04	5.23	3.76	20.54	8.99	2.29	11.55	2.49	12.27
38	1	0	0	1	0	1	16.25	7.21	9.63	3.29	23.46	12.91	1.82	10.55	3.01	15.67
39	1	0	0	1	1	0	12.50	6.19	9.63	3.65	18.69	13.27	1.41	5.42	2.29	10.55
40	1	0	0	1	1	1	16.25	5.37	9.63	3.17	21.62	12.80	1.69	8.82	2.82	13.94
41	1	0	1	0	0	0	−2.00	7.45	5.23	3.36	5.45	8.59	0.63	−3.14	1.10	0.51
42	1	0	1	0	0	1	1.75	6.62	9.63	2.89	8.37	12.51	0.67	−4.14	1.13	0.98
43	1	0	1	0	1	0	−2.00	5.60	9.63	3.25	3.60	12.87	0.28	−9.27	0.47	−4.15
44	1	0	1	0	1	1	1.75	4.78	9.63	2.78	6.53	12.40	0.53	−5.87	0.90	−0.75
45	1	0	1	1	0	0	23.50	7.08	5.23	2.69	30.58	7.91	3.86	22.67	4.25	23.39
46	1	0	1	1	0	1	27.25	6.25	9.63	2.21	33.50	11.84	2.83	21.66	4.99	26.79
47	1	0	1	1	1	0	23.50	5.24	9.63	2.57	28.74	12.20	2.36	16.54	4.06	21.66
48	1	0	1	1	1	1	27.25	4.41	9.63	2.10	31.66	11.73	2.70	19.94	4.80	25.06

49	1	1	0	0	0	0	−5.50	3.99	4.40	2.34	−1.51	6.74	−0.22	−8.24	−0.39	−5.42
50	1	1	0	0	0	1	−1.75	3.17	8.80	1.86	1.42	10.66	0.13	−9.25	0.22	−4.95
51	1	1	0	0	1	0	−5.50	2.15	8.80	2.22	−3.35	11.02	−0.30	−14.37	−0.50	−10.07
52	1	1	0	0	1	1	−1.75	1.32	8.80	1.75	−0.43	10.55	−0.04	−10.97	−0.07	−6.67
53	1	1	0	1	0	0	20.00	3.63	5.23	1.66	23.63	6.89	3.43	16.74	3.84	17.46
54	1	1	0	1	0	1	23.75	2.80	9.63	1.19	26.55	10.81	2.46	15.74	4.67	20.86
55	1	1	0	1	1	0	20.00	1.78	9.63	1.55	21.78	11.17	1.95	10.61	3.60	15.74
56	1	1	0	1	1	1	23.75	0.96	9.63	1.07	24.71	10.70	2.31	14.01	4.43	19.13
57	1	1	1	0	0	0	5.50	3.04	5.23	1.26	8.54	6.49	1.32	2.05	3.01	5.70
58	1	1	1	0	0	1	9.25	2.21	9.63	0.79	11.46	10.41	1.10	1.05	2.17	6.17
59	1	1	1	0	1	0	5.50	1.19	9.63	1.15	6.69	10.77	0.62	−4.08	1.18	1.05
60	1	1	1	0	1	1	9.25	0.37	9.63	0.68	9.62	10.30	0.93	−0.68	1.86	4.44
61	1	1	1	1	0	0	31.00	2.67	5.23	0.59	33.67	5.81	5.79	27.86	6.62	28.58
62	1	1	1	1	0	1	34.75	1.84	9.63	0.11	36.59	9.74	3.76	26.85	7.93	31.98
63	1	1	1	1	1	0	31.00	0.83	9.63	0.47	31.83	10.10	3.15	21.73	6.40	26.85
64	1	1	1	1	1	1	34.75	0.00	9.63	0.00	34.75	9.63	3.61	25.13	7.72	30.25

TABLE 5.2 Analysis of designs *DX*1 and *DX*2.

Disassembly Time			
Design *DX1*		Design *DX2*	
Subassembly	Time	Subassembly	Time
$Root_{DX1}$	3	$Root_{DX2}$	3.5
$Sub_{DX1,1}$	5	$Sub_{DX2,1}$	6.5
$Sub_{DX1,2}$	1.5		
$Sub_{DX1,3}$	8		

TABLE 5.3 Disassembly times for designs *DX1* and *DX2*

The total recycling revenue can be obtained from recycling components that are not reused (i.e., P_1, P_5, and P_6 in combination 29). From Eq. (5.3),

$$\begin{aligned} TCR &= ((7.5)(4.5)(0.85)(2)(1) + (7.5)(2.7)(0.65)(1)(1) \\ &\quad + (3.5)(3.75)(0.45)(1)(1)) \cdot (0.14) \\ &= \$10.70 \end{aligned}$$

Subassembly module $Root_{DX1}$, $Sub_{DX1,1}$ and $Sub_{DX1,2}$ must be disassembled to obtain components P_2, P_3, and P_4. With the disassembly times provided in Table 5.3, we can calculate the total processing cost with Eqs. (5.4) and (5.5) as shown below

$$\begin{aligned} TPC &= (T(\mathrm{Root}_{DX1}) + T(\mathrm{Sub}_{DX1,1}) + T(\mathrm{Sub}_{DX1,2})) \cdot (0.55) \\ &= \$5.23 \end{aligned}$$

From Eq. (5.6), the total disposal cost can be calculated as follows:

$$\begin{aligned} TDC &= ((3)(4.5)(1 - 0.85)(2)(1) + (1.2)(2.7)(1 - 0.65)(1)(1) \\ &\quad + (2.3)(3.75)(1 - 0.45)(1)(1)) \cdot (0.1) \\ &= \$0.99 \end{aligned}$$

The resulting values *TRR*, *TCR*, *TPC*, and *TDC* of every mutually exclusive combination is shown in columns (K), (L), (M), and (N), respectively, in Table 5.2.

Step 4. In Table 5.2, for each combination, the total benefit (*TRR* + *TCR*), and the total cost (*TPC* + *TDC*), the DfDI [calculated by dividing column (O) by (P)] and the net benefit [calculated by subtracting column (P) from column (O)] are shown in columns (O), (P), (Q), and (R), respectively.

For design *DX1*, the maximum value of the net benefit is $35.48 (combination number 29). By following a similar procedure, we can show that the maximum value of the net benefit for design $DX2$ is \$39.61 (Table 5.2) (for combination number 30). Since design $DX2$ has a higher value of the net benefit than design $DX1$, the design $DX2$ is preferred. Thus, with respect to the optimal design ($DX2$) the list of components recommended for recovery and reuse are P_2, P_3, P_4, and P_6, which correspond to the power supply, printed circuit boards (PCBs), motherboard, and hard disk drive, respectively. The remaining components (P_1, housing assembly; and P_5, floppy disk drive) can be pulverized and recycled and/or processed for environmentally benign disposal.

The DfDI calculation procedure performs an exhaustive search over all possible EOL alternatives and then chooses the best among them. This procedure provides insight into different combinations of reuse, recycling, and disposal alternatives. However, this procedure comes at a price. As the number of components grows, the time to perform the procedure also increases. The computational complexity of the procedure is of the order of $O(2^m)$, where m is the number of components in the product. An alternative to the procedure is the use of integer programming to directly determine the net benefit of a particular design. By finding out the net benefit of every design under consideration, the best one can be chosen. In the next section, we formulate this problem as in an integer programming problem.

5.2.5 The Optimization Model

The integer programming formulation for calculating the net benefit is as discussed in the following.

The Objective Function

The objective of the model is to maximize the net benefit of a design. Thus, the objective function can be written as:

$$\text{Max } Z = TRR + TCR - TPC - TDC \tag{5.7}$$

where TRR, TCR, TPC, and TDC need to be slightly modified from the way they were defined before. They are as follows:

$$TRR = \sum_{j \exists P_j \in LS^S(\mathbf{Root}_i)} (RV_j \cdot \{R_{ij}\} \cdot \{X_{ij}\}) - TC_i \tag{5.8}$$

$$TCR = \sum_{j \exists P_j \in LS^S(\mathbf{Root}_i)} (CI_j \cdot DW_j \cdot CRP_j \cdot (\{Q_{ij}\} - \{R_{ij}\} \cdot \{X_{ij}\})) \cdot CF \tag{5.9}$$

$$\begin{aligned} TPC = \Bigg[& \left(\operatorname*{Max}_{\forall P_j \in LS^s(\mathbf{Root}_i)} \lceil \{X_{ij}\} \rceil \right) \left(T(\mathbf{Root}_i) \right) \\ & + \sum_{k=1}^{si} \left\{ \left(\operatorname*{Max}_{\forall P j \in LS^s(A_{ik})} \lceil \{X_{ij}\} \rceil \right) \left(T(A_{ik}) \right) \right\} \Bigg] \cdot PC \end{aligned} \tag{5.10}$$

$$TDC = \sum_{j \exists P_j \in LS^S(\mathbf{Root}_i)} (DI_j \cdot DW_j \cdot (1 - CRP_j) \cdot (\{Q_{ij}\} - \{R_{ij}\} \cdot \{X_{ij}\})) \cdot DF \qquad (5.11)$$

The Constraints

The following set of constraints needs to be considered:

1. **Product Structure Constraints**
   The number of the selected components to retrieve must not exceed the number of each component's multiplicity.

$$\{R_{ij}\} \leq \{Q_{ij}\}; \text{ for all } j \exists P_j \in LS^S(\mathbf{Root}_i) \qquad (5.12)$$

2. **Integer and Non-Negativity Constraints**
   The number of selected components to retrieve must be non-negative integer values.

$$\{R_{ij}\} \geq 0 \text{ and Integer; for all } i \text{ and } j \qquad (5.13)$$

3. **0/1 Constraints**
   The selection of components are represented by non-negative 0 or 1 integer values.

$$\{X_{ij}\} = 0 \text{ or } 1; \text{ for all } i \text{ and } j \qquad (5.14)$$

4. **Additional Constraints**
   Additional constraints may be needed to represent potential interrelationships among selected components. A few examples are given below:
   a. If components p, q, and r (in product i) are mutually exclusive, then

$$\{X_{ip}\} + \{X_{iq}\} + \{X_{ir}\} \leq 1; \qquad (5.15)$$

   b. If, in product i, component r can be retrieved only if component s is also retrieved, then

$$\{X_{ir}\} \leq \{X_{is}\} (\text{or } \{X_{ir}\} - \{X_{is}\} \leq 0); \qquad (5.16)$$

   c. If components u and v (in product i) are mutually exclusive, and component r is dependent on the acceptance of either u or v, then

$$\{X_{iu}\} + \{X_{iv}\} \leq 1; \quad \text{and } \{X_{ir}\} \leq \{X_{iu}\} + \{X_{iv}\}; \qquad (5.17)$$

d. If components y and z (in product i) are not mutually exclusive and component r is strictly dependent on the acceptance of both u and v, then

$$2 \cdot \{X_{ir}\} \leq \{X_{iy}\} + \{X_{iz}\}; \tag{5.18}$$

e. If, in product i, component u has a minimum mandatory retrieval requirement and/or if component v has a maximum mandatory retrieval requirement, then

$$\begin{aligned} \{R_{iu}\} \cdot \{X_{iu}\} &\geq \lceil RPN_{iu} \cdot \{Q_{iu}\} \rceil; \\ \text{and/or } \{R_{iv}\} \cdot \{X_{iv}\} &\leq \lfloor RPX_{iv} \cdot \{Q_{iv}\} \rfloor. \end{aligned} \tag{5.19}$$

The Mathematical Formulation

The following integer programming formulation will maximize the net benefit of a design. The output of the model will provide optimal value of the net benefit of a design and the parts to reuse, recycle, and/or dispose of.

$$\text{Max } Z = TRR + TCR - TPC - TDC$$

subject to:

$\{R_{ij}\} \leq \{Q_{ij}\}$; for all $j \exists P_j \in LS^S(\mathbf{Root}_i)$

$\{R_{ij}\} \geq 0$ and Integer; for all i and j

$\{X_{ij}\} = 0$ or 1; for all i and j

plus additional constraints, if any, where:

$$TRR = \sum_{j \exists P_j \in LS^S(\mathbf{Root}_i)} (RV_j \cdot \{R_{ij}\} \cdot \{X_{ij}\}) - TC_i$$

$$TCR = \sum_{j \exists P_j \in LS^S(\mathbf{Root}_i)} (CI_j \cdot DW_j \cdot CRP_j \cdot (\{Q_{ij}\} - \{R_{ij}\} \cdot \{X_{ij}\})) \cdot CF$$

$$\begin{aligned} TPC = \Bigg[& \left(\max_{\forall P_j \in LS^s(\mathbf{Root}_i)} \lceil \{X_{ij}\} \rceil \right) \left(T(\mathbf{Root}_i) \right) \\ & + \sum_{k=1}^{si} \left\{ \left(\max_{\forall P_j \in LS^s(A_{ik})} \lceil \{X_{ij}\} \rceil \right) \left(T(A_{ik}) \right) \right\} \Bigg] \cdot PC \end{aligned}$$

$$TDC = \sum_{j \exists P_j \in LS^S(\mathbf{Root}_i)} (DI_j \cdot DW_j \cdot (1 - CRP_j) \cdot (\{Q_{ij}\} - \{R_{ij}\} \cdot \{X_{ij}\})) \cdot DF$$

5.2.6 An Application of the Optimization Procedure

Consider the example investigated in Sec. 5.2.4. Using the optimization model of the previous section, we can formulate the example for product design $DX1$ as follows:

$$\text{Max } Z = TRR + TCR - TPC - TDC;$$

subject to:

$$\{R_{1,1}\} \leq 2;$$
$$\{R_{1,2}\} \leq 1;$$
$$\{R_{1,3}\} \leq 4;$$
$$\{R_{1,4}\} \leq 1;$$
$$\{R_{1,5}\} \leq 1;$$
$$\{R_{1,6}\} \leq 1;$$

$$\{R_{1,1}\}, \{R_{1,2}\}, \{R_{1,3}\}, \{R_{1,4}\}, \{R_{1,5}\}, \{R_{1,6}\} \geq 0 \text{ and Integer}$$
$$\{X_{1,1}\}, \{X_{1,2}\}, \{X_{1,3}\}, \{X_{1,4}\}, \{X_{1,5}\}, \{X_{1,6}\} = 0 \text{ or } 1$$

where:

$$\begin{aligned}
TRR = {} & (0^*\{R_{1,1}\}^*\{X_{1,1}\} + 7.50^*\{R_{1,2}\}^*\{X_{1,2}\} + 2.75^*\{R_{1,3}\}^*\{X_{1,3}\} \\
& + 25.50^*\{R_{1,4}\}^*\{X_{1,4}\} + 0^*\{R_{1,5}\}^*\{X_{1,5}\} + 3.75^*\{R_{1,6}\} \\
& ^*\{X_{1,6}\}) - 12; \\
TCR = {} & (7.5^*4.5^*0.85^*(2 - \{R_{1,1}\}^*\{X_{1,1}\}) + 5^*10.5^*0.6^*(1 - \{R_{1,2}\} \\
& ^*\{X_{1,2}\}) + 6.5^*0.8^*0.35^*(4 - \{R_{1,3}\}^*\{X_{1,3}\}) + 7^*1.5^*0.25 \\
& ^*(1 - \{R_{1,4}\}^*\{X_{1,4}\}) + 7.5^*2.7^*0.65^*(1 - \{R_{1,5}\}^*\{X_{1,5}\}) \\
& + 3.5^*3.8^*0.45^*(1 - \{R_{1,6}\}^*\{X_{1,6}\}))^*0.14; \\
TPC = {} & (Max(\{X_{1,1}\}, \{X_{1,2}\}, \{X_{1,3}\}, \{X_{1,4}\}, \{X_{1,5}\}, \{X_{1,6}\})^*3 \\
& + Max(\{X_{1,2}\}, \{X_{1,3}\}, \{X_{1,4}\}, \{X_{1,5}\}, \{X_{1,6}\})^*5 \\
& + Max(\{X_{1,3}\}, \{X_{1,4}\})^*1.5 \\
& + Max(\{X_{1,5}\}, \{X_{1,6}\})^*8)^*0.55; \\
TDC = {} & (3^*4.5^*(1 - 0.85)^*(2 - \{R_{1,1}\}^*\{X_{1,1}\}) \\
& + 5^*10.5^*(1 - 0.6)^*(1 - \{R_{1,2}\}^*\{X_{1,2}\}) \\
& + 5.5^*0.8^*(1 - 0.35)^*(4 - \{R_{1,3}\}^*\{X_{1,3}\}) \\
& + 6^*1.5^*(1 - 0.25)^*(1 - \{R_{1,4}\}^*\{X_{1,4}\}) \\
& + 1.2^*2.7^*(1 - 0.65)^*(1 - \{R_{1,5}\}^*\{X_{1,5}\}) \\
& + 2.3^*3.8^*(1 - 0.45)^*(1 - \{R_{1,6}\}^*\{X_{1,6}\}))^*0.1;
\end{aligned}$$

When the above integer program was solved, the optimal net benefit Z^* was found to be \$35.48. This was achieved by retrieving components P_2 (power supply), P_3 (PCBs), and P_4 (motherboard) for reuse. Other components were partially recycled (according to the percentages given) and the remains were discarded. Note that this result is identical to the one obtained in Sec. 5.2.4. Similar formulation and calculations can be carried out for product design $DX2$.

To demonstrate an application of additional constraints [Eqs. (5.15)–(5.19)] described in Sec. 5.2.5, let us introduce a new product design under a more stringent regulation on the reuse and recycling of electronic products.

Let the new design, *DX*3, have a same set of characteristics (i.e., disassembly times, cost and benefit variables) and the exact product structure as design *DX*1 (see Sec. 5.2.4). However, for *DX*3, only one of either component P_2 or P_3 may be selected for retrieval, and the retrieval of component P_4 is dependent on the retrieval of at least one of the former components. Furthermore, suppose the legislation imposes a special restriction requiring a minimum mandatory amount of retrieval of component P_3 (RPN_{33}) to be 45%. Similarly, the company's policy allows no more than 80% of component P_3 (RPX_{33}), to be retrieved. We can formulate the additional constraints as follows [see Eqs. (5.15), (5.17), and (5.19)}:

$\{X_{3,2}\} + \{X_{3,3}\} \leq 1;$	(mutually exclusive constraint)
$\{X_{3,4}\} \leq \{X_{3,2}\} + \{X_{3,3}\};$	(dependency constraint)
$\{R_{3,3}\}^*\{X_{3,3}\} \geq \lceil(0.45^*4)\rceil$	(minimum requirement constraint)
$\{R_{3,3}\}^*\{X_{3,3}\} \leq \lfloor(0.8^*4)\rfloor$	(maximum requirement constraint)

When the previous integer program is solved by adding the above constraints, the result shows an optimal DfDI of $27.52 for design *DX*3 with the selection of three units of P_3 and one unit of P_4. The result demonstrates that, although the net benefit has been reduced by $7.96 because of the stringent requirements imposed on *DX*3, the overall benefit confirms that the design is still suitable for EOL component retrieval.

5.3 Use of Sensor Embedded Products

In this section, we present a framework that interconnects sensor-embedded products (SEPs), a remote monitoring center (RMC), maintenance centers, disassembly centers, recycling centers, disposal centers, and remanufacturing centers; see Fig. 5.2. The framework integrates several functions: (a) record data from the point products are manufactured until they are disposed of or are recycled, and (b) provide access to the recorded data and product-specific information to the maintenance, disassembly, recycling, disposal, and remanufacturing centers (Vadde et al., 2008). The functions and characteristics of each entity in the framework are presented below.

5.3.1 Sensor-Embedded Products (SEPs)

SEPs contain sensors implanted at the time of their production to monitor their critical components. By facilitating data collection during

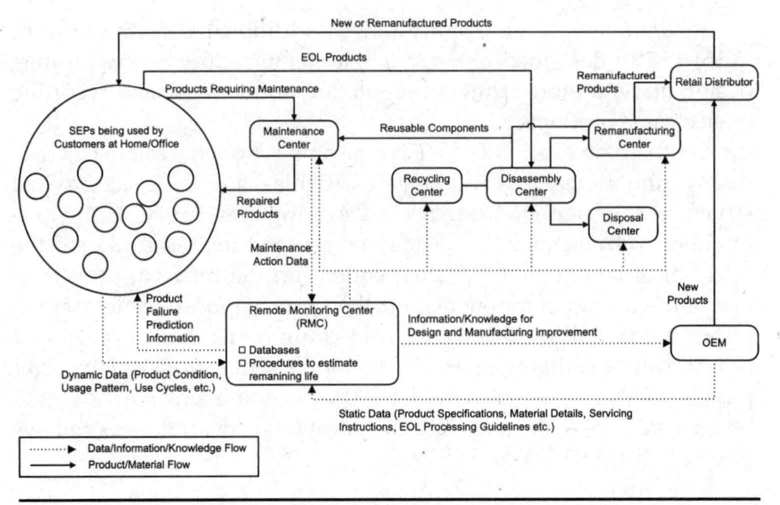

FIGURE 5.2 Use of SEP for effective life cycle management of products.

product usage, these embedded sensors enable one to predict product/component failures and estimate the remaining life of components as the products reach their end-of-life.

Sensors that are embedded in products contain (a) a sensing element to register environmental parameters (e.g., temperature, vibration), (b) a microprocessor to extract features from the signals to perform local data processing, (c) memory with limited capacity to store sensor data, (d) data transceiver, (e) unique identification code (UIC) which can be read using RFID tag readers, and (e) onboard power supply from the product power source or a separate battery or an energy harvesting mechanism (energy scavenging sensors). The sensors can either be wired or wireless depending upon the product characteristics.

5.3.2 Remote Monitoring Center (RMC)

The RMC is the nerve center of the product life cycle monitoring framework. It documents the static data/information of products compiled by the *original equipment manufacturer* (OEM) during the product manufacturing stage and the dynamic data/information chronicled by sensors in SEPs during their use phase. Static data/information is comprised of the bill of materials, component suppliers, configuration options, servicing instructions, and EOL guidelines such as the disassembly sequence. Dynamic data/information generated during the use of a product consists of sensor data, patterns of usage, number of use cycles, runtime in each use cycle, and environmental conditions. Dynamic data also includes service history on inspections, and parts replaced and repaired.

Much of the static and dynamic data/information is archived at the RMC in a database against a product's UIC. Sensors can use an onboard radio-frequency transmitter, telephone line, local area network, or the Internet to transmit the registered data/information to the RMC. Dynamic data can be transmitted to the RMC via a communication network, either continuously or intermittently on an hourly, daily, weekly, or monthly basis, depending on the product usage pattern. In intermittent transmission mode, sensors can either send entire data generated during each cycle of the product's operation. or statistics of the measured data during the cycle, or an approximate signal form representing the data in each cycle.

The RMC contains procedures to estimate the remaining life and condition of the product/component and to predict the product/component failures in the immediate future. It stores product condition data and remaining life information against the product's UIC. When product failure is either detected or predicted, the RMC alerts the user of the product so that precautionary measures are taken to avert unexpected failures. The RMC facilitates better life cycle management of products by providing the archived data/information to the maintenance, disassembly, recycling, disposal, and remanufacturing centers. Wireless communication technologies, satellite links, and the Internet act as enablers for the RMC to receive and transmit data/information from geographically remote products.

5.3.3 Maintenance Center

Customers bring their products to the maintenance center when the products are predicted to fail or if they are malfunctioning. For example, when sensors predict that a product is going to fail in a given time window, the RMC sends an electronic message to the owner of the product and the maintenance center, giving notification of the imminent failure. Before the customer sends the product to the maintenance center, the maintenance personnel can access the static and dynamic data of the product from the RMC to analyze its current condition, diagnose the reasons for failure, and prepare a maintenance plan. This prediagnosis, before the product arrives at the maintenance center, can save maintenance time and cost. Depending on the reasons for failure, the service personnel can repair the defective components or replace them with either remanufactured or new components. Servicing information such as the reason(s) for a product's arrival to the maintenance center, the problem as diagnosed by maintenance personnel, any action taken to rectify the problem, and new or refurbished components used to fix the problem are all communicated to the RMC for record keeping.

5.3.4 Disassembly Center

After products have been sold to customers, they take different journeys before they arrive at the disassembly center, thus making it difficult to determine the exact usage conditions, current state, and remaining life of products. Lack of such information on product life history is a major barrier that renders current EOL practices inefficient and ineffective.

5.3.5 Recycling Center

Details regarding the composition of materials (static data) used in the product/component are accessed from the RMC to determine the appropriate recycling processes.

5.3.6 Remanufacturing Center

Remanufactured products can be built with components that have similar remaining lives. The remaining life information is obtained from the RMC. Remanufactured components are either sent to the maintenance center for use as spares or are sold to retailers.

5.3.7 Disposal Center

The recovered components from the disassembly center that are either physically damaged or have negligible remaining lives are sent to disposal centers where they are disposed of based on the static data/information obtained from the RMC.

5.3.8 Sensor Data Mining

Data mining techniques can be run on data/information stored at the RMC to extract information/knowledge such as patterns of product usage, components causing frequent failures, environmental working conditions of the product, and any information that can provide the OEM valuable feedback. The extracted information/knowledge can be incorporated by the OEM to improve the design and/or manufacturing of the product.

5.4 Benefits from SEP and Product Monitoring Framework

Sensors can have a significant positive impact on the life cycle management of products. A marketing advantage can be gained by extracting patterns of consumer use from the stored data at the RMC. Embedded sensors can enhance reliability, maintainability, serviceability, and recyclability, and promote design improvements in each subsequent generation of products.

References

Vadde, S., Kamarthi, S.V., Gupta, S.M. and Zeid, I., 2008, Product life cycle monitoring via embedded sensors, in *Environment Conscious Manufacturing*, S.M. Gupta and A.J.D. Lambert (eds.), Boca Raton, FL: CRC Press, Chapter 3, pp. 91–104.

Veerakamolmal, P., Gupta, S.M., 1999. Analysis of design efficiency for the disassembly of modular electronic products. *Journal of Electronics Manufacturing* 9, 79–95.

PART 3

Green Value Chain Management

CHAPTER 6
Green Procurement
CHAPTER 7
Green Production
CHAPTER 8
Green Logistics
CHAPTER 9
Green Customers
CHAPTER 10
End-of-Life Management

CHAPTER 6

Green Procurement

Vendor Selection with Risk Analysis

Due to the rising consciousness in recent decades about environmental protection, many issues related to *green supply chain management* (GSCM) have been discussed by businesses and governments at all levels. Different from traditional supply chain management, GSCM concerns itself with environmental impacts and material utilization issues, which make the selection of suppliers a more complicated decision than usual.

The mindset of "prevention prior to cure" and the legislative restrictions of the European Union have made consideration of green materials and components critical issues from the very beginning of production.

According to the regulations of the EU and the criteria of supplier selection, vendor selection is a typical multi-attribute problem. Thus, in this chapter, we first discuss the criteria for selecting a supplier and then analyze their relation by their hierarchical structure. Then, by employing the AHP (analytic hierarchy process) and FMEA (failure mode and effects analysis), their relative weights of importance can be derived.

As part of the process of selection, apart from the costs, both the risk from the environmental impact as well as the decision maker's (DM) perception are taken into consideration. In addition, to support the decision, a mechanism to calibrate the results of decisions is introduced so that decision makers can establish a benchmark to identify qualified suppliers. Sensitivity analysis is provided for determining the critical factors for risk management and control.

6.1 Introduction

There have been various directives and regulations stipulated for green products, such as the Waste Electrical and Electronic Equipment (WEEE) directive (2003), the Restriction of Hazardous Substances (RoHS) directive (2010), and the Energy-Using Products (EuP) directive (2010).

In the past, traditional supply chain management focused on information flows from vendors to customers; whereas present-day supply chain management must also fulfill the four R's of green factor regulations: (1) *reduction*, (2) *redesign*, (3) *remanufacturing*, and (4) *reuse*.

Purchasing is one of the most critical stages in a green supply chain. Hazardous and harmful substances will accumulate throughout the processes, and the purchasing activities of a green supply chain seek green suppliers to provide cleaning materials and components so that the end-product can conform to the green regulations. This chapter introduces a method of risk estimation and selection for green suppliers as a guide for a manufacturer, so that for the alliance of suppliers, a manufacturer is not only able to select qualified vendors, but also, by improving the quality of vendors, is able to ensure the entire green supply chain.

Referring to the procedure of *analytic hierarchy process* (AHP) (Saaty, 1980) that was introduced in Sec. 2.3 of Chap. 2, we shall show in the following how to perform vendor selection with risk analysis. A numerical example will be provided to demonstrate the whole process with sensitivity analysis.

6.2 Risk Analysis of Green Vendor Selection

Because green materials and components play a key role in a supply chain, violating a regulation carries a high risk for both a manufacturer's finances and reputation. Therefore, selecting reliable suppliers who do not present a risk is very important for a manufacturer.

To measure the objective factors of risk, three dimensions of occurrence (O), severity (S), and detection (D) are provided by the FMEA method (Teoh and Case, 2004). In contrast, to measure the subjective factors of risk, the attitude of a manufacturer—employing a utility function described in Sec. 2.2 of Chap. 2 to represent possible approaches of risk aversion, neutral risk, and risk-prone—will be defined. The objective risk measures are integrated into the procedure of AHP to provide objective weights of importance for the attributes, while the level of risk attitude will contribute to the subjective weights of importance. By combining the objective weights measured by AHP and the subjective weights of utility values, the overall risk measure of a vendor can be estimated. Then, accompanied by the additional factor of cost, a complete evaluation of suppliers can be illustrated by way of a two-dimensional risk map (risk, cost) for ranking and selecting suppliers.

Since the weights from AHP and the preference information from utility theory provide different aspects of a DM's objective and subjective viewpoints, their relative and individual impacts towards final evaluation are essential for analysis and adjustment. Therefore, sensitivity analysis will be carried out. The whole procedure will be introduced in the following discussion.

6.2.1 Criteria of Selection with Their Hierarchical Relations

When evaluating a green vendor, the green improvement of this evaluation structure is the consideration of environmental impact and *life cycle analysis* (LCA), which are different from traditional evaluation criteria. There are four major criteria as specified below:

1. *Income quality control (IQC), denoted by q* The content of environmentally sensitive substances are controlled or forbidden; thus, this chapter aims at IQC of green suppliers in order to distinguish whether or not suppliers conform with RoHS (Directive, 2010).
2. *Disassembly effort index (DEI), denoted by d* This criterion measures the effects of a product's life cycle. DEI helps us to know the influence of the materials/components in the future.
3. *Vendor management, denoted by vm* This criterion is meant to assess the performance of management from five resources below:
   a. Self-inspection reports, where each supplier is compelled to initiate basic improvements in management performance.
   b. Actual inspection reports, where the experts who specialize in inspections conduct the assessment.
   c. Warranty, where suppliers make a commitment, by vouching for their material/component contents as conformed to law.
   d. Official certifications, which come from international certification organizations.
   e. Private certifications, which come from well-known enterprises.
4. *Logistic Level, denoted by l* This criterion measures the performance of delivery in terms of time, ordered quantity, and quality; also, the ability of the suppliers to work cooperatively with all partners in a supply chain, which results in increased efficiency and profits for all members.

The above-mentioned criteria with their individual content can be drawn as a hierarchical structure, as shown in Fig. 6.1. The measurement method of each criterion is discussed below.

6.2.2 Weighting of the Criteria

Subjective Weighting: Risk Utility Function

Referring to the introduction of utility function in Sec. 2.2 of Chap. 2, let us consider a simple risk utility function in the form of Eq. (6.1). The risk attitude of each criterion for a decision maker depends on the

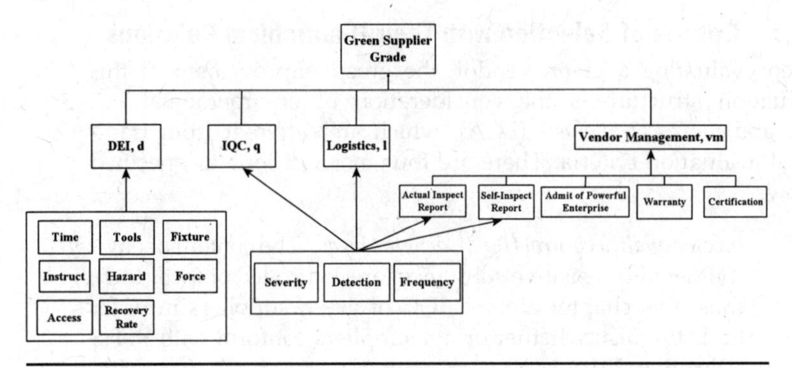

FIGURE 6.1 Hierarchical analysis structure of green supplier selection.

value of the exponent parameter a_v as described below:

$$f_v(x) = x^{a_v} \quad v \in \{d, q, vm, l\}. \tag{6.1}$$

where

if $0 < a_v < 1$: the convex form indicates that the DM is conservative regarding the criterion, in other words, the DM is risk-averse and has a low tolerance in the criterion;
if $a_v > 1$: the concave form indicates that the DM is radical regarding the criterion; in other words, the DM is risk-prone and has a high tolerance in the criterion;
if $a_v = 1$: the linear form indicates that the DM is risk-neutral towards certain criteria.

Objective Weighting: Analytical Hierarchical Process

While the utility function shows the subjective measure of a DM's risk attitude towards each attribute, the objective measure of the weights of importance is estimated by conducting AHP with sequential aggregation of pairwise evaluation of criteria in each hierarchy. The pairwise evaluation is based on a 9-level scale described in Table 2.1 of Chap. 2.

Let A be the pairwise matrix given by a DM, based on Eq. (6.2) of an eigensystem; the maximum eigenvalue, $\lambda_{\max}$, is computed to obtain the respective eigenvector, w, which is used as the weights of importance of the criteria:

$$Aw = \lambda_{\max} w \tag{6.2}$$

Although pairwise comparison is a valid method for obtaining weights, possible inconsistencies can occur and should be examined.

6.2.3 Measures of the Attributes

Now we introduce the methods proposed to quantify these criteria, including the four major criteria: IQC, DEI, vendor management, and logistics.

Income Quality Control (IQC, *q*)

Following the RoHS directive as the most well-known regulation for hazardous materials, based on the feedstock inspection report, IQC is used to evaluate the risk of hazardous materials from three dimensions of FMEA: (1) severity, (2) frequency, and (3) detection.

(a) Severity, S_q S_q defines the probability that the content of hazardous materials will not pass the required level. It can use the request of RoHS as the upper bound of normalization. The lower bound is dependent on technical development. Therefore, by Eq. (6.3), the data of the inspection report can be normalized into a value between 0 and 1:

$$x_{ijk} = \frac{\textbf{Inspected Value} - \textbf{LB}}{UB - LB}, i = 1, \ldots, n; j = 1, 2, 3; \quad k = 1, \ldots, 6. \tag{6.3}$$

where i is the number of inspection reports, j indicates the risk level expressed by linguistic terms such as $j = 1$ for *high* ($\tilde{H}$), $j = 2$ for *medium* ($\tilde{M}$), and $j = 3$ for *low* ($\tilde{L}$). These linguistic terms are defined by the fuzzy numbers (Definition 2.1) with the corresponding membership functions defined in Eq. (6.4) and shown in Fig. 6.2. Finally, k refers to 6 hazardous materials indicated in Table 6.1.

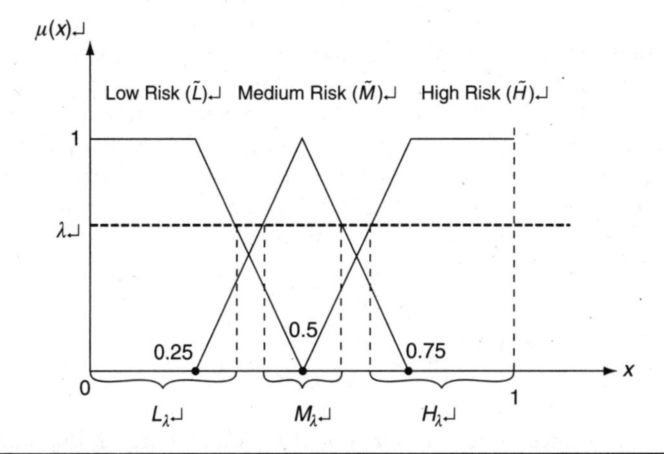

Figure 6.2 The fuzzy number of risk level.

k =	1	2	3	4	5	6
	Hg	Cd	Pb	Cr(VI)	PBBs	PBDEs

TABLE 6.1 Hazardous materials

$$\left.\begin{aligned}
\mu_{\tilde{L}}(x) &= \begin{cases} 1, & 0 \le x \le 0.25 \\ 4(0.5 - x), & 0.25 \le x \le 0.5 \\ 0, & x \ge 0.5 \\ 0, & x \le 0.25 \end{cases} \\
\mu_{\tilde{M}}(x) &= \begin{cases} 4(x - 0.25) & 0.25 \le x \le 0.5 \\ 4(0.75 - x) & 0.5 \le x \le 0.75 \\ 0, & x \ge 0.75 \end{cases} \\
\mu_{\tilde{H}}(x) &= \begin{cases} 0, & x \le 0.5 \\ 4(x - 0.5), & 0.5 \le x \le 0.75 \\ 1, & x \ge 0.75 \end{cases}
\end{aligned}\right\} \tag{6.4}$$

Because the result of the assessment is according to experience of the inspector, each risk level is represented by a fuzzy number, defined by Eq. (6.4) and shown in Fig. 6.2. On the premise that the categorization's reliability is [I?, the fuzzy numbers $\tilde{L}$, $\tilde{M}$, and $\tilde{H}$ will become the crisp intervals of L_λ $L_{I?}$, M_λ $M_{I?}$, and H_λ $H_{I?}$, how to remain the original mathematical form?] respectively, by λ-level cut defined in Definition 2.1. Formulation (6.5) shows how to integrate each part of a component's risk estimation to denote the severity of the component.

$$S_q = \frac{L_\lambda}{L_\lambda + M_\lambda + H_\lambda}\left(\sum_{i=1}^{n}\sum_{j,k=1}^{3}\frac{x_{ijk}}{n}\right) \tag{6.5}$$

(b) Frequency, F_q The total number of failed inspection reports are recorded. The frequency, F_q, refers to the ratio of total inspection reports to failed inspection reports. When a failed report appears in the inspection process, it means the supplier's product contains a certain risk, as denoted by h. Then the failed inspection reports' ratio cannot excess a level h' assigned by the DM. If a failed report appears, F_q will be counted from h with $h + h' = 1$, which is calculated by Eq. (6.6):

$$F_q = \begin{cases} 0 & \text{if} \quad x_q = 0 \\ h + \dfrac{x_q}{n} & \text{if} \quad 0 < \dfrac{x_q}{n} \le h' \\ M & \text{otherwise} \end{cases} \tag{6.6}$$

where n denotes the total number of inspection reports, x_q denotes the number of failed inspection reports, and M represents a large positive number. If the ratio of the failed inspection reports exceeds h', then the supplier will not be considered at all.

(c) Detection, D_q Upon inspection of an entire component, the concentration of hazardous material would be diluted, and thus decrease the accuracy of the inspection. So the risk of detection is according to the purity degree of material. This leads us to adopt a mixed ratio to denote the risk of detection. D_q^ρ examined by an auditor, which is expressed in fuzzy terms as high ($\tilde{H}$), medium ($\tilde{M}$), and low ($\tilde{L}$), as is defined similarly in Eq. (6.5) with confidence level ρ.

After obtaining the three dimensions of risk estimates, according to the relative weights obtained from AHP, it becomes the IQC's risk estimate as shown in Eq. (6.7):

$$Risk_{\text{IQC}} = w_{S_q} S_q + w_{F_q} F_q + w_{D_q} D_q^\rho \tag{6.7}$$

Index of Disassembly Effort (DEI)

The ideal green supply chain is a closed loop. All components/materials can be reused, remanufactured, or recycled, and will not produce any waste. Therefore, at the beginning stage, preconceptions of the other stage situations requires pre-action. In addition to the *disassembly effort index* (DEI) scorecard (Kuo, 2006) used to analyze the impacts of a supplier's components/materials during the disassembly stage, this study considers additional recovery rate as a criterion that evaluates the components/materials to conform with the EU's directive on the restriction of WEEE (2002).

Each criterion score ranges between 0 and 25—the smaller the better. Each score is then normalized into a unit interval between 0 and 1 to present the risk estimation. After obtaining the risk measures of all subcriteria, the DEI's risk level is estimated by the weighted sum of these subcriteria with the weights calculated from AHP as follows:

$$Risk_{\text{DEI}} = \sum_{i \in \text{DEI}} w_i y_i \tag{6.8}$$

where DEI = {time, tool, fixture, instruct hazard, force, access, recovery rate}.

Vendor Management (VM)

According to the practice of AVECTEC (2009), the criteria of vendor management include two major evaluation items and three bonus items. The two major evaluated resources are (1) self-inspection reports and (2) actual-inspection reports. There are 12 main items in the inspection reports, and each main item has a different number of subitems, which are summarized in Table 6.2.

i =	1	2	3	4	5	6	7	8	9	10	11	12
j =	19	9	9	21	12	24	76	15	14	6	8	8

TABLE 6.2 Amount of Subitems j in a Main Item i (AVECTEC, 2009)

1. Green product's quality certification system: 19 subitems.
2. Documents and records management: 9 subitems.
3. RoHS contract accreditation: 9 subitems.
4. Design management: 21 subitems.
5. Altered management: 12 subitems.
6. Purchase and supplier management: 24 subitems.
7. Manufacture management: 76 subitems.
8. Test and determination: 15 subitems.
9. Failure management and tracing: 14 subitems.
10. Correct treatment: 6 subitems.
11. Training program: 8 subitems.
12. Risk management: 8 subitems.

Rule of Report Evaluation

There are rules for audition on the different sources of reports. First, we shall transform semantic items into scores for quantified measure. Then, the threshold of pass or fail will be defined. Finally, integrating the scores with respect to the two sources of report—self-inspection and actual inspection—will be carried out based on the relative weights of importance derived from AHP. The details are illustrated below:

Recording Method Indicate "0" as totally unsatisfied, "1" as seldom satisfied, "3" as partly satisfied, and "9" as totally satisfied; the score of each subitem can be given. Therefore, when the subitem's score is lower than, or equal to 1, the subitem is recorded as "1" for a failure.

$$f_r(\mathbf{item}_{ij}) = \begin{cases} 1, & \textbf{if} \quad \mathbf{item}_{ij} \leq 1 \\ 0, & \textbf{if} \quad \mathbf{item}_{ij} \geq 3 \end{cases} \qquad i \in \{1, \ldots, 12\} \tag{6.9}$$

Evaluation Method Each main item uses the average score of the subitems to represent the performance level shown in Eq. (6.10):

$$\mathbf{item}_i = f_s(\mathbf{item}_{ij}) = \frac{\sum_j \mathbf{item}_{ij}}{\# j}, \qquad i \in \{1, \ldots, 12\} \tag{6.10}$$

Standard for Pass Adopting a general rule of thumb, when the average score of the main items is over 7, then the supplier passes the inspection standards. If a supplier's average score is lower than 7, then the performance of vendor management will show a large number "*M*" to denote the high risk.

Inspection Reports

Since there are two types of inspection reports: self-inspection report denoted by s, and the actual inspection report denoted by a, their risk levels are evaluated by three dimensions of FMEA.

Severity, S_s and S_a When a supplier has a good performance in management, then the component's/material's reliability will also be high. Because "1" denotes the worst case in risk evaluation, this study uses a "1 minus average score of the main items" in order to represent the severity, as shown in Eq. (6.11).

$$S_\varepsilon = 1 - \frac{\sum_i \textbf{item}_i}{12 \times 9}, \quad \varepsilon \in \{s, a\} \tag{6.11}$$

Frequency, F_s F_s and F_a Both these values use the ratio of the total number of subitems to failure items in order to express frequency. The severity S_s and S_a represent the risk of degree, and frequency F_s F_s and F_a represent the risk of occurrence. The computation method is as shown in Eq. (6.12):

$$F_\varepsilon = \frac{\sum_{i=1}^{12} \sum_j f_r(\textbf{item}_{ij})}{\sum_j} \quad \varepsilon \in \{s, a\} \tag{6.12}$$

Detection, D_s D_s and D_a This rating is according to an auditor's subjective opinion, and can result in a three-level rating inspection process. Since the rating is in a linguistic form, this study uses three fuzzy numbers, namely, $\tilde{E}$, $\tilde{N}$, and $\tilde{H}$ to represent *easy*, *normal*, and *hard*, respectively, with a similar definition to Eq. (6.4). On the premise that the auditor's reliability is μ, then, the fuzzy numbers $\tilde{E}$, $\tilde{N}$, and $\tilde{H}$ will become the crisp set, E_μ, N_μ, and H_μ.

After obtaining the three dimensions of risk estimations, their relative weights of importance with respect to the self-inspection and actual inspection of risk will be derived from AHP. Because these two evaluation items' forms are the same, the only difference is the auditor's opinion. The risk estimation of a self-inspection and actual inspection can be obtained by following Eqs. (6.13) and (6.14):

$$Risk_{\text{self-inspect}} = w_{S_s} S_s + w_{F_s} F_s + w_{D_s} D_s^\mu \tag{6.13}$$

$$Risk_{\text{actual-inspect}} = w_{S_a} S_a + w_{F_a} F_a + w_{D_a} D_a^\mu \tag{6.14}$$

Recognition and Certification

The warranty, self-declaration of an influential enterprise, and certification are each items of exception for vendor management. The warranty is a basic guarantee of a supplier's product and is quantified by a binary variable, P, where "1" means the supplier having a warranty; otherwise, this is denoted by "0". Standardized certification from international organizations and the declaration of an influential enterprise are considered to be better than a warranty, so using an increasing half-normal distribution to express the levels of satisfaction is shown in Eq. (6.15).

Based on different standards and effects of the products, the declaration of powerful enterprise will enable the classification for these enterprises. There are three level of classification, namely, α, β, and γ. The α class is the most notable, held by powerful international enterprises. After obtaining a risk estimation from both self-inspection and actual inspection reports, and any items of exception data, the next step is to aggregate by respective weights, which are obtained from AHP.

$$\mu_{\Delta}(x) = \begin{cases} 1 - e^{-kx^2}, & \textbf{when} \quad x > 0 \\ 0, & \textbf{otherwise} \end{cases} \qquad \Delta \in \{C, Y_{\alpha}, Y_{\beta}, Y_{\gamma}\} \tag{6.15}$$

$$\begin{aligned} Risk_{\text{management}} &= Max\{(w_a Risk_{\text{actual-inspect}} + w_s Risk_{\text{self-inspect}} \\ &\quad - (w_p P + w_c C + \sum_{i=\alpha,\beta,\gamma} w_i Y_i)), 0\}, \\ &\quad \text{where} P, C \in \{0, 1\}, Y \in Z^{+}. \end{aligned} \tag{6.16}$$

Note that while risk estimation is computed, the range of a risk value is between 0 and 1. Therefore, in Eq. (6.16), we take the maximum between the computed result and 0, and thus obtain a non-negative value.

Logistics (*L*)

Severity, S_l The expression S_l measures the time period x_t indicating the number of weeks the supplier was out-of-stock in the past two years. The unit between the two time points is one week, with 104 weeks over two years. A power function, $f_l(x_t)$, will express the severity of the time period, x_t. Because the larger the time period x_t, the better, using total weeks minus time period x_t represents the risk to be out-of-stock. Parameter a represents the influence of out-of-stock, which is decided by the DM.

$$S_l = f_l(x_t) = \left(\frac{104 - x_t}{104}\right)^a \quad 0 \le x_t \le 104 \; 0 \le a \le 1 \tag{6.17}$$

Frequency, F_l The expression F_l represents the number of orders, x_N, that have been out-of-stock among the total orders N, as shown in Eq. (6.18) with a similar definition as F_q.

$$F_l = \begin{cases} 0 & \text{if} \quad x_N = 0 \\ G + \dfrac{x_N}{N} & \text{if} \quad 0 < \dfrac{x_N}{N} \leq G' \\ M & \text{otherwise} \end{cases} \tag{6.18}$$

Detection, D_1 Because S_l and F_l use the records of transactions to evaluate a supplier's logistics level, an auditor's assessment will provide a linguistic statement to describe the status of the transaction records. There are three levels of status: Clear $\tilde{C}$, Normal $\tilde{N}$, and Unclear $\tilde{U}$, with definitions similar to Eq. (6.4). On the premise that the auditor's reliability is η, the fuzzy numbers $\tilde{C}$, $\tilde{N}$, and $\tilde{U}$ will become a crisp set C_η, N_η, and U_η.

After obtaining the three dimensions of risk estimation, and according to each of their weight, they are aggregated according to their relative weights of importance obtained from AHP to become the logistics' risk estimation in Eq. (6.19):

$$Risk_{\text{logistic}} = w_{S_l} S_l + w_{F_l} F_l + w_{D_l} D_l^\eta \tag{6.19}$$

6.3 Vendor Evaluation and Selection

In this section, the risk aggregation method and supplier ranking method are introduced.

6.3.1 Risk Aggregation Method

Prior to quantification, the calculated utility functions of the four major criteria are based on the risk attitude of the DM. Since the risk estimation of each criterion is an interval with a certain required level, to measure its corresponding risk utility value, the transformation method through the utility function is based on Eq. (6.20) and is illustrated in Fig. 6.3 below.

$$f_v(x) = \begin{cases} f_v(x) & \textbf{if} \quad UB = LB \\ \displaystyle\int_{LB}^{UB} f_v(x)dx \Big/ (UB-LB)\times 1 & \textbf{if} \quad UB \neq LB \end{cases} \tag{6.20}$$

Finally, these risk utility values will be integrated with objective weights to obtain the final risk level as in Eq. (6.21), of which the relative weights, w, of four main criteria—IQC(q), DEI(d), vendor

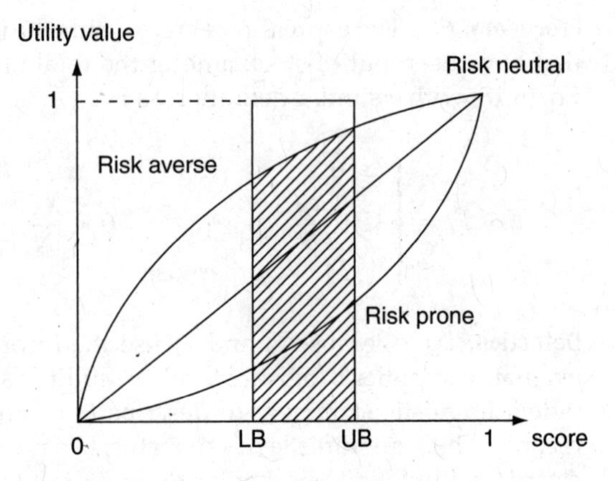

FIGURE 6.3 Risk level *wrt* interval score.

management (*vm*), and logistics (*l*)—of vendor selection are obtained from AHP.

$$R_s = w_q f_q(Risk_{\mathrm{IQC}}) + w_d f_d(Risk_{\mathrm{DEI}}) + w_{\mathrm{vm}} f_{\mathrm{vm}}(Risk_{\mathrm{management}}) + w_l f_l(Risk_{\mathrm{logistic}}) \quad (6.21)$$

6.3.2 Ranking Method

Each supplier will have a risk estimation, R_s. By incorporating the cost of each supplied component or material, a risk map can be constructed by using this information as two coordinate axes. Then, according to the minimum allowed values with respect to risk and cost, two thresholds can be drawn on the map to divide it into four sectors that correspond to four classes of suppliers, as shown in Fig. 6.4.

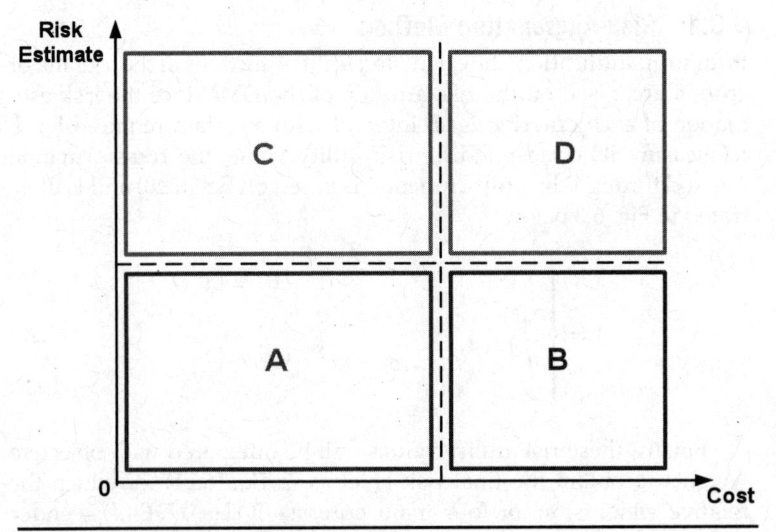

FIGURE 6.4 Risk map of suppliers.

Class A is the first priority because the risk estimations and costs there are all lower than the allowed levels assigned by the DM. Class B is the second priority, followed by Class C. These priorities are derived from international practice, whereby if an enterprise has any record of violating green legislation, the damages from risk are greater than the costs. Finally, the suppliers classified in Class D would be considered to have failed based on their combined high risk and high cost.

Based on this risk map, the DM may guide the supply partners to improve their performance, which will not only ensure the enterprise to use green suppliers, but also maintain a long-standing partnership with upper-stream suppliers. To achieve these purposes, factors for effective control and management need to be identified, which in turn rely on sensitivity analysis. Therefore, in the following section, we conduct such analysis.

6.4 Sensitivity Analysis and Alliance Development

There are two types of decision support methods adopted in this study to help the DM throughout the decision process: (1) AHP and (3) the utility theory. The two methods are independent of each other with respect to objective and subjective judgments. After developing a risk map, the analysis focuses on assistance to either the suppliers in Class C to cross the risk threshold, or to the suppliers in Class A to maintain a desirable performance. Therefore, sensitivity analysis is conducted to determine the most sensitive parameters, which can help both suppliers and the company to expand their goals to reach lower costs and achieve a higher level of greenness.

6.4.1 Issues of Sensitivity Analysis

There are two issues investigated by sensitivity analysis:

1. With respective to objective and subjective risk measures, which is more sensitive in maintaining their classes in the risk map?
2. With respective to the risk estimates of the criteria, which criterion would be most sensitive in maintaining their relative weights of importance?

To answer both questions, and to maintain the conditions, we investigate the allowed range for the parameters of a_v in the utility function, and the exponential parameter ϕ defined below, and determine that the parameter with the smaller allowable range has the higher degree of sensitivity.

Because subjective risk measures reflect the experience and knowledge domain of a DM, they are often more stable than objective risk measures. Therefore, we discuss how to perform sensitivity analysis on the weights derived from AHP, suggested by Hurley (2001).

6.4.2 Sensitivity Analysis on AHP

It has been noted that decision makers are often certain regarding the ranking order of attributes for a particular pairwise comparison matrix, but uncertain about the precise numerical weights that AHP produces for that matrix. Hurley (2001) has proposed a method to vary the weights derived from AHP by giving each element of a pairwise matrix an exponential parameter ϕ, defined below, but preserving the ranked order of the attributes in a matrix, which not only results in confidence for the AHP recommendations, but also provides the DM some flexibility in adjusting the weights. We may apply this method to our sensitivity analysis:

$\phi = 1$: represents the original matrix and will not affect the consistency ratio of the pairwise matrix;

$\phi > 1$: the order of weights will be the same as the original, but the largest weight will increase while the other weights decrease, and the consistency ratio will increase;

$0 < \phi < 1$: the order of weights will not change, but the largest weight will decrease while the other weights will increase, and the consistency ratio will decrease;

$\phi = 0$: all the elements in the comparison pairwise matrix are equal to 1; thus, all the weights remain the same. The matrix becomes the most consistent matrix, so the consistency ratio is 0.

6.5 Summary of the Selection Procedure

In summary, the risk estimation of a green supplier can be computed by the following procedure:

Step 1. Construct a hierarchical structure based on the relation of the considered criteria.

Step 2. Derive the subjective risk estimates from the utility functions of the main criteria and measure the weights of criteria by AHP.

Step 3. Quantify the risk level of each subcriterion by following the procedure of Sec. 2.2.

Step 4. Aggregate the risk estimations in Eq. (6.21) to obtain the overall risk level.

Step 5. Given the thresholds of allowed risk and costs to establish a risk map, plot the risk score of each supplier on the map for ranking.

Step 6. Use sensitivity analysis to determine the most sensitive parameters and establish an alliance strategy.

Different DMs have different preferences and attitudes toward risk for each criterion, which lead to different risk estimations. According to sensitivity analysis, the DM understands the allowed variations and also the key factors for management. A numerical example is provided in the following example.

6.6 Numerical Example

We consider an example of backlight suppliers who are considered by a TFT-LCD (thin film transistor liquid crystal display) original equipment manufacturer (OEM). According to the EU regulations and directives, two suppliers have gone through five steps of evaluation as shown in Sec. 6.6.1. The sixth step of sensitivity analysis is illustrated in Sec. 6.6.2. Finally, the conclusion of the example is given in Sec. 6.6.3.

6.6.1 Estimation of Risks and Ranking of Two Suppliers

There are six steps in the proposed selection process. The details of the process are illustrated in the Appendix, and the results are summarized below:

Step1: Constructing Hierarchical Structure

The hierarchical analysis structure is shown in Fig. 6.1.

Step 2: Establishing the Utility Function and Measuring the Weights of Criteria

A *purchase-in-charge* is a DM of the OEM who was asked four questions to establish his risk attitude in four major criteria. The risk utility functions are shown as follows:

$$f_d(x) = x \tag{6.22}$$

$$f_q(x) = x^{0.515} \tag{6.23}$$

$$f_{vm}(x) = x^{1.152} \tag{6.24}$$

$$f_l(x) = x^{1.322} \tag{6.25}$$

The four utility functions above represent the attitude of the DM, respectively: risk neutral toward DEI (d) [Eq. (6.22)]; risk aversion towards IQC (q) [Eq. (6.23)]; and risk prone towards vendor management (vm [Eq. (6.24)] and logistics (l) [Eq. (6.25)].

Level	Weights
1st	$W\{\text{DEI, IQC}, vm, I\} = \{0.166, 0.454, 0.263, 0.117\}$
2nd	$W_{\text{DEI}}\{\text{TI, TO, FI, A, I, H, FO}\} = \{0.078, 0.215, 0.108, 0.072, 0.101, 0.312, 0.114\}$; $W_{\text{IQC}}\{\text{S, F, D}\} = \{0.359, 0.517, 0.124\}$; $W_{vm}\{\text{S, A, W}, a\alpha, a\beta, a\gamma, \text{C}\} = \{0.163, 0.341, 0.063, 0.179, 0.098, 0.063, 0.093\}$; $W_I\{\text{S, F, D}\} = \{0.429, 0.429, 0.142\}$.

TABLE 6.3 Summary of the Relative Weights Derived from AHP

The relative weights of importance of the criteria were derived from the pair-wise matrices of each hierarchy based on AHP. Details can be found in the Appendix, and the weights are summarized in Table 6.3. All weights conform to consistent levels (<0.1) measured by *consistent index* (CI) and *consistent ratio* (CR).

Step 3: Measures of Risks

This step quantifies the suppliers' data. The substances used in a backlight must conform with international green regulations. According to expert experience, the risks of restricted substances are classified according to the volume used in the backlight. The risk level of each restricted substance is shown in Table 6.4.

Due to improvements in the technology, the Hg used in backlights can be exempted. The reliability of the risk-level classification, set by experts, is 0.9. Under this reliability as the α-level, the fuzzy number of each risk level, as defined in Eq. (6.4), will become a crisp set, as shown below:

$$L_{0.9} = [0, 0.275] \tag{6.26}$$

$$M_{0.9} = [0.475, 0.525] \tag{6.27}$$

$$H_{0.9} = [0.725, 1] \tag{6.28}$$

Supplier 1 has provided 10 inspection reports, with a mixed inspection ratio of 25%. Through normalization, as shown in Sec. 2.3, Hg can be exempted in the backlight, thus, the lower bound is 0; and the lower and upper bounds of the other substances are determined according to the rules. Each inspection report's information and result

Material/component	Hg	Cd	Pb	Cr(VI)	PBBs	PBDEs
Lamps, backlight	H	L	H	M	N/A	N/A

TABLE 6.4 Risk Level of Restricted Substances in the Backlight

Before [After] (*LB, UB*)	Hg (0,30)	Cd (3,10)	Pb (35,100)	Cr (20,55)
1	15 [0.500]	7 [0.571]	50 [0.231]	25 [0.143]
2	17 [0.567]	4 [0.143]	70 [0.538]	30 [0.286]
3	17 [0.567]	5 [0.286]	65 [0.462]	25 [0.143]
4	13 [0.433]	7 [0.571]	55 [0.308]	45 [0.714]
5	15 [0.500]	5 [0.286]	85 [0.769]	50 [0.857]
6	16 [0.533]	5 [0.286]	80 [0.692]	40 [0.571]
7	14 [0.467]	8 [0.714]	75 [0.615]	30 [0.286]
8	13 [0.433]	7 [0.571]	70 [0.538]	35 [0.429]
9	19 [0.633]	8 [0.714]	55 [0.308]	25 [0.143]
10	17 [0.567]	8 [0.714]	65 [0.462]	40 [0.571]
Total	156 [5.200]	64 [4.857]	670 [4.923]	340 [4.143]

TABLE 6.5 Inspection Report Before and After Normalization

of normalization are shown in Table 6.5 with normalization measures. It shows that all the inspection reports are under the restrictions, so the frequency F_q is 0.

IQC and risk estimation are shown as follows:

$$S_q = \frac{L_\lambda}{L_\lambda + M_\lambda + H_\lambda}\left(\sum_{i=1}^{n}\sum_{j,k=1}^{3}\frac{x_{ijk}}{n}\right)$$
$$= [0, 0.23]\cdot\frac{4.86}{10} + [0.26, 0.44]\cdot\frac{4.14}{10} + [0.40, 0.83]\cdot\frac{2.60+2.46}{10}$$
$$= [0.31, 0.71] \quad (6.29)$$
$$F_q = 0 \quad D_q = 0.25 \quad (6.30)$$
$$R_q = w_{S_q}S_q + w_{F_q}F_q + w_{D_q}D_q$$
$$= 0.359\cdot[0.313, 0.714] + 0.517\cdot 0 + 0.124\cdot 0.25 = [0.143, 0.287] \quad (6.31)$$

The DEI scorecard has been completed by an auditor. The score of each item, and the normalization results, are shown in Table 6.6. The normalization results will be integrated according to each weight,

Item	Time	Tool	Fixture	Access	Instruct	Hazard	Force
Score	5	10	5	0	0	10	5
$f_d(x_{DEI})$	0.2	0.4	0.2	0	0	0.4	0.2

TABLE 6.6 DEI Score Card

y \ Subitems score	0	1	3	9	Subitems number	Average score
1	1/1	2/2	6/6	10/10	19	5.789/5.789
2	0/0	0/0	1/2	8/7	9	8.333/7.667
3	0/0	0/0	3/3	6/6	9	7/7
4	1/2	4/3	1/1	15/15	21	6.762/6.714
5	1/1	0/0	2/3	9/8	12	7.250/6.750
6	1/1	1/1	3/4	19/18	24	7.542/7.292
7	0/2	2/2	4/4	70/68	76	8.474/8.237
8	0/0	0/0	2/3	13/12	15	8.200/7.800
9	0/0	0/1	1/1	13/12	14	8.571/8.000
10	1/1	0/0	0/0	5/5	6	7.500/7.500
11	0/0	1/1	1/1	6/6	8	7.250/7.250
12	1/2	1/0	0/0	6/6	8	6.875/6.750
Total	6/10	11/10	24/28	180/173	221	89.546/86.749

TABLE 6.7 Reports of Self-Inspection and Actual Inspection of Supplier 1 (Self/Actual)

which is computed by AHP, as shown in Eq. (6.32) as below.

$$\begin{aligned} R_d(x) &= \sum_{\text{DEI}} w_{\text{DEI}} y_{\text{DEI}} \\ &= 0.08 \cdot 0.20 + 0.22 \cdot 0.40 + 0.11 \cdot 0.2 + 0.07 \cdot 0 + 0.10 \cdot 0 \\ &\quad + 0.31 \cdot 0.40 + 0.11 \cdot 0.20 \\ &= 0.27 \end{aligned} \tag{6.32}$$

Supplier 1's reports of self-inspection and actual inspection are shown in Table 6.7.

Each report records the different scores of all subitems. The scoring method is shown in Sec. 6.2.3. The detection methods of inspection reports are similar to the risk levels of restricted substances, with an auditor's rating of normal and a reliability inspection of 0.9.

The risk estimations of the three dimensions are calculated respectively to self - and actual inspection reports as below:

$$S_s = 1 - \frac{\sum_{y=1}^{12} \text{item}_y}{12 \cdot 9} = 1 - \frac{89.546}{12 \cdot 9} = 0.171 \tag{6.33}$$

$$F_s = \frac{\sum_{y=1}^{12} \sum_{z} f_r(\text{item}_{yz})}{\sum^{z}} = \frac{6+11}{221} = 0.077 \tag{6.34}$$

$$D_s = [0.475, 0.525] \tag{6.35}$$

$$
\begin{aligned}
R_{\text{self-inspect}} &= w_{S_i} S_s + w_{F_i} F_s + w_{D_i} D_s^{\mu} \\
&= 0.558 \cdot 0.171 + 0.320 \cdot 0.077 + 0.122 \cdot [0.475,0.525] \\
&= [0.78,0.184] \qquad (6.36)
\end{aligned}
$$

$$
S_a = 1 - \frac{\sum_{y=1}^{12} \text{item}_y}{12 \cdot 9} = 1 - \frac{89.749}{12 \cdot 9} = 0.197 \qquad (6.37)
$$

$$
F_a = \frac{\sum_{y=1}^{12} \sum_{z} f_r(\text{item}_{yz})}{\sum^{z}} = \frac{10 + 10}{221} = 0.090 \qquad (6.38)
$$

$$
D_a = [0.475,0.525] \qquad (6.39)
$$

$$
\begin{aligned}
R_{\text{actual inspect}} &= w_{S_i} S_a + w_{F_i} F_a + w_{D_i} D_a^{\mu} \\
&= 0.558 \cdot 0.197 + 0.320 \cdot 0.090 + 0.122 \cdot [0.475,0.525] \\
&= [0.201,0.207] \qquad (6.40)
\end{aligned}
$$

Since Supplier 1 has certification of ISO and QC080000, then the risk estimation of vendor management is shown as follows:

$$
\begin{aligned}
R_{vm} &= Max\{w_A R_{\text{actual inspect}} + w_S R_{\text{self-inspect}} - (w_P P + w_C C + w_\alpha Y_\alpha \\
&\quad + w_\beta Y_\beta + w_\gamma Y_\gamma), 0\} \\
&= [0.163 \cdot [0.178,0.184] + 0.341 \cdot [0.201,0.207] - 0.093 \qquad (6.41)
\end{aligned}
$$

In logistics, Supplier 1 has one out-of-stock record during the past two years, occurring about 36 weeks ago. Following the rule of Sec. 6.3.1, 36 weeks is a time point in estimated severity and the out-of-stock record would be a consideration in estimated frequency. The acceptable ratio of out-of-stock is 0.25, as established by the DM. The risk estimation of logistics is shown as the following:

$$
S_l = f_l(x_t) = \left(\frac{104 - x_t}{104}\right)^a = \left(\frac{104 - 36}{104}\right)^{0.8} = 0.712 \qquad (6.42)
$$

$$
F_l = 0.75 + \frac{1}{30} = 0.783 \qquad (6.43)
$$

$$
D_l = [0,0.275] \qquad (6.44)
$$

$$
\begin{aligned}
R_l &= w_{S_l} S_l + w_{F_l} F_l + w_{D_l} D_l^{\eta} \\
&= 0.429 \cdot 0.712 + 0.429 \cdot 0.783 + 0.142 \cdot [0,0.275] \\
&= [0.641,0.680] \qquad (6.45)
\end{aligned}
$$

The details of the operational processes of Supplier 2 are omitted. Table 6.8 summarizes the objective information of two suppliers with the weights and the costs.

	Supplier 1	Supplier 2
DEI	0.272	0.326
IQC	[0.143,0.287]	[0.205,0.313]
Vendor management	[0.005,0.008]	[0.174,0.180]
Logistic	[0.641,0.680]	[0.718,0.766]
Cost (dollars/unit)	195	160

Table 6.8 Summary of the Risk Measures and Costs of Two Suppliers

Step 4: Risk Aggregation

This step aggregates risk estimations by the utility function of Eqs, (6.20) and (6.21). The result represents the risk estimation for Supplier 1 and Supplier 2, respectively, as shown below:

$$\begin{aligned} R_1 &= w_q f_q(R_q) + w_d f_d(R_d) + w_{vm} f_{vm}(R_{vm}) + w_l f_l(R_l) \\ &= 0.454 \cdot 0.451 + 0.166 \cdot 0.272 + 0.263 \cdot 0.003 + 0.117 \cdot 0.578 \\ &= 0.318 \end{aligned} \tag{6.46}$$

$$\begin{aligned} R_2 &= w_q f_q(R_q) + w_d f_d(R_d) + w_{vm} f_{vm}(R_{vm}) + w_l f_l(R_l) \\ &= 0.454 \cdot 0.498 + 0.166 \cdot 0.326 + 0.263 \cdot 0.136 + 0.117 \cdot 0.674 \\ &= 0.394 \end{aligned} \tag{6.47}$$

Step 5: Establish a Supplier Risk Map

According to the method of establishing thresholds for a risk map, the threshold of risk level is 0.380 and the threshold of cost is 200. The risk map is shown in Fig. 6.5. Supplier 1 in Category A is preferred.

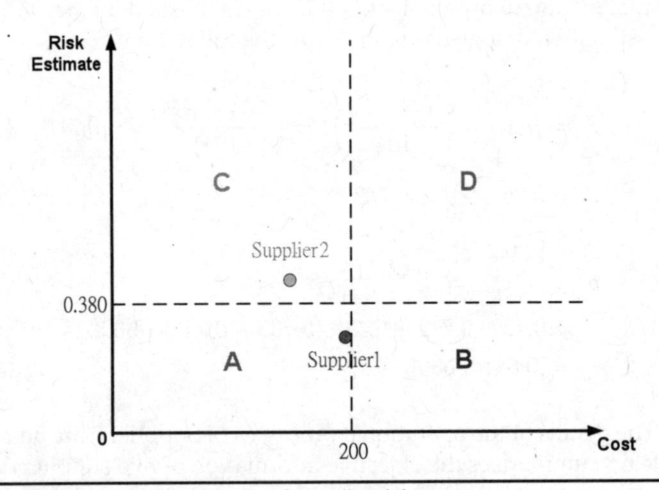

Figure 6.5 Numerical example of supplier risk map.

6.6.2 Sensitivity Analysis

Although Supplier 2 is not preferred, in order to maintain the partnership along the supply chain, it is the DM of this OEM who is responsible to supervise Supplier 2 so that their risk level can be reduced. To achieve this, the DM has to know which factors affect the risk level the most.

Recall that during the vendor selection processes, two measures of risks were derived from AHP and the utility function. AHP is an objective method using pairwise comparison to obtain the relationship between each criterion; whereas, the utility function is a subjective method to measure the utility value of each risk based on the attitude of the DM. Therefore, the sensitivity analysis of AHP and the utility function were examined first and then compared. The results show that the parameters of the utility function have lower sensitivity than the parameters ψ of AHP, which suggests that the parameters ψ of AHP have greater influence on risk measure than those of the utility function. The result is consistent with the analysis by Wang (2004) in which the objective weights are more important than the subjective weights if robust decisions are expected.

Thus, further analysis on the parameters, ψ, of AHP is conducted, and the results are shown in Table 6.9.

ϕ	CR	Supplier 1	Supplier 2
0	0	0.326	0.408
0.1	0.0003	0.324	0.406
0.2	0.001	0.321	0.403
0.3	0.002	0.320	0.400
0.4	0.004	0.318	0.399
0.5	0.006	0.318	0.397
0.6	0.009	0.317	0.396
0.7	0.013	0.317	0.396
0.8	0.016	0.317	0.395
0.9	0.021	0.318	0.395
1	0.026	0.318	0.394
1.1	0.033	0.320	0.396
1.2	0.038	0.322	0.397
1.3	0.045	0.324	0.398
1.4	0.052	0.326	0.399
1.5	0.060	0.328	0.401
1.6	0.068	0.331	0.403
1.7	0.077	0.333	0.404
1.8	0.086	0.337	0.407
1.9	0.097	0.340	0.409
2	0.108		

TABLE 6.9 Results of Sensitivity Analysis

The parameter ϕ ranges from 0 to 1.9. The DM can amend risk estimations by varying the parameter ϕ. In this example, when parameter ϕ increases, it means the order of the weights is the same as the original, but the largest IQC weight is increased, while the other weights are decreased, and thus, the risk estimation is increased. The DM can establish a set of weights that are most consistent with personal opinion.

According to the hierarchy structure, the weights of each subcriterion could be computed, as shown in Table 6.10.

Subcriterion	Weight (Rank)
DEI-Time	0.013 (21)
DEI-Tools	0.036 (9)
DEI-Fixture	0.018 (15)
DEI-Instruct	0.017 (16)
DEI-Hazard	0.052 (4)
DEI-Force	0.024 (12)
DEI-Access	0.012 (22)
IQC-Severity	0.163 (2)
IQC-Frequency	0.235 (1)
IQC-Detection	0.056 (3)
VM-Actual inspect-severity	0.050 (5)
VM-Actual inspect-frequency	0.029 (10)
VM-Actual inspect-detection	0.011 (23)
VM-Self inspect-severity	0.024 (12)
VM-Self inspect-frequency	0.014 (20)
VM-Self inspect-detection	0.005 (24)
VM-Warranty	0.017 (16)
VM-Admit α	0.047 (8)
VM-Admit β	0.026 (11)
VM-Admit γ	0.017 (16)
VM-Certification	0.024 (12)
Logistic-Severity	0.050 (5)
Logistic-Frequency	0.050 (5)
Logistic-Detection	0.017 (16)

Table 6.10 Weights of Subcriteria

6.6.3 Conclusion of the Example

In this example, based on Table 6.10, IQC is the most important criterion and should be improved to effectively reduce the risk level. Furthermore, among the factors of IQC, it can be noted that if Supplier 2 wants to advance to a higher class, the first improvement should be taken first, to decrease the number of failed inspection reports. That is, they need to check the content of hazardous substances, which includes the improvement of their production technology and their manufacturing quality level. Second, they need to increase the accuracy of their inspection reports by disassembling components so that purer material can be presented for inspection.

6.7 Summary and Conclusion

This chapter proposes an evaluation procedure to support the decision of green supplier selection. Different from traditional supplier selection, the criteria of green supplier selection must consider worldwide regulations for environmental protection. Based on these green criteria, this chapter constructs a hierarchical structure to conduct a complete analysis of green suppliers. Through risk analysis of green suppliers, each supplier will be classified according to their risk estimation and the cost, tracked on a supplier risk map. The supplier risk map provides the required reference for the DM to make a decision.

In summary:

- By engaging in discussions with numerous experts in related fields, the green criteria of this chapter completely considers the regulations of the EU, such as RoHS, WEEE, and EuP, and are suitable for instant application.
- Integrating FMEA with AHP renders risk evaluation more systematic, and the integration of the utility theory reflects the risk attitude of a DM more honestly.
- The supplier risk map provides instinctive information, which can help the DM to comprehend the interrelations between suppliers.

Future work that can be extended from this chapter is suggested as follows:

- Application of this methodology to more real problems and more case studies is required for further verification.
- Using risk estimations and costs as input parameters, construct a mathematical programming model to calculate the purchase quantity of each supplier.

References

AVECTEC (2009). Methodology and prototype development for green RoHS risk control module, *AVECTEC Report*, 1/5/2008–4/30/2009.

Chan, F. T. S., and N. Kumar (2007). Global supplier development considering risk factors using fuzzy extended AHP-based approach, *Omega*, 35, 417–431

Dickson, G. W. (1966). Analysis of vendor selection systems and decisions, *Journal of Purchasing*, 2(1), 5–17.

Directive 2002/96/EC of the European Parliament and of the Council of 27 January 2003 on Waste Electrical and Electrical Equipment (WEEE), *Official Journal of the European Union* 13.2.2003, 24–37.

EUP Directives, 25 April 2010, http://www.eup-network.de.

Hatush, Z., and M. Skitmore (1998). Contractor selection using multicriteria utility theory: An additive model, *Building and Environment*, 33, 105–115.

Ho, W. (2008). Integrated analytic hierarchy process and its applications: A literature review, *European Journal of Operational Research*, 186, 211–228.

Ho, W., X. Xu, and P. K. Dey (2010). Multicriteria decision-making approaches for supplier evaluation and selection: A literature review, *European Journal of Operational Research*, 202, 16–24.

Hu, A. H., C-W. Hsu, T-C. Kuo, W-C. Wu (2009). Risk evaluation of green components to hazardous substance using FMEA and FAHP, *Expert Systems with Applications*, 36, 7142–7147.

Hurley, W. J. (2001). The analytic hierarchy process: A note on an approach to sensitivity which preserves rank order, *Computers & Operations Research*, 28, 185–188.

Kainuma, Y., and N. Tawara (2006). A multiple-attribute utility theory approach to lean and green supply chain management, *Internal Journal of Production Economics*, 101, 99–108.

Kuo, T. C. (2006). Enhancing disassembly and recycling planning using life-cycle analysis, *Robotics and Computer-Integrated Manufacturing*, 22, 420–428.

ROHS Directives, 25 April 2010, http://www.rohs.gov.uk.

Saaty, T. L. (1980). *The Analytic Hierarchical Process*, New York: McGraw-Hill.

Teoh, P. C., and K. Case (2004). Failure modes and effects analysis through knowledge modelling, *Journal of Materials Processing Technology*, 154, 253–260.

Vaidya, O. S., and S. Kumar (2006). Analytic hierarchy process: An overview of applications, *European Journal of Operational Research*, 169, 1–29.

Wang, H. F. (2004). *Multicriteria Decision Analysis: From Certainty to Uncertainty*, Taiwan: Ting Lung Book Co.

Weber, C. A., J. R. Current, and W. C Benton, Vendor selection criteria and methods, *European Journal of Operational Research*, 50, 2–18, 1991.

Appendix: Pairwise Matrices Given for AHP of the Numerical Example

The DM is asked to perform a pairwise comparison of six matrices. The first pairwise matrix compares the four major criteria, and the results are shown in Table 6A.1.

The second pairwise matrix compares seven DEI attributes. The other pairwise matrices are shown in Tables 6A.2–6A.6. When the consistency ratio is less than 0.1, the pairwise matrix is acceptable.

	d	*q*	*vm*	*l*	Weight
d	1	1/3	1/2	2	0.166
q	3	1	2	3	0.454
vm	2	1/2	1	2	0.263
l	1/2	1/3	1/2	1	0.117
$\lambda_{max} = 4.071$, CI = 0.024, CR= 0.026					

Table 6A.1 The Pairwise Matrix of the Four Major Criteria

	TI	TO	FI	A	I	H	FO	Weight
TI	1	1/3	1/2	2	1	1/4	1/2	0.078
TO	3	1	2	2	3	1/2	3	0.215
FI	2	1/2	1	1	2	1/3	1/2	0.108
A	1/2	1/2	1	1	1/2	1/3	1/2	0.072
I	1	1/3	1/2	2	1	1/4	2	0.101
H	4	2	3	3	4	1	3	0.312
FO	2	1/3	2	2	1/2	1/3	1	0.114
$\lambda_{max} = 7.085$, CI = 0.081 CR = 0.061								

Table 6A.2 The Pairwise Matrix of DEI

	Severity	Frequency	Detection	Weight
Severity	1	1/2	4	0.359
Frequency	2	1	3	0.517
Detection	1/4	1/3	1	0.124
$\lambda_{max} = 3.108$, CI = 0.054, CR = 0.093				

Table 6A.3 The Pairwise Matrix of IQC

	S	A	W	$a\alpha$	$a\beta$	$a\gamma$	C	Weight
S	1	1/3	2	1	2	3	2	0.163
A	3	1	4	2	4	5	3	0.341
W	1/2	1/4	1	1/3	1/2	1	1/2	0.063
$a\alpha$	1	1/2	3	1	2	3	2	0.179
$a\beta$	1/2	1/4	2	1/2	1	2	1	0.098
$a\gamma$	1/3	1/5	1	1/3	1/2	1	1	0.063
C	1/2	1/3	2	1/2	1	1	1	0.093
$\lambda_{max} = 7.119$, CI = 0.020, CR = 0.015								

TABLE 6A.4 The Pairwise Matrix of Vendor Management

	Severity	Frequency	Detection	Weight
Severity	1	2	4	0.558
Frequency	1/2	1	3	0.320
Detection	1/4	1/3	1	0.122
$\lambda_{max} = 3.018$, CI = 0.009, CR = 0.016				

TABLE 6A.5 The Pairwise Matrix of Self- and Actual Inspection Reports

	Severity	Frequency	Detection	Weight
Severity	1	1	3	0.429
Frequency	1	1	3	0.429
Detection	1/3	1/3	1	0.142
$\lambda_{max} = 3$, CI = 0, CR = 0				

TABLE 6A.6 The Pairwise Matrix of Logistics

CHAPTER 7

Green Production

Manufacture and Remanufacture in Certain and Uncertain Environments

7.1 Introduction

Due to the increasing development of technology and the enhancement of living standards worldwide, natural resources have often been overused, leading to damage to the environment. Only in recent decades have people begun to give serious consideration to protecting the environment by reusing and recycling resources.

Due to the fact that pollution can spread from one region to another, and to the entire world by way of air and water, concern about pollution is no longer just a local issue but has become a global issue. As a result, many countries have sought to reach agreements and form global organizations with the goal of maintaining the ecological balance of the environment.

For example, in 1993, most countries joined in signing an international treaty—the United Nations Framework Convention on Climate Change (UNFCCC, http://unfccc.int/)—the goal of which is worldwide commitment to reducing global warming. To understand, quantify, and manage greenhouse gas emissions as part of this commitment, government and business leaders may use the accounting tool known as the Greenhouse Gas Protocol (GHG Protocol) (http://www.ghgprotocol.org/). In addition, the European Union strictly monitors that manufacturers adhere to legislation such as:

1. RoHS (http://www.rohs.gov.uk/): Restriction of the Use of Certain Hazardous Substances in Electrical and Electronic Equipment;
2. WEEE (http://en.wikipedia.org/wiki/Waste_Electrical_and_Electronic_Equipment_Directive): Imposes on manufacturers the responsibility for proper disposal of waste electrical and electronic equipment;

3. EuP (http://en.wikipedia.org/wiki/Eco-Design_of_Energy-Using_Products_Directive): Sets mandatory ecological requirements for energy-using and energy-related products.

Nowadays, under pressure from international treaties and environmental legislation, manufacturers need to produce products not only to meet the customer needs but also to engage in product recovery. In other words, manufacturers need now to be responsible for the entire life cycle of a product. As a consequence, manufacturers may need to take actions on returned products such as repair, refurbishing, remanufacturing, cannibalization, recycling or disposal. That is, traditional production planning systems need now to take into account both product recovery and greenhouse gas emission.

In general, production planning is a planning activity concerned with the overall operations of an organization over a specified period of time. In order to achieve its organizational mission, a company's planner must determine an efficient way of responding to market conditions and effectively allocate organizational resources over a designated period. In other words, the planner must consider anticipated sales opportunities, production requirements, and manpower in order to design a suitable plan before beginning production. According to different time ranges, Karimi et al. (2003) have classified three types of production planning, which are shown in Fig. 7.1.

In the long term, strategic planning usually focuses on market analysis and forecasts. By considering these factors, managers can make decisions about product design, product or service choices, strategic capacity planning, process selection, and facility layout. Under strategic planning, managers can obtain information such as (1) the resources and facilities available, (2) demand forecast for the period, (3) the costs for various alternatives (e.g., subcontracting, backordering), and (4) the costs for various resources (e.g., the cost of maintaining inventory, the cost of ordering, production costs, etc.). In addition, the information obtained will be used as input for aggregate planning.

Aggregate planning is concerned with meeting demand requirements and making good use of organizational resources over the medium time range in order to optimize integral production performance. A manager must decide what quantity of products is to be

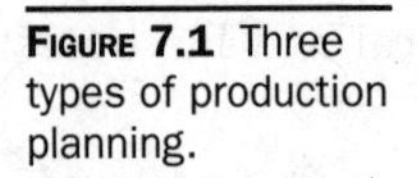
FIGURE 7.1 Three types of production planning.

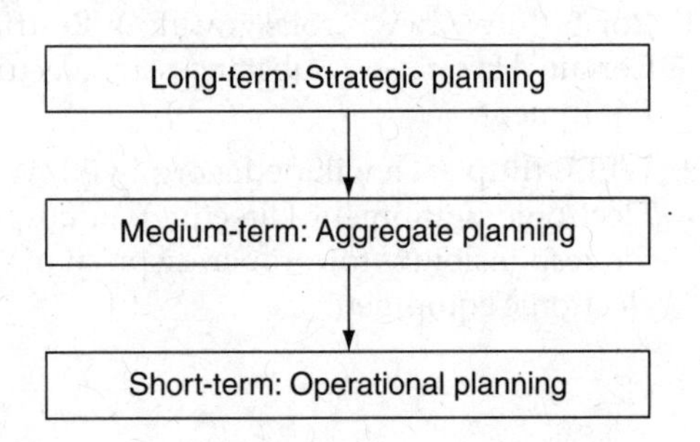

produced or outsourced, how many laborers are to be hired, and the amount of inventory to be held in stock and to be backlogged for each period. Finally, *operational planning* is needed to support aggregate planning by making more detailed plans that involve day-to-day scheduling of operations.

In order to reduce carbon emissions and lower total costs, a single-item capacitated dynamic lot-sizing problem is used, with batch manufacturing and remanufacturing in the closed-loop supply chain when carbon emissions for each period are taken into account. That is, if the manufacturer has exceeded the mandated limit for carbon emissions, then the manufacturer must pay the penalty cost.

In addition, because the return rate of the used products is very difficult to estimated, a fuzzy model is defined in this chapter to cope with this uncertainty.

The remainder of this chapter is organized as follows. In Sec. 7.2, we introduce current developments in production planning. In Sec. 7.3, mathematical models in the form of the mixed-integer programming model for both certain and uncertain environment are defined respectively is with analysis and numerical illustration. Finally, a summary and conclusion are provided in Sec. 7.4.

7.2 Current Development

In this section we focus on the lot-sizing model.

7.2.1 Elements of the Lot-Sizing Model

The deterministic production planning and inventory control model can be subdivided into *static* and *dynamic* models. The static model corresponds to the classical *economic order quantity* (EOQ), which seeks an optimal tradeoff between fixed setup and variable holding costs. The dynamic model corresponds to lot sizing. Li et al. (2007) defined the lot-sizing problem as identifying when and how much of a product to produce such that setup, production, and holding costs can all be minimized. Therefore, it is a type of production problem with the following elements.

Planning Horizon

The *planning horizon* sets a finite time frame for the lot-sizing problem. In aggregate planning, the planning horizon ranges between about 2 months and 12 months. That is, it is to assume the finite-horizon in lot-sizing problem.

Capacity

In long-term planning, managers must consider the constraints of manpower, equipment, machines, budget, etc., in a production system. These constraints can also be applied to the upper bounds of manufacturing or remanufacturing for each period. Both Lia et al. (2007)

and Pan et al. (2009) have considered the concept of *capacity* for a single item and have assumed that manufacturing and remanufacturing are independent during the production process. However, the outputs for manufacturing and remanufacturing are the same and may in fact overlap during production. Therefore, the capacity relationship between manufacturing and remanufacturing may be considered as a lot-sizing problem.

Periodic Demand

Periodic demand considers the input data in a lot-sizing problem and the demand change over time. Managers can obtain this input from a forecast derived from one of two methods—qualitative and quantitative. The *qualitative method* is based on soft information such as experts' comments, consumer market surveys, and a jury of executive opinions. The *quantitative method* is based on historical sales data or current data from test markets and is used to forecast demand, for example, by way of time-series forecasts and associative forecasts.

Inventory Shortage

In order to satisfy anticipated demand or reduce the variation within each demand period, a manufacturer usually produces finished goods or work-in-process ahead of time and then stores them as *inventory*. However, insufficient capacity may lead to an *inventory shortage*. Lia et al. (2007) considered the option of emergency procurement to satisfy demand in the current period when the shortage occurs. But there is another choice to be considered when such a shortage occurs, that is, backlogging, in which a manufacturer delays delivery of currently ordered goods.

7.2.2 Current Lot-Sizing Models

Due to environmental concerns, the lot-sizing model has started to integrate product recovery as part of its criteria. One of the important steps in product recovery is remanufacturing, which involves activities that make remanufactured products or major modules—potentially as good as new—to be marketable and sold again.

Richter et al. (2000) proposed the reverse Wagner–Whitin's dynamic production planning and inventory control model. In this reverse model, single-item production planning is considered in a discrete-time finite horizon with minimum total cost and there is no difference between the manufactured product and the remanufactured product. Relevant information about demand and the number of returned products for the entire planning horizon were known, and backlogging was not permitted.

As a result of varying periodic demand, returned products may not need to be remanufactured completely. Golany et al. (2001) consider three options for returned products: (1) get rid of the returned product; (2) keep the returned product for a future period; (3) remanufacture the returned product. Due to the constraint of capacity, the production system may incur a shortage, and the manager must then take action to avoid the shortage by substitutions or emergency procurement. Therefore, by considering the capacity of manufacturing and remanufacturing, Li et al. (2007) proposed the capacitated dynamic lot-sizing problem with substitutions and return products, where both manufacturing and remanufacturing are both considered as production. Tang et al. (2009) generalized the model to cope with the capacitated dynamic lot-sizing problem arising in a closed-loop supply chain by combining the capacities of manufacturing and remanufacturing, disposal, and the impermissibility of shortage and backlogging.

These models have offered a good framework for single-item production planning in a closed-loop supply for product recovery. However, there is a lack of consideration for carbon emissions from manufacturing and remanufacturing.

7.2.3 Conclusion

We introduce here a capacitated dynamic lot-sizing problem model that additionally considers three important issues in green production:

1. Because remanufactured products are considered to be as good as new, the processes of manufacturing and remanufacturing may overlap in utilizing system capacity. Therefore, the interaction between manufacturing and remanufacturing should be taken into account.
2. In a capacitated production system, a shortage may incur. In order to prevent such a shortage, two options in the lot-sizing problem can be considered. The first is emergency procurement to satisfy the insufficiency of current demand. The other is backlogging, wherein unsatisfied demand can be provided in future periods.
3. In a green production plan, not only should total costs be minimized, but excessive carbon emissions should also be avoided. Therefore, there is a need to trade off production modes between manufacturing and remanufacturing.

Table 7.1 summarizes the properties of these models with the model we shall introduce below. Therefore, the model we are going to introduce is to cope with a capacitated dynamic lot-sizing problem of which the batch manufacturing and remanufacturing, the relationship between manufacturing and remanufacturing in capacity, and the two options for preventing shortage are considered for reducing carbon emission and minimum total cost.

Features	BOAZ et al. (2001)	Teunter et al. (2006)	Yongjian Lia et al. (2007)	Absi et al. (2009)	Pan et al. (2009)	The Proposed Model
No. of product	Single	Single	Two	Multiple	Single	Single
No of periods	Multiple	Multiple	Multiple	Multiple	Multiple	Multiple
Manufacturing	Considered	Considered	Considered (batch production)	Considered	Considered	Considered (batch production)
Remanufacturing	Considered	Considered	Considered (batch production)	Not considered	Considered	Considered (batch production)
Disposal	Not considered	Not considered	Not considered	Not considered	Considered	Not considered
Capacity	Noncapacitated	Noncapacitated	Capacitated	Capacitated	Capacitated	Capacitated
The relation of capacity between manufacturing and remanufacturing	Not considered	Not considered	Independent	Independent	Independent	Dependent
Set-up cost	Not considered	Considered	Considered	Considered	Not considered	Considered
Backlogging	Not permitted	Not permitted	Not permitted	Not permitted	Not permitted	Permitted
Shortage	Not permitted	Not permitted	Permitted	Permitted	Not permitted	Permitted
Carbon emissions	Not considered	Not considered	Not considered	Not considered	Not considered	Considered

TABLE 7.1 Comparison of existing models with the proposed model

7.3 The Green Lot-Sizing Production Model

In the following, we introduce the structure of the closed-loop supply chain system in Sec. 7.3.1, and then detail the problem description with a mixed-integer programming model of the lot-sizing problem in a certain environment is formulated in Sec. 7.3.2. Due to the return rate is commonly uncertain, in Sec. 7.3.3 a fuzzy model is developed to cope with such uncertainty; and a simple example is presented as an illustration in Sec. 7.3.4. Finally, concluding remarks are provided in Sec. 7.3.5.

7.3.1 Framework of the Periodic Closed-Loop Production System

A factory produces a certain kind of product, for which demand is deterministic but time-varying during a finite planning horizon and which can also be satisfied by emergency procurement and backlogging. But the cost of emergency procurement is much higher than batch manufacturing and remanufacturing and backlogging will induce a penalty cost. Therefore, the manufacturer should keep batch production as normal as possible

The factory is also responsible for processing used products returned from customers. At the beginning of each period, there is no inventory of either returned or new products. The amount of the periodic returned products is fixed proportional ro the quantities of new product that are provided to customers before the current period over the planning horizon. There are two options available for these returned products: (1) remanufacturing and (2) disposal. Remanufactured products can be sold as new ones. In addition, the capacities of manufacturing and remanufacturing are limited, and may overlap when utilizing system capacity. According to their different production methods, manufacturing and remanufacturing have different levels of carbon emissions. If a manufacturer exceeds the mandated level of carbon emissions for any period, it will be required to pay a penalty. The framework for this periodic closed-loop production system is shown in Fig. 7.2.

7.3.2 Modeling in a Certain Environment

Referring to Fig. 7.2, we introduce in this section a deterministic, capacitated lot-sizing problem for a single item by batch manufacture and remanufacture in a finite planning horizon, so that the demand in each period along the planning horizon can be satisfied with the minimum total cost. Total costs involves production costs (batch manufacturing, batch remanufacturing, and disposal), holding costs (new product, returned product), emergency procurement costs, backlogging costs, and the penalty for excessive carbon emissions.

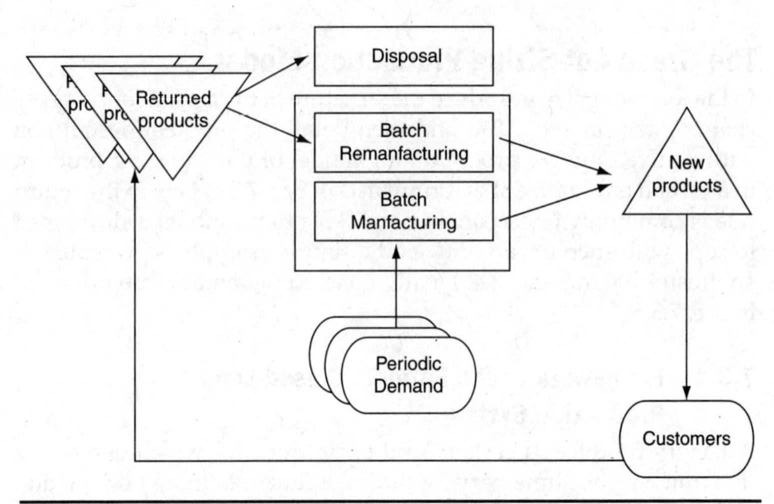

FIGURE 7.2 The framework of the periodic closed-loop production system.

We first define the notation to be used:

Parameters

T	Planning horizon
t	Index for periods in the planning horizon, $t = 1, 2, \ldots, T$
K	Resources for manufacturing or remanufacturing
k	Index for resources $k = 1, 2, \ldots, K$
U	The utilization period of the new product
D_t	The demand of new products in period t
p	The rate of returned product collected from each period available for remanufacturing
MC:	The cost of manufacturing per batch
RC:	The cost of remanufacturing per batch
DC:	The cost of disposal per unit
MS:	The set-up cost of manufacturing per period
RS:	The set-up cost of remanufacturing per period
MN:	The number of manufactured items per batch
RN:	The number of remanufactured items per batch
C_k :	The amount of resource k available for each period
CM_k	The quantity of resource k required per batch of manufacturing

CR_k The quantity of resource k required per batch of remanufacturing

SC The emergency procurement cost of new product per unit

BC_t The cost of backlogging after t periods per unit

HS The holding cost of new product per unit

HR The holding cost of returned product per unit

CE The limit of carbon emissions per period

CC The penalty cost for over-the-limit carbon emissions per unit

ME The carbon emissions of manufacturing per batch

RE The carbon emissions of remanufacturing per batch

M Large number

R The recyclable rate for each period

Variables

x_t The quantity of batches manufactured in period t

y_t The quantity of batches remanufactured in period t

z_t The quantity of return products disposed of in period t

r_t The quantity of return products in period t

is_t The quantity of new products held in inventory at the end of period t

ir_t The quantity of returned products held in inventory at the end of period t

s_t The quantity of shortage of new products in period t

b_{ij} The quantity of the ith demand supplied by manufacturing or remanufacturing in period j

$$1 \leq i < T \text{ and } i < j \leq T$$

m_t 0–1 binary variable for manufacturing set-up in period t

rm_t 0–1binary variable for remanufacturing set-up in period t

o_t The unit of carbon emissions over the limit in period t

Then a deterministic model is defined as follows.

Crisp Mixed-Integer Programming Model (CMIP)

Minimize

$$\sum_{t=1}^{T}[(MSm_t + MCx_t) + (RCy_t + RSy_t) + (DCz_t) + (HSis_t + HRir_t) + SCs_t + CCo_t] + \sum_{i=1}^{T-1}\sum_{j=i+1}^{T} BC_t b_{ij} \tag{7.1}$$

subject to

$$is_0 = 0 \quad (7.2a)$$

$$ir_0 = 0 \quad (7.2b)$$

$$r_t = 0 \quad t \in T \quad \text{and} \quad t \leq U \quad (7.3a)$$

$$r_t = R[D_{t-U} + \sum_{i=1}^{t-U-1} b_{i,t-U} - s_{t-U}] \quad t \in T \quad \text{and} \quad t > U \quad (7.3b)$$

$$MNx_t + RNy_t + is_{t-1} + s_t + \sum_{i=t+1}^{T} b_{ti} = D_t + \sum_{i=1}^{t-1} b_{it} + is_t \quad t \in T \quad (7.4)$$

$$ir_t - ir_{t-1} + RNy_t + z_t = pr_t \quad t \in T \quad (7.5)$$

$$CM_k x_t + CR_k y_t \leq C_k \quad t \in T, k \in K \quad (7.6)$$

$$x_t \leq Mm_t \quad t \in T \quad (7.7)$$

$$y_t \leq Mrm_t \quad t \in T \quad (7.8)$$

$$MEMNx_t + RERNy_t - CE \leq o_t \quad t \in T \quad (7.9)$$

$$o_t, r_t, z_t, ir_t, \geq 0 \quad (7.10)$$

$$x_t, y_t, is_t, s_t \geq 0 \text{ and integer } t \in T \quad (7.11)$$

$$m_t, rm_t \text{ binary variable} \quad (7.12)$$

$$b_{ij} \geq 0 \text{ and integer } 1 \leq i < T \text{ and } i < j \leq T \quad (7.13)$$

According to the constraints in Eqs. (7.3a) and (7.3b), the constraint in Eq. (7.5) can be rewritten as the following:

$$ir_t - ir_{t-1} + RNy_t + z_t = 0 \quad t \in T \quad \text{and} \quad t \leq U \quad (7.14a)$$

$$ir_t - ir_{t-1} + RNy_t + z_t - pR \sum_{i=1}^{t-U-1} b_{i,t-U} + pRs_{t-U}$$

$$= pRD_{t-U} \quad t \in T, \quad \text{and} \quad t > U \quad (7.14b)$$

In this model, we propose a crisp mixed-integer programming model (CMIP) of the lot-sizing problem with many parameters given a priori. Some of parameters can be estimated by historical data, but most of them are difficult for the decision maker to assign the precise value for the parameters because of unavailability or incompleteness of data in real-world situations, especially the quantity of returned products in the reverse supply chains. Therefore, the returned rate was often given by the experts based on their experience, which is normally expressed in linguistic terms. Thus, in the CMIP, the returned rate (R) will be considered as a fuzzy number as developed in the following sections.

7.3.3 Modeling in an Uncertain Environment

Referring to the definition of fuzzy number in Sec. 2.1 of Ch. 2, let us assume that the returned rate is a trapezoidal fuzzy number $\widetilde{R} = (r_2, r_3, r_2 - r_1, r_4 - r_3)$for ease of computation, and that it obeys the law of convolution. Its membership function is defined in Eq. (7.15):

$$\mu_{\widetilde{R}}(x) = \begin{cases} 0 & \forall x \in (-\infty, r_1] \\ \dfrac{x - r_1}{r_2 - r_1} & \forall x \in [r_1, r_2] \\ 1 & \forall x \in [r_2, r_3] \\ \dfrac{x - r_3}{r_4 - r_3} & \forall x \in [r_3, r_4] \\ 0 & \forall x \in [r_4, \infty) \end{cases} \tag{7.15}$$

Then the fuzzy mixed-integer programming model (FMIP) is proposed as below:

The Fuzzy Mixed Integer Programming Model (FMIP)

Minimize

$$\sum_{t=1}^{T} [(MSm_t + MCx_t) + (RCy_t + RSy_t) + (DCz_t) + (HSis_t + HRir_t) + SCs_t + CCo_t] + \sum_{i=1}^{T-1} \sum_{j=i+1}^{T} BC_t b_{ij} \tag{7.16}$$

subject to

$$is_0 = 0 \tag{7.17}$$

$$ir_0 = 0 \tag{7.18}$$

$$MNx_t + RNy_t + is_{t-1} + s_t + \sum_{i=t+1}^{T} b_{ti} = D_t + \sum_{i=1}^{t-1} b_{it} + is_t \quad t \in T \tag{7.19}$$

$$ir_t - ir_{t-1} + RNy_t + z_t = 0 \quad t \in T \quad \text{and} \quad t \le U \tag{7.20a}$$

$$ir_t - ir_{t-1} + RNy_t + z_t - p\widetilde{R} \sum_{i=1}^{t-U-1} b_{i,t-U} + p\widetilde{R}s_{t-U} = p\widetilde{R}D_{t-U} \quad t \in T, \ \text{and} \ \ t > U \tag{7.20b}$$

$$CM_k x_t + CR_k y_t \le C_k \quad t \in T, k \in K \tag{7.21}$$

$$x_t \le Mm_t \quad t \in T \tag{7.22}$$

$$y_t \le Mrm_t \quad t \in T \tag{7.23}$$

$$MEMNx_t + RERNy_t - CE \le o_t \quad t \in T \tag{7.24}$$

$$o_t, r_t, z_t, ir_t, \geq 0 \tag{7.25}$$

$$x_t, y_t, is_t, s_t \geq 0 \text{ and integer } t \in T \tag{7.26}$$

$$m_t, rm_t \text{ binary variable} \tag{7.27}$$

$$b_{ij} \geq 0 \text{ and integer } 1 \leq i < T \text{ and } i < j \leq T \tag{7.28}$$

Jiménez et al. (2007) have developed a method to convert the FMIP model into an equivalent auxiliary crisp model in the form of a possibility model involving imprecise coefficients. That is, consider the membership function of a fuzzy number, $\tilde{a}$, as defined in Eq. (7.29); the concepts of "expected value" and "expected interval" of a fuzzy number can then be described by Definition 7.1 below:

$$\mu_{\tilde{a}}(x) = \begin{cases} 0 & \forall x \in (-\infty, a_1], \\ & f_a(x) \text{ increasing on } [a_1, a_2] \\ 1 & \forall x \in [a_2, a_3], \\ & g_a(x) \text{ decreasing on } [a_3, a_4], \\ 0 & \forall x \in [a_4, \infty). \end{cases} \tag{7.29}$$

where $f_a(x)$ and $g_a(x)$ are continuous functions, $f_a(x)$ is increasing from 0 to 1, and $g_a(x)$ is decreasing from 1 to 0.

Definition 7.1: (Jiménez et al., 2007) The expected interval (EI) and expected value (EV) of fuzzy number $\tilde{a}$ are defined as follow:

$$EI(\tilde{a}) = [E_1^a, E_2^a] = \left[\int_0^1 f_a^{-1}(x)dx, \int_0^1 g_a^{-1}(x)dx \right] \tag{7.30}$$

$$EV(\tilde{a}) = \frac{E_1^a + E_2^a}{2} \tag{7.31}$$

When a fuzzy number is linear trapezoidal, as shown in Fig. 7.3 below, its expected interval and its expected value are as follows:

$$EI(\tilde{a}) = \left(\frac{1}{2}(a_1 + a_2), \frac{1}{2}(a_3 + a_4) \right) \tag{7.32}$$

$$EV(\tilde{a}) = \frac{1}{4}(a_1 + a_2 + a_3 + a_4) \tag{7.33}$$

Jiménez et al. (2007) have proposed a ranking method for any pair of fuzzy numbers $\tilde{a}$ and $\tilde{b}$, and the degree in which $\tilde{a}$ is bigger than $\tilde{b}$ is as follows:

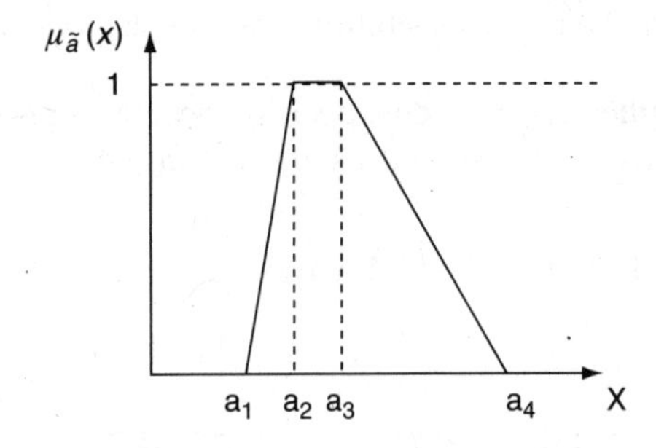

FIGURE 7.3 Trapezoidal fuzzy number $\tilde{a}$.

Definition 7.2 Given any pair of fuzzy numbers $\tilde{a}$ and $\tilde{b}$ as defined in Eq. (7.15), the degree that $\tilde{a}$ is bigger than $\tilde{b}$ is defined by

$$\mu_M(\tilde{a},\tilde{b}) = \begin{cases} 0 & \text{if } E_2^a - E_1^b < 0, \\ \dfrac{E_2^a - E_1^b}{E_2^a - E_1^b - (E_1^a - E_2^b)} & \text{if } 0 \in [E_1^a - E_2^b, E_2^a - E_1^b], \\ 1 & \text{if } E_1^a - E_2^b > 0, \end{cases} \tag{7.34}$$

where $[E_1^a, E_2^a]$ and $[E_1^b, E_2^b]$ are the expected intervals of $\tilde{a}$ and $\tilde{b}$, respectively.

When $\mu_M(\tilde{a},\tilde{b}) = 0.5$, we say that $\tilde{a}$ and $\tilde{b}$ are indifferent. When $\mu_M(\tilde{a},\tilde{b}) \geq \alpha$, we say that $\tilde{a}$ is bigger than, or equal to, $\tilde{b}$ at least at a degree α, and we will represent it by $\tilde{a} \geq {}_\alpha \tilde{b}$.

Since Definition 7.2 cannot handle indifference properly, Parra et al. (2005) proposed the α-indifference between $\tilde{a}$ and $\tilde{b}$ by a semantic as: "$\mu_M(\tilde{a},\tilde{b})$ is approximately ½."

Definition 7.3 According to Eq. (7.34), giving any pair of fuzzy numbers $\tilde{a}$ and $\tilde{b}$, $\tilde{a}$ is indifferent from $\tilde{b}$ to the degree α, $0\,\alpha\,1$, denoted by $\tilde{a} \approx {}_\alpha \tilde{b}$ if the following relationships hold simultaneously: $\tilde{a} \leq {}_{\alpha/2} \tilde{b}$ and $\tilde{b} \leq {}_{\alpha/2} \tilde{a}$, that is, $\tilde{a}$ is indifferent to $\tilde{b}$ in a degree α if $\frac{\alpha}{2} \leq \mu_M(\tilde{a},\tilde{b}) \leq 1 - \frac{\alpha}{2}$.

Now, we consider the following fuzzy mathematical programming model in which the constraints involve fuzzy numbers.

Minimize

$$z = cx$$

subject to

$$\begin{aligned} \tilde{a}_i x &\geq {}_\alpha \tilde{b}_i, \; i = 1, \ldots, l \\ \tilde{a}_i x &= {}_\alpha \tilde{b}_i, \; i = l+1, \ldots, m \quad x \geq 0 \end{aligned} \tag{7.35}$$

Then, the fuzzy feasibility is defined as follows.

Definition 7.4 Given a decision vector x, it is feasible in degree α if $\min_{i=1,\ldots,m} \{\mu_M(\widetilde{a}_i x, \widetilde{b}_{\partial i})\} = \alpha$ where $\widetilde{a}_i = (\widetilde{a}_{i1}, \widetilde{a}_{i2}, \ldots, \widetilde{a}_{in})$.

Based on Definition 7.2, we know $\widetilde{a}_i x \geq {}_\alpha \widetilde{b}_i \; i = 1, \ldots, l$ are equivalent to:

$$\frac{E_2{}^{a_i x} - E_2{}^{b_i}}{E_2{}^{a_i x} - E_1{}^{a_i x} + E_2{}^{b_i} - E_1{}^{b_i}} \geq \alpha, i = 1, \ldots, l \tag{7.36}$$

or

$$[(1-\alpha)E_2{}^{a_i} + \alpha E_1{}^{a_i}]x \geq \alpha E_2{}^{b_i} + (1-\alpha)E_1{}^{b_i}, \; i = 1, \ldots, l \tag{7.37}$$

Based on Definition 7.3 we know $\widetilde{a}_i x = {}_\alpha \widetilde{b}_i, i = l+1, \ldots, m$ are equivalent to:

$$\left(\left(1-\frac{\alpha}{2}\right) E_2{}^{a_i} + \frac{\alpha}{2} E_1{}^{a_i}\right) x \geq \frac{\alpha}{2} E_2{}^{b_i} + \left(1-\frac{\alpha}{2}\right) E_1{}^{b_i}, \; i = l+1, \ldots, m \tag{7.38a}$$

$$\left(\frac{\alpha}{2} E_2{}^{a_i} + \left(1-\frac{\alpha}{2}\right) E_1{}^{a_i}\right) x \leq \left(1-\frac{\alpha}{2}\right) E_2{}^{b_i} + \frac{\alpha}{2} E_1{}^{b_i}, \; i = l+1, \ldots, m \tag{7.38b}$$

Consequently, based on Definition 7.4, the equivalent auxiliary crisp α-parametric model of the model (7.35) can be written as follows:

Minimize

$$z = cx$$

subject to

$$\begin{gathered} ((1-\alpha)E_2{}^{a_i} + \alpha E_1{}^{a_i})x \geq \alpha E_2{}^{b_i} + (1-\alpha)E_1{}^{b_i}, \; i = 1, \ldots, m \\ \left(\left(1-\tfrac{\alpha}{2}\right) E_2{}^{a_i} + \tfrac{\alpha}{2} E_1{}^{a_i}\right)x \geq \tfrac{\alpha}{2} E_2{}^{b_i} + \left(1-\tfrac{\alpha}{2}\right) E_1{}^{b_i}, \; i = l+1, \ldots, m \\ \left(\tfrac{\alpha}{2} E_2{}^{a_i} + \left(1-\tfrac{\alpha}{2}\right)E_1{}^{a_i}\right)x \leq \left(1-\tfrac{\alpha}{2}\right) E_2{}^{b_i} + \tfrac{\alpha}{2} E_1{}^{b_i}, \; i = l+1, \ldots, m \quad x \geq 0 \end{gathered} \tag{7.39}$$

According to above descriptions, the FMIP can transform into the equivalent auxiliary crisp model (ACM) as described in the following.

The Equivalent Auxiliary Crisp Model (ACM)

Minimize

$$\sum_{t=1}^{T}\left[(MSm_t + MCx_t) + (RCy_t + RSy_t) + (DCz_t) + (HSis_t + HRir_t)\right] + SCs_t + CCo_t + \sum_{i=1}^{T-1}\sum_{j=i+1}^{T} BC_t b_{ij} \tag{7.40}$$

subject to

$$is_0 = 0 \tag{7.41}$$

$$ir_0 = 0 \tag{7.42}$$

$$MNx_t + RNy_t + is_{t-1} + s_t + \sum_{i=t+1}^{T} b_{ti} = D_t + \sum_{i=1}^{t-1} b_{it} + is_t \quad t \in T \tag{7.43}$$

$$ir_t - ir_{t-1} + RNy_t + z_t = 0 \quad t \in T, \quad \text{and} \quad t \le U \tag{7.44}$$

$$ir_t - ir_{t-1} + RNy_t + z_t - p\left(\left(1-\frac{\alpha}{2}\right)E_2{}^R + \frac{\alpha}{2}E_1{}^R\right)\sum_{i=1}^{t-U-1} b_{i,t-U} + p\left(\left(1-\frac{\alpha}{2}\right)E_2{}^R + \frac{\alpha}{2}E_1{}^R\right)s_{t-U} \ge p\left(\frac{\alpha}{2}E_2{}^R + \left(1-\frac{\alpha}{2}\right)E_1{}^R\right)D_{t-U}$$

$$t \in T, \text{ and } t > U \tag{7.45a}$$

$$ir_t - ir_{t-1} + RNy_t + z_t - p\left(\frac{\alpha}{2}E_2{}^R + \left(1-\frac{\alpha}{2}\right)E_1{}^R\right)\sum_{i=1}^{t-U-1} b_{i,t-U} + p\left(\frac{\alpha}{2}E_2{}^R + \left(1-\frac{\alpha}{2}\right)E_1{}^R\right)s_{t-U} \le p\left(\left(1-\frac{\alpha}{2}\right)E_2{}^R + \frac{\alpha}{2}E_1{}^R\right)D_{t-U}$$

$$t \in T, \text{ and } t > U \tag{7.45b}$$

$$CM_k x_t + CR_k y_t \le C_k \quad t \in T, k \in K \tag{7.46}$$

$$x_t \le Mm_t \quad t \in T \tag{7.47}$$

$$y_t \le Mrm_t \quad t \in T \tag{7.48}$$

$$MEMNx_t + RERNy_t - CE \le o_t \quad t \in T \tag{7.49}$$

$$o_t, r_t \ge 0 \tag{7.50}$$

$$o_t, r_t, z_t, ir_t, \ge 0 \tag{7.51}$$

$$x_t, y_t, is_t, s_t \ge 0 \text{ and integer } t \in T \tag{7.52}$$

$$b_{ij} \ge 0 \text{ and integer } 1 \le i < T \text{ and } i < j \le T \tag{7.53}$$

There are $T(6+K) - U$ constraints and $(T^2+19T)/2$ variables in this model. Comparison on the scale between CMIPM and ACM can be referred to Table 7.2:

	CMIPM	ACM	Comparison of These Two Model
Variables	$(T^2 + 19T)/2$	$(T^2 + 19T)/2$	0
Constraints	$T(5 + K)$	$T(6 + K) - U$	$T - U$

TABLE 7.2 The comparison of two models

7.4 Numerical Illustration

In this section, we use an example to illustrate the proposed model. The input data is as follows:

- $\widetilde{R} = (0.7, 0.7, 0.3, 0.2)$, the returned rate for product per period
- Time period: 8
- D_t, the number of new products in period t

t	1	2	3	4	5	6	7	8
Demand (unit)	450	850	531	348	500	700	350	670

- $MC = 90$ (price), the cost of manufacturing per batch
- $RC = 60$ (price), the cost of remanufacturing per batch
- $DC = 0.15$ (price), the cost of disposal per unit
- $MS = 100$ (price), the set up cost of manufacturing per period
- $RS = 60$ (price), the set up cost of remanufacturing per period
- $MN = 80$ (unit), the number of manufacturing per batch
- $RN = 60$ (unit), the number of remanufacturing per batch
- $SC = 2.3$ (price), the shortage cost of new product per unit
- $HS = 0.3$ (price), the holding cost of new product per unit
- $HR = 0.1$ (price), the holding cost of returned product per unit
- $CE = 400$ (unit), the limit of carbon emission per period
- $CC = 2.5$ (price), the penalty cost of over the limit of carbon emission per unit
- $ME = 50$ (price), the carbon emission of manufacturing per batch
- $RE = 10$ (price) the carbon emission of remanufacturing per batch
- $p = 0.9$, the rate of returned product that can be remanufactured

- Resources and capacities

K (resource)	1	2	3
C_k	350	400	90
CM_k	20	50	10
CR_k	35	5	10

- The cost of backlogging after t period per unit

t	1	2	3	4	5	6	7
BC_t	0.1	0.2	0.3	0.4	0.5	0.6	0.7

Then the optimization package ILOG OPL Studio 3.5 has been used to solve the example on a Pentium IV 1.6 GHz PC. Under the different feasibility degrees the optimal values of different cost items are shown in Table 7.3.

In Fig. 7.4, we can observe the positive correlation between the total cost and the feasibility degree α in the fuzzy model. This is due to the fact that, the higher the degree of satisfaction of constraints (feasibility degree, α), the smaller the feasible region becomes; and the worse is the optimal value.

With this information, the decision maker is able to trade off between the degree of satisfaction of the constraints and the objective

Feasibility Degree, α	Manufacturing Cost	Remanufacturing Cost	Disposal Cost
0	2460	2760	7.911
0.1	2460	2760	7.7409
0.2	2460	2760	7.5708
0.3	2460	2760	7.4007
0.4	2460	2760	0
0.5	2460	2760	0
0.6	2650	2700	0.6318
0.7	2650	2700	0.5852
0.8	2650	2700	6.4517
0.9	2650	2700	6.2836
1	2650	2700	6.1875
Deterministic model	2650	2700	6.1875

TABLE 7.3 The level costs of different items (*continued*)

Feasibility Degree, α	Inventory Cost		Shortage		Total Cost
	New Item	Returned Item	Emergency Procurement	Backlogging	
0	96	38.961	43.7	35.2	5441.772
0.1	96	42.7059	43.7	35.2	5445.3468
0.2	101.1	44.676	43.7	36.9	5453.9468
0.3	114.3	43.6626	43.7	41.3	5470.3633
0.4	114.6	58.5891	43.7	40.9	5477.7891
0.5	114.6	62.7458	43.7	40.9	5481.9458
0.6	55.2	38.7516	0	41	5485.5834
0.7	55.2	40.4405	0	41	5487.2257
0.8	39.9	43.878	0	63.9	5504.1297
0.9	39.9	46.119	0	63.9	5506.2026
1	39.9	48.312	0	63.9	5508.2995
Deterministic model	39.9	48.312	0	63.9	5508.2995

Table 7.3 *(Continued)*

value, and thereby choose the best one for production planning. In terms of the bases, Table 7.4 provides a different product-mix with respect to different levels of satisfaction on the constraints.

By increasing the feasibility degree of the constraints [Eqs. (7.45a) and (7.45b)], the returned product become stricter, and the variation

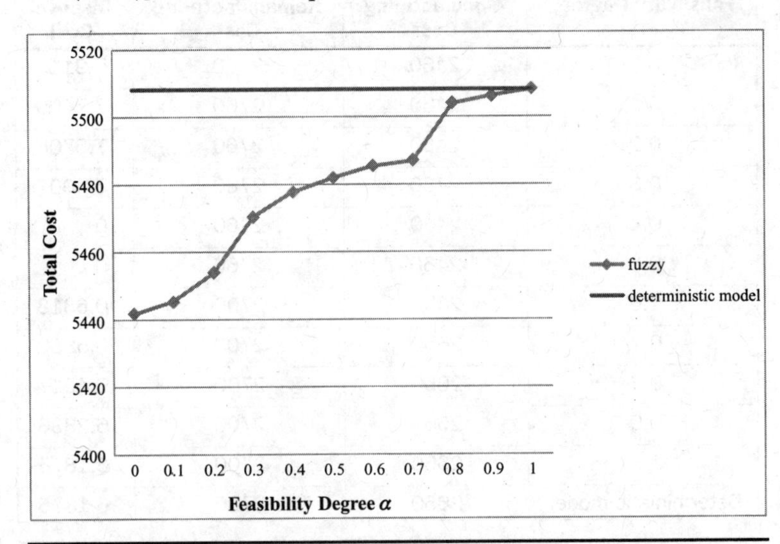

Figure 7.4 Total costs of deterministic and fuzzy models.

Feasibility degree, α	x								y							
	1	2	3	4	5	6	7	8	1	2	3	4	5	6	7	8
0	8	8	0	0	0	8	0	0	0	0	5	9	9	0	9	9
0.1	8	8	0	0	0	8	0	0	0	0	5	9	9	0	9	9
0.2	8	8	0	0	0	8	0	0	0	0	5	9	9	0	9	9
0.3	8	8	0	0	0	8	0	0	0	0	5	9	9	0	9	9
0.4	8	8	0	0	0	8	0	0	0	0	5	9	9	0	9	9
0.5	8	8	0	0	0	8	0	0	0	0	5	9	9	0	9	9
0.6	6	8	3	0	0	8	0	0	0	0	5	9	8	0	9	9
0.7	6	8	3	0	0	8	0	0	0	0	5	9	8	0	9	9
0.8	5	8	4	0	0	8	0	0	0	0	4	9	9	0	9	9
0.9	5	8	4	0	0	8	0	0	0	0	4	9	9	0	9	9
1	5	8	4	0	0	8	0	0	0	0	4	9	9	0	9	9
Deterministic model	5	8	4	0	0	8	0	0	0	0	4	9	9	0	9	9

Table 7.4 The optimal production mixes under different feasibility degrees

region, which is the batch of available remanufacturing in every period, also becomes smaller. According to a different interval feasibility degree, we can classify three types of production planning with respect to levels of satisfaction : (1) 0~0.5, (2) 0.6~0.7, (3) 0.8~1.

7.5 Summary and Conclusion

In this chapter, we consider a lot-sizing green production problem in both certain and uncertain environments. By considering the reuse of the recycled products for remanufacture and also the reduction of the carbon emissions, the corresponding models in basic mixed-integer programs were introduced in the basic form of mixed-integer programs. Because the uncertain model arises from an uncertain return rate and was developed by communicating with a decision maker, the linguistic data are transformed into a fuzzy number and incorporated into the modeling. This enables the decision maker to make decisions by trading off the levels of satisfaction on the resource constraints with the total costs.

From the analysis, we also can realize the flexibility of the uncertain model developed in the form of a fuzzy model, so that the deterministic model becomes its special case. Solving the fuzzy model by transforming into its auxiliary crisp model, crisp decisions with respect to different levels of satisfaction can be obtained.

References

Absia, N., and S. Kedad-Sidhoumb (2009). The multi-item capacitated lot-sizing problem with safety stocks and demand shortage costs, *Computers & Operations Research*, 36, 2926–2936.

Golany, B., J. Yang, and G. Yu (2001). Economic lot-sizing with remanufacturing options, *IIE Transactions* 33, 995–1003.

Jiménez, M., M. Arenas, A. Bilbao, and M. V. Rodríguez (2007). Linear programming with fuzzy parameters: An interactive method resolution, *European Journal of Operational Research* 177, 1599–1609.

Karimi, B., S. M. T. F. Ghomi, and J. M. Wilson (2003). The capacitated lot-sizing problem: A review of models and algorithms, *Omega* 31, 365–378.

Parra, M.A., A. Bilbao-Terol, B. Pérez-Gladish, M. V. Rodríguez Uría (2005). Solving a multiobjective possibilistic problem through compromise programming, *European Journal of Operational Research* 164, 748–759.

Li, Y., J. Chen, and X. Cai (2007). Heuristic genetic algorithm for capacitated production planning problems with batch processing and remanufacturing, *International Journal of Production Economics* 105, 301–317.

Pan, Z., J. Tang, and O. Liu (2009). Capacitated dynamic lot-sizing problems in closed-loop supply chain, *European Journal of Operational Research* 198, 810–821.

Richter, K. (2000). Remanufacturing planning for the reverse Wagner/Whitin models, *European Journal of Operational Research* 121, 304–315.

Teunter, R.H., Z. P. Bayindir, and W. van den Heuvel (2006). Dynamic lot-sizing with product returns and remanufacturing, *International Journal of Production Research*, 44(20), 4377–4400.

CHAPTER 8

Green Logistics

Recycling with Certain and Uncertain Situations

8.1 Introduction

Given the limited energy and resources, sustainability has become an important topic in environmental protection in recent years. The *green supply chain* (GSC) management is now suggested as an efficient tactic to achieve this goal. On the basis of the three Rs—recycling, reuse, and recovery (3R)—in GSC management, a green company puts in effort to prevent any waste of materials from the life cycle of a product. Therefore, closed-loop logistics are required to facilitate the 3R processes.

Closed-loop logistics for a green company consists of two parts: forward logistics and reverse logistics. Apart from the conventional logistics, GSC has an additional role called *dismantlers*, which allow green logistics to operate with the additional functions of recovery and recycling. Figure 8.1 shows such a structure.

The uncertainty embedded in reverse logistics has been a challenge for GSC managers. For the conventional supply chain, uncertain demand has affected inventory level, production amounts, and logistics. The uncertain factors of the reverse supply chain are more complex than those of the forward supply chain. Apart from the uncertain demand, the values of recovery rate and landfilling rate pose difficulties in estimation, yet both contribute to the major factors of reverse logistics management (Kongar, 2004; Ovidiu, 2007; Salema et al., 2007). To incorporate logistics, the differences between forward supply and reverse supply chains were studied by Kongar (2004), and a comparison was summarized as shown in Table 8.1. In addition, the table shows the difficulties in the high level of uncertainty embedded in a reverse supply chain.

Incorrect estimation and judgment based on uncertain information will cause a high risk of loss. Thus, in this chapter, after a basic model based on Hsu and Wang (2009) for deterministic closed-loop

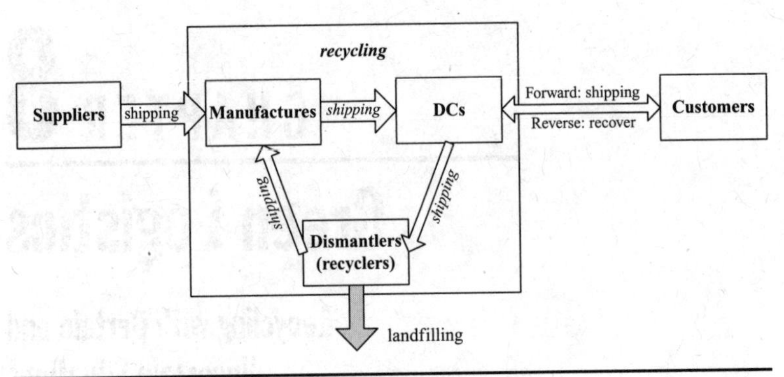

FIGURE 8.1 Framework of green supply chain logistics.

logistics is introduced, the uncertain factors of demand, rates of recovery, and landfilling along the reverse logistics will be considered to support more realistic decisions of logistics and facility locations. Owing to the capability of fuzzy presentation to engage uncertain patterns, fuzzy numbers will be used to describe these uncertain factors.

Forward Supply Chain	Reverse Supply Chain
Easier to forecast	Harder to forecast
Profit benefit orientated	Environmentally benign product benefit orientated
Distribution to multiple locations from a single source	Distribution to a single location from multiple sources
Stable product quality	Unstable product quality
Stable product packaging	Unstable product packaging
Stable product structure	Unstable product structure
Route of distribution is known/determined	Route of distribution is unknown/undetermined
Known main characteristics	Unknown main characteristics
More or less stable pricing	Pricing is effected by various factors/less stable
Speed is important	Speed is not a factor
Easily visualized cost factors	Hard to determine the cost
Stable inventory management	Unstable inventory management
Manageable product life cycle	More complicated product life cycle
Well known marketing techniques	Marketing techniques involve more complicated factors
Clearly observed processes	Less visible processes

TABLE 8.1 Comparison of Forward Supply and Reverse Supply Chains (Kongar, 2004)

Thus, based on Wang and Hsu (2010), fuzzy mathematical programming will be introduced for uncertain modeling.

Fuzzy programming has been discussed with different viewpoints in the literature. Among these, the statistical approach towards possibility is a way to synthesize fuzzy information. In the framework of fuzzy programming, possibilistic mean and variance will be formulated in this chapter to transform the proposed fuzzy mathematical model into a crisp form to facilitate efficient computation and analysis. Furthermore, because of such uncertainty, there is a risk caused by violating the estimated resource constraints. Therefore, risk analysis will be conducted so that *decision makers* (DMs) can trade off between the expected cost saving and resource utilization based on their perceptions towards the GSC management in question.

After reviewing the literature on the analysis of GSC management in uncertain environments, including modeling and defuzzification procedures discussed in Sec. 8.2, a mathematical programming model of the green supply chain logistics will be introduced in Sec. 8.3. Then, the solution process and risk method of interval programming after transferring fuzzy numbers into interval numbers will be proposed with specifications in Sec. 8.4. A numerical example of an uncertain GSC will be presented in Sec. 8.5. Finally, in Sec. 8.6, the conclusion will be drawn.

8.2 Deterministic Modeling of Closed-Loop Logistics

From the concepts we described above, we know that the closed-loop supply chain is different from a conventional supply chain. The problems involved are more complex, and need more than double the effort to analyze both forward and reverse logistics simultaneously.

To measure the effectiveness of the logistics in a closed-loop network, the cost is normally considered by a company. Besides, in a multistage supply chain network problem, the following conditions should be satisfied in modeling:

1. The demand of each customer must be satisfied.
2. The flow is only allowed to be transferred between two consecutive stages.
3. Both the number of facilities that can be opened and their capacities are limited.

Because they are also the basic conditions for closed-loop logistics, we shall consider them as our assumptions in modeling.

Note that there are essentially five stages along a green logistic network: suppliers, manufacturers, DCs, customers, and dismantlers. Apart from the common conditions of the satisfied demand in (1), and limited capacities in (3); from assumption (2) it can be noted that there are no flows between the facilities at the same stage.

One special issue of closed-loop logistics is the recycling rate, including the recovery and landfilling rates. Laan et al. (1999) point out that in the recovery systems, a common assumption is that the amounts of the returned products depend on the demand of the products. To adopt this assumption, the recovery amount is assumed to be a percentage of the customer demand in our model. This leads to the fourth assumption as

4. The recovery and landfilling rates are given.

The goal of this study is to design a closed-loop supply chain logistics system that can minimize the total transportation and the operation costs by determining locations of the facilities and the flows of the operation units along each capacity-constrained stage when the demand of customers and the recycling rates are given. This closed-loop system is meant to support long-term steady-state logistics decisions. Therefore, from the economic point of view, we can suggest the minimal cost flows and opening facilities in the system.

8.2.1 The Deterministic Closed-Loop Logistics Model (DCLL)

Consider the integer-valued basic logistics units in our system. In this section, based on four assumptions and the network structure, we shall define a mathematical model to describe such logistic system.

Before modeling, we define the related parameters and notations as follows:

Indices

I: The number of suppliers with $i = 1, 2, \ldots, I$

J: The number of manufactories with $j = 1, 2, \ldots, J$

K: The number of DCs with $k = 1, 2, \ldots, K$

L: The number of customers with $l = 1, 2, \ldots, L$

M: The number of dismantlers with $m = 1, 2, \ldots, M$

Parameters

a_i: Capacity of supplier i

b_j: Capacity of manufactory j

Sc_k: Total capacity of forward and reverse logistics in the DC k

pd_k: The percentage of total capacity for reverse logistics in DC k

pc_l: Recovery percentage of customer l

pl_m: The landfilling rate of dismantler m

d_l: Demand of the customer l

e_m: Capacity of dismantler m

s_{ij}: Unit cost of production in manufactory j using materials from supplier i

t_{jk}: Unit cost of transportation from each manufactory j to each DC k

u_{kl}: Unit cost of transportation from DC k to customer l

v_{km}: Unit cost of transportation from DC k to dismantler m

w_{mj}: Unit cost of transportation from dismantler m to manufactory j

Ru_{lk}: Unit cost of recovery in DC k from customer l

f_j: Fixed cost for operating manufactory j

g_k: Fixed cost for operating DC k

h_m: Fixed cost for operating dismantler m

φ: Fixed cost for landfilling per unit.

Variables

x_{ij}: Quantity produced at manufactory j using raw materials from supply i

y_{jk}: Amount shipped from manufactory j to DC k

z_{kl}: Amount shipped from DC k to customer l

o_{km}: Amount shipped from DC k to dismantler m

Rd_{mj}: Amount shipped from dismantler m to manufactory j

Rz_{lk}: Quantity recovered at DC k from customer l

$$\alpha_j = \begin{cases} 1, & \text{if production takes place at manufactory } j \\ 0, & \text{otherwise} \end{cases}$$

$$\beta_k = \begin{cases} 1, & \text{if DC } k \text{ is opened} \\ 0, & \text{otherwise} \end{cases}$$

$$\delta_m = \begin{cases} 1, & \text{if dismantler } m \text{ is opened} \\ 0, & \text{otherwise} \end{cases}$$

Because the recovery and landfilling rates are the estimated proportional values of the demand and the recovery amount, they are nonintegral. Similarly, the percentage of the capacity for the reverse logistics in DC is also estimated proportionally with real numbers. Therefore, in order to maintain integral properties, the Gauss symbol is used in our mathematical model and the model is shown below with TC as the total cost:

Object function:

$$\min TC = \sum_i \sum_j s_{ij} x_{ij} + \sum_j \sum_k t_{jk} y_{jk} + \sum_k \sum_l u_{kl} z_{kl} + \sum_k \sum_m v_{km} o_{km} + \sum_m \sum_j w_{mj} Rd_{mj} + \sum_l \sum_k Ru_{lk} Rz_{lk} + \sum_j f_i \alpha_j + \sum_k g_k \beta_k + \sum_m h_m \delta_m + \varphi \sum_m \left\lfloor pl_m \sum_k o_{km} \right\rfloor \quad (8.1)$$

Subject to:

$$\sum_j x_{ij} \leq a_i, \quad \forall i \quad (8.2)$$

$$\sum_k y_{jk} \leq b_j \alpha_j, \quad \forall j \quad (8.3)$$

$$\sum_i x_{ij} + \sum_m Rd_{mj} = \sum_k y_{jk}, \quad \forall j \quad (8.4)$$

$$\sum_l z_{kl} + \sum_m o_{km} \leq Sc_k \beta_k, \quad \forall k \quad (8.5)$$

$$\sum_j y_{jk} = \sum_l z_{kl}, \quad \forall k \quad (8.6)$$

$$\sum_m o_{km} \leq \lfloor pd_k Sc_k \beta_k \rfloor, \quad \forall k \quad \lfloor\ \rfloor: \text{floor for Gauss' symbol} \quad (8.7)$$

$$\sum_l Rz_{lk} = \sum_m o_{km}, \quad \forall k \quad (8.8)$$

$$\sum_k Rz_{lk} \geq \left\lceil pc_l \sum_k z_{kl} \right\rceil, \quad \forall l \quad \lceil\ \rceil: \text{ceiling for Gauss' symbol} \quad (8.9)$$

$$\sum_k z_{kl} \geq d_l, \quad \forall l \quad (8.10)$$

$$\sum_j Rd_{mj} + \left\lfloor pl_m \sum_k o_{km} \right\rfloor \leq e_m \delta_m, \quad \forall m \quad \lfloor\ \rfloor: \text{floor for Gauss' symbol} \quad (8.11)$$

$$\sum_k o_{km} = \sum_j Rd_{mj} + \left\lfloor pl_m \sum_k o_{km} \right\rfloor, \quad \forall m \quad \lfloor\ \rfloor: \text{floor for Gauss' symbol} \quad (8.12)$$

$$\alpha_j, \beta_k, \delta_m \in \{0, 1\}, \quad \forall j, k, m \quad (8.13)$$

$$x_{ij}, y_{jk}, z_{kl}, o_{km}, Rd_{mj}, Rz_{lk} \in \mathrm{N} \cup \{0\} \quad \forall i; j, k, l, m \quad (8.14)$$

The objective is to minimize the total cost of the transportation and the operations, and the objective function (8.1) represents this goal. The constraints mainly contain two types: one is for limited capacities and the other is for the law of the flow conservation. Constraints (8.2) and (8.3) represent the limit of the capacity for suppliers and manufactories in forward logistics. Constraint (8.5) shows that the total flows of forward and backward can not exceed the total capacity of DC. Constraints (8.7) and (8.11) mean the reverse limit of the capacity for DCs and dismantlers, respectively. Constraint (8.9) describes the customer recovery relationship with the recovery rate. Constraints (8.4), (8.6), (8.8), and (8.12) satisfy the law of the flow conservation by inflow equal to out-flow. Constraint (8.10) satisfies customer demand. Constraint (8.13) denotes the binary variables, and constraint (8.14) is the non-negative, integral condition in our model.

The parameter of pd_k is used to describe the role of DC k. If $pd_k = 0$, DC k has a sole duty for distribution in the forward logistics; and when $pd_k = 1$, DC k may play a singular role for the collection center. If pd_k falls between zero and one, it means DC k not only can be a distribution center, but also a collection center. The concept is similar for manufactories. If a manufactory accepts the resources from the dismantlers for reuses, it acts as both manufacturer and remanufacturer; otherwise it only uses raw materials for manufactory. These concepts have been presented in our general closed-loop logistics model.

Because the variables denote the basic units of logistics, apart from 0–1 decision variables, all other variables are all integers and thus Gauss's symbols are introduced in the model. The floor and ceiling of the Gaussian are defined below.

Definition 8.1 The floor or ceiling of a real number x is an integer denoted by $\lfloor x \rfloor$ and $\lceil x \rceil$, respectively, and defined, respectively, as below:

$$\lfloor x \rfloor = \{\underline{x} | x \geq \underline{x}, \underline{x} \geq y, \forall y, \underline{x} \in I\}$$
$$\lceil x \rceil = \{\overline{x} | x \leq \overline{x}, \overline{x} \leq y, \forall y, \overline{x} \in I\}$$

Because of the Gaussian, the model will be transformed further into a linear model.

8.2.2 The Transformed Integer Linear Programming Model

To transform into a computable model, we propose the following process to transform two rates into linear forms as below:

Let us consider the constraints (8.7) and (8.9) first. Since $\lfloor pd_k Sc_k \beta_k \rfloor$ and $\lceil pc_l \sum_k z_{kl} \rceil$ are the joint given input values of all parameters: pd_k, pc_l, and Sc_k; and $\sum_k z_{kl}$ equals to the demand of customer l, because customer demand must be satisfied by constraint (8.10) and the optimal solution exists if and only if constraint (8.10) equals to the lower bound. Therefore, by rewriting $SP_k = \lfloor pd_k Sc_k \rfloor$ and $ZP_l = \lceil pc_l \sum_k z_{kl} \rceil$, constraints (8.7), (8.9), and (8.10) can be transformed respectively into (8.7a), (8.9a), and (8.10a) as below:

$$\sum_m o_{km} \le SP_k \beta_k, \quad \forall k \tag{8.7a}$$

$$\sum_k Rz_{lk} \ge ZP_l, \quad \forall l \tag{8.9a}$$

$$\sum_k z_{kl} = d_l, \quad \forall l \tag{8.10a}$$

With regard to constraints (8.11) and (8.12) which are different from the situation above with decision value, $\sum_k o_{km}$ are not known in advance, and also pl_m are the parameters related to the estimated landfilling amounts. To transform these two constraints with Gauss' symbol, three additional inequalities are needed as defined below.

$$OP_m \le pl_m \sum_k o_{km}, \quad \forall m \tag{8.15}$$

$$OP_m \ge pl_m \sum_k o_{km} - \varepsilon, \quad \forall m \quad \text{where } \varepsilon \to 1^- \tag{8.16}$$

$$OP_m \in \mathrm{N} \cup \{0\} \quad \forall m \tag{8.17}$$

$$\sum_j Rd_{mj} + OP_m \le e_m \delta_m, \quad \forall m \tag{8.11a}$$

$$\sum_k o_{km} = \sum_j Rd_{mj} + OP_m, \quad \forall m \tag{8.12a}$$

where a very small real number $\varepsilon > 0$ is given to ensure inequality holds for integer solutions.

Then, the original model with Gauss's symbol can be transformed into an *integer linear program* (ILP) which is summarized as below:

Deterministic Closed-Loop Logistics Model (DCLL Model)

$$\begin{aligned} \min TC = &\sum_i \sum_j s_{ij} x_{ij} + \sum_j \sum_k t_{jk} y_{jk} + \sum_k \sum_l u_{kl} z_{kl} + \sum_k \sum_m v_{km} o_{km} \\ &+ \sum_m \sum_j w_{mj} Rd_{mj} + \sum_l \sum_k Ru_{lk} Rz_{lk} + \sum_j f_i \alpha_j + \sum_k g_k \beta_k \\ &+ \sum_m h_m \delta_m + \varphi \sum_m OP_m \end{aligned} \tag{8.1}$$

Subject to:

$$\sum_j x_{ij} \le a_i, \quad \forall i \tag{8.2}$$

$$\sum_{k} y_{jk} \leq b_j \alpha_j, \quad \forall j \tag{8.3}$$

$$\sum_{i} x_{ij} + \sum_{m} Rd_{mj} = \sum_{k} y_{jk}, \quad \forall j \tag{8.4}$$

$$\sum_{l} z_{kl} + \sum_{m} o_{km} \leq Sc_k \beta_k, \quad \forall k \tag{8.5}$$

$$\sum_{j} y_{jk} = \sum_{l} z_{kl}, \quad \forall k \tag{8.6}$$

$$\sum_{m} o_{km} \leq SP_k \beta_k, \quad \forall k \tag{8.7a}$$

$$\sum_{l} Rz_{lk} = \sum_{m} o_{km}, \quad \forall k \tag{8.8}$$

$$\sum_{k} Rz_{lk} \geq ZP_l, \quad \forall l \tag{8.9a}$$

$$\sum_{k} z_{kl} = d_l, \quad \forall l \tag{8.10a}$$

$$OP_m \leq pl_m \sum_{k} o_{km}, \quad \forall m \tag{8.15}$$

$$OP_m \geq pl_m \sum_{k} o_{km} - \varepsilon \quad \forall m \tag{8.16}$$

$$\sum_{j} Rd_{mj} + OP_m \leq e_m \delta_m, \quad \forall m \tag{8.11a}$$

$$\sum_{k} o_{km} = \sum_{j} Rd_{mj} + OP_m, \quad \forall m \tag{8.12a}$$

$$\alpha_j, \beta_k, \delta_m \in \{0, 1\}, \quad \forall j, k, m \tag{8.13}$$

$$x_{ij}, y_{jk}, z_{kl}, o_{km}, Rd_{mj}, Rz_{lk}, OP_m \in \mathrm{N} \cup \{0\}, \varepsilon \to 1^- \quad \forall i, j, k, l, m$$

(8.14) and (8.17)

In this closed-loop logistics model, there are $(I + 2J + 4K + 2L + 4M)$ constraints (8.18), and $(I \times J + J \times K + K \times L + L \times K + K \times M + M \times J + J + K + 2M)$ variables (8.19) including $(J + K + M)$ binary variables (8.20), in which additional $2M$ variables and M constraints are derived from transformation. With this structure, the number of variables is always more than that of constraints, and thus the model is always feasible.

Suppliers	Manufactories	DCs	Customers	Dismantlers	pd_k	pc_l	pl_m	φ
3	5	3	4	2	10%	10%	10%	5

TABLE 8.2 The Size and Estimated Constants of the Example

8.2.3 An Illustrative Example

In this paragraph, we shall use a small example to illustrate the properties of the problem and the model.

Tables 8.2 and 8.3 are the given data. The example contains thee suppliers, five manufactories, three distribution centers, four customers, and two dismantlers. Five types of roles are involved with the respective numbers (recovery, landfilling, and percentage of capacity for reverse in DC) as shown in Table 8.2, and three rates are assumed to be equal with respect to each customer l, dismantler m, and DC k, respectively. Tables 8.3 and 8.4 list all the unit costs of operation and transportation respectively.

In this example, with $I = 3$, $J = 5$, $K = 3$, $L = 4$ and $M = 2$, there are 41 constraints and 82 variables. Using both LINGO 8.0 and ILOG-CPLEX 7.0 with at most 1 s elapsed time, we obtained the optimal solution as shown in Table 8.5 and Fig. 8.2.

From this numerical example, it can be seen that with the conservation law, all logistic units were reserved in the system of the forward and backward flows. Also, from this solution, it can be seen that only three manufactory sites out of five, two distribution centers out of three, and one dismantler out of two are needed to meet the overall demands of four customers and recycling. Thus, the model is able to serve optimal green supply chain management from an economic viewpoint.

In the closed-loop logistics planning problem, the recovery and landfilling rates are the most critical yet uncertain factors. In this numerical example, they are assumed to be 10%. Hsu and Wang (2009)

	Manufactory		DC			Dismantler	
Supplier Capacity	Capacity	Fixed Cost	Capacity	Fixed Cost	Customer Demand	Capacity	Fixed Cost
500	400	1800	870	1000	500	540	900
650	550	900	890	900	300	380	800
390	490	2100	600	1600	400	—	—
—	300	1100	—	—	300	—	—
—	500	900	—	—	—	—	—

TABLE 8.3 Capacity, Demand (in units) and Fixed Cost (US $)

Supplier	Manufactory				
	1	2	3	4	5
1	5	6	4	7	5
2	6	5	6	6	8
3	7	6	3	9	6
Manufactory	**DC**				
	1	**2**	**3**	**—**	**—**
1	5	8	5	—	—
2	8	7	8	—	—
3	4	7	4	—	—
4	3	5	3	—	—
5	5	6	6	—	—
DC	**Customer**				
	1	**2**	**3**	**4**	**—**
1	7	4	5	6	—
2	5	4	6	7	—
3	7	5	3	6	—
Customer	**DC**				
	1	**2**	**3**	**—**	**—**
1	3	7	4	—	—
2	8	5	5	—	—
3	4	3	4	—	—
4	3	2	5	—	—
DC	**Dismantler**				
	1	**2**	**—**	**—**	**—**
1	3	2	—	—	—
2	2	5	—	—	—
3	3	3	—	—	—
Dismantler	**Manufactory**				
	1	**2**	**3**	**4**	**5**
1	2	3	4	2	5
2	3	4	6	3	4

TABLE 8.4 Unit Shipping Cost for Each Stage (US $)

	Objective value: 29848.00				
x_{ij}	$x_{13} = 40$	$x_{15} = 460$	$x_{22} = 415$	$x_{23} = 60$	$x_{33} = 390$
y_{jk}	$y_{22} = 550$	$y_{31} = 490$	$y_{51} = 319$	$y_{52} = 141$	—
z_{kl}	$z_{12} = 109$	$z_{13} = 400$	$z_{14} = 300$	$z_{21} = 500$	$z_{22} = 191$
o_{km}	$o_{11} = 61$	$o_{21} = 89$	—	—	—
Rd_{mj}	$Rd_{12} = 135$	—	—	—	—
Rz_{lk}	$Rz_{11} = 50$	$Rz_{22} = 30$	$Rz_{32} = 40$	$Rz_{41} = 11$	$Rz_{42} = 19$
α_j	$\alpha_2 = 1$	$\alpha_3 = 1$	$\alpha_5 = 1$	—	—
β_k	$\beta_1 = 1$	$\beta_2 = 1$	—	—	—
δ_m	$\delta_1 = 1$	—	—	—	—
OP_m	$OP_1 = 15$	—	—	—	—

TABLE 8.5 The Optimal Solution of the Numerical Example

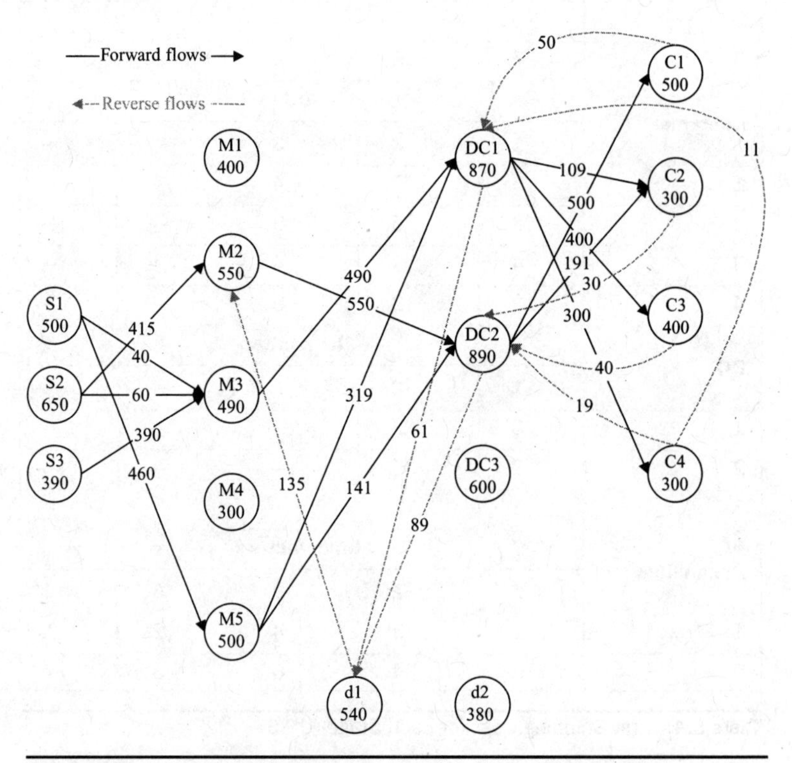

FIGURE 8.2 Optimal distribution pattern of the illustrative example.

have carried out parameter analysis to obtain the ranges of landfilling and recovery rates with the same optimal logistics pattern. Also, they pointed out that because of the recovery rate is more sensitive than the landfilling rate, therefore, with a smaller tolerance range, any change of recovery rate should be given more attention in control and management.

8.3 Closed-Loop Modeling for Uncertain Logistic

This section focuses on the uncertain issues of GSC management raised in literature. After the basic knowledge of uncertain environment, the resolution methods will be discussed.

The high complexity and necessity in application have become a challenge. From Table 8.1, it can be concluded that the basic difference between the conventional and green supply chains is the uncertainty. Among them, demand, landfilling, and recovery rates are the basic factors that contribute to the uncertainty in management (e.g., Biehl et al., 2007; Christos and George, 2007; Kongar, 2004). Based on the *closed-loop* model introduced in Sec. 8.2, fuzzy programming will be adopted in the following to cope with the three major uncertain factors.

8.3.1 The Fuzzy Programming Model

As a result of the uncertainty embedded in the customer demand, recycling rate, and landfill rate, apart from the facility location and the flows between each pair of facilities found in the proposed model, the capacity in the reverse channel, which is shared with the forward channel, is also considered to be an issue at each facility. All of the uncertain parameters are first described as fuzzy numbers to realize their possibilistic patterns. Then, an uncertain mathematical programming model will be proposed accordingly of which the notations are defined below.

Indices

I: The number of suppliers with $i = 1, 2, \ldots, I$

J: The number of manufactories with $j = 1, 2, \ldots, J$

K: The number of distribution centers (DCs) with $k = 1, 2, \ldots, K$

L: The number of customers with $l = 1, 2, \ldots, L$

M: The number of dismantlers with $m = 1, 2, \ldots, M$

Parameters

a_i: Capacity of supplier i

b_j: Capacity of manufactory j

Sc_k: Capacity of the DC k

$\widetilde{pc}_l$: Uncertain recovery percentage of customer l

$\widetilde{pl}_m$: The uncertain landfilling rate of dismantler m

$\widetilde{d}_l$: Uncertain demand of the customer l

e_m: Capacity of dismantler m

s_{ij}: Unit cost of production in manufactory j using materials from supplier i

t_{jk}: Unit cost of transportation from each manufactory j to each DC k

u_{kl}: Unit cost of transportation from DC k to customer l

v_{km}: Unit cost of transportation from DC k to dismantler m

w_{mj}: Unit cost of transportation from dismantler m to manufactory j

Ru_{lk}: Unit cost of recovery in DC k from customer l

f_j: Fixed cost for operating manufactory j

g_k: Fixed cost for operating DC k

h_m: Fixed cost for operating dismantler m

LC: Fixed cost for landfilling per unit.

Decision Variables

$\widetilde{x}_{ij}$: Quantity produced at manufactory j using raw materials from supply i

$\widetilde{y}_{jk}$: Amount shipped from manufactory j to DC k

$\widetilde{z}_{kl}$: Amount shipped from DC k to customer l

$\widetilde{o}_{km}$: Amount shipped from DC k to dismantler m

$\widetilde{Rd}_{mj}$: Amount shipped from dismantler m to manufactory j

$\widetilde{Rz}_{lk}$: Quantity recovered at DC k from customer l

$$\alpha_j = \begin{cases} 1, & \text{if production takes place at manufactory } j \\ 0, & \text{otherwise} \end{cases}$$

$$\beta_k = \begin{cases} 1, & \text{if DC } k \text{ is opened} \\ 0, & \text{otherwise} \end{cases}$$

$$\delta_m = \begin{cases} 1, & \text{if dismantler } m \text{ is opened} \\ 0, & \text{otherwise} \end{cases}$$

Then, the fuzzy green closed-loop logistics model is defined below.

Fuzzy-GCLL Model
Object function:

$$\min \widetilde{TC} = \sum_i \sum_j s_{ij} x_{ij} + \sum_j \sum_k t_{jk} y_{jk} + \sum_k \sum_l u_{kl} z_{kl} + \sum_k \sum_m v_{km} o_{km} + \sum_m \sum_j w_{mj} Rd_{mj} + \sum_l \sum_k Ru_{lk} Rz_{lk} + \sum_j f_i \alpha_j + \sum_k g_k \beta_k + \sum_m h_m \delta_m + LC \sum_m \widetilde{pl}_m \sum_k o_{km} \tag{8.21}$$

Subject to:

$$\sum_j x_{ij} \leq a_i, \quad \forall i \tag{8.22}$$

$$\sum_k y_{jk} - b_j \alpha_j \leq 0, \quad \forall j \tag{8.23}$$

$$\sum_i x_{ij} + \sum_m Rd_{mj} - \sum_k y_{jk} = 0, \quad \forall j \tag{8.24}$$

$$\sum_l z_{kl} + \sum_m o_{km} - Sc_k \beta_k \leq 0, \quad \forall k \tag{8.25}$$

$$\sum_j y_{jk} - \sum_l z_{kl} = 0, \quad \forall k \tag{8.26}$$

$$\sum_l Rz_{lk} - \sum_m o_{km} = 0, \quad \forall k \tag{8.27}$$

$$\sum_k Rz_{lk} - \widetilde{pc}_l \sum_k z_{kl} \gtrsim 0, \quad \forall l \tag{8.28}$$

$$\sum_k z_{kl} \gtrsim \widetilde{d}_l, \quad \forall l \tag{8.29}$$

$$\sum_j Rd_{mj} + \widetilde{pl}_m \sum_k o_{km} - e_m \delta_m \lesssim 0, \quad \forall m \tag{8.30}$$

$$\sum_k o_{km} - \sum_j Rd_{mj} - \widetilde{pl}_m \sum_k o_{km} \cong 0, \quad \forall m \tag{8.31}$$

$$\alpha_j, \beta_k, \delta_m \in \{0, 1\}, \quad \forall j, k, m \tag{8.32}$$

$$x_{ij}, y_{jk}, z_{kl}, o_{km}, Rd_{mj}, Rz_{lk} \in R^+ \quad \forall i, j, k, l, m \tag{8.33}$$

The objective is to minimize the possible total cost (TC) of transportation and operations. The constraints contain mainly two types: one refers to the limited capacities and the other refers to the law of flow conservation. Constraints (8.22) and (8.23) represent the possible amounts that could be provided by suppliers and manufacturers in forward logistics, respectively. Constraint (8.25) is the joint-capacity limit between forward and reverse in DCs. Constraint (8.30) is the uncertain reverse capacity limits of the dismantlers. Constraint (8.28) describes the possible recovery amount of the customers from an uncertain recovery rate. Constraints (8.24), (8.26), (8.27), and (8.31) ensure the law of flow conservation through the possible in/out flows. Constraint (8.9) is the uncertain customer demand to be satisfied. Constraint (8.32) denotes the binary decision variables, and constraint (8.33) is the non-negative requirement in our model.

The closed-loop green logistics contains two parts, the forward and reverse logistics. The fuzziness in the sign is derived from three fuzzy parameters. Constraint (8.28) is the recovery amount that is uncertain under an uncertain recovery rate. Reverse logistics is an activity pushed by the uncertain recovery rate, which in turn causes uncertainty in the reverse chain. Constraint (8.29) refers to the uncertain customer demand that causes uncertainty in the pull system and forward chain. Customer demand is uncertain, and the recovery amount depends on the uncertain recovery percentage of an uncertain demand. Thus, the derived degree of uncertainty in the reverse chain is more than that in the forward chain. Constraints (8.30) and (8.31) pertain to the uncertain landfilling rate that reverts the reuse amount back to the forward chain. The three uncertain factors have a very close relation, which causes the whole closed-loop chain to become a highly uncertain environment.

8.3.2 Resolution of Uncertainly

Based on their intrinsic uncertainty, the uncertain factors are assumed to be fuzzy numbers described by fuzzy membership functions. Since the final decision is always crisp with the degree of possibility, we shall develop a procedure to transform the fuzzy numbers into crisp numbers without losing any information. Thus, in this section, the details of transforming a fuzzy membership function into intervals through all level cuts will be described. Moreover, the solution with its risk level determined from the interval programming model will be introduced. The resolution procedure can be seen in Fig. 8.3, in which the indicated section has the chapter number deleted for simplification.

First, a solution procedure in general terms to cope with the uncertain environment is developed. To facilitate the general development of the procedure, a fuzzy linear model is considered to determine the n decision variables, **x**, subjective to M constraints with m "$\geq$" constraints, and M-m "$\leq$" constraints. This canonical form is presented below.

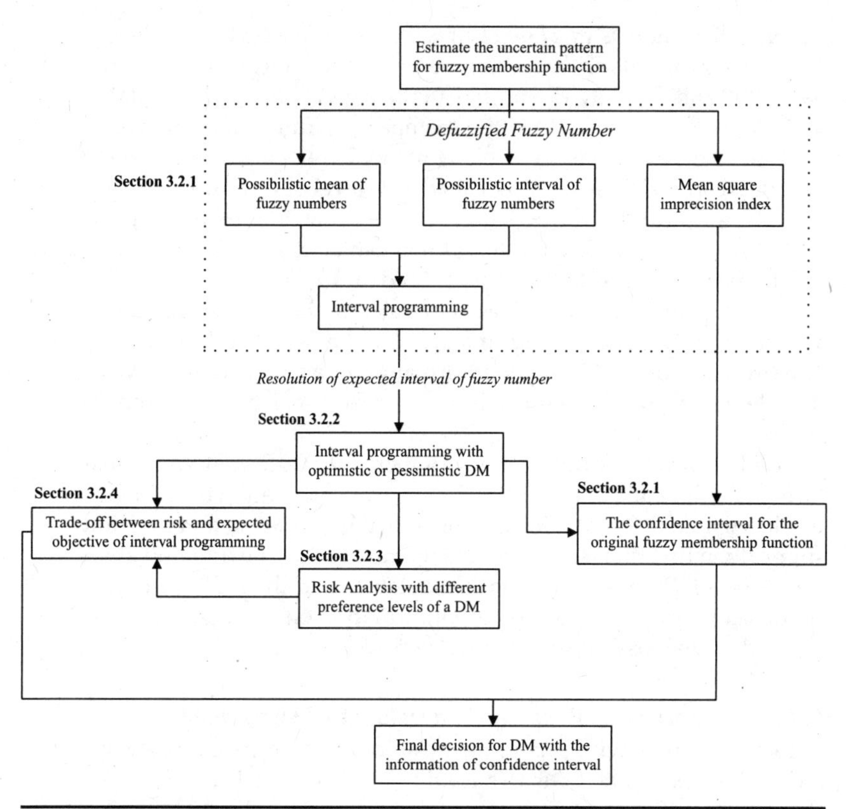

FIGURE 8.3 Analysis and resolution under an uncertain environment.

General Fuzzy Programming Model (GFPM)

$$\begin{aligned}
&\text{Min } \tilde{\mathbf{c}}^{\mathrm{T}}\mathbf{x} \\
&\text{s.t. } \tilde{\mathbf{D}}\mathbf{x} \tilde{\geq} \tilde{\mathbf{F}} \\
&\tilde{\mathbf{E}}\mathbf{x} \tilde{\leq} \tilde{\mathbf{G}} \\
&\mathbf{x} \geq 0
\end{aligned}$$

where

$$\begin{aligned}
\tilde{\mathbf{D}} &= \{(d_{ij}, \mu_{\tilde{\mathbf{D}}}(d_{ij}))|\mathbf{D} = [d_{ij}] \in \mathbf{R}^{m\times n}, \mu_{\tilde{\mathbf{D}}}(d_{ij}) \in [0,1]\}, \\
&\quad i = 1, 2, \ldots, m,\ j = 1, 2, \ldots, n \\
\tilde{\mathbf{E}} &= \{(e_{ij}, \mu_{\tilde{\mathbf{E}}}(e_{ij}))|\mathbf{E} = [e_{ij}] \in \mathbf{R}^{(M-m)\times n}, \mu_{\tilde{\mathbf{E}}}(e_{ij}) \in [0,1]\}, \\
&\quad i = m+1, m+2, \ldots, M,\ j = 1, 2, \ldots, n \\
\tilde{\mathbf{F}} &= \{(f_i, \mu_{\tilde{\mathbf{F}}}(f_i))|\mathbf{F} \in \mathbf{R}^{m}, \mu_{\tilde{\mathbf{F}}}(f_i) \in [0,1]\}, i = 1, 2, \ldots, m \\
\tilde{\mathbf{G}} &= \{(g_i, \mu_{\tilde{\mathbf{G}}}(g_i))|\mathbf{G} \in \mathbf{R}^{M-m}, \mu_{\tilde{\mathbf{G}}}(g_i) \in [0,1]\}, \\
&\quad i = m+1, m+2, \ldots, M \\
\tilde{\mathbf{c}} &= \{(c_j, \mu_{\tilde{\mathbf{c}}}(c_j))|\mathbf{c} \in \mathbf{R}^{n}, \mu_{\tilde{\mathbf{c}}}(c_j) \in [0,1]\}, j = 1, 2, \ldots, n
\end{aligned}$$

Note that there is no constraint with "=" in the GFPM model because all constraints with "=" can be transferred into two constraints with "≥" and "≤". By defining a fuzzy set $\widetilde{A} = \{(x, \mu_{\widetilde{A}}(x)) | \mu_{\widetilde{A}}(x) \in [0,1], \forall x \in R\}$ where $\mu_{\widetilde{A}}(x)$ is the membership function or degree of truth of x in A. A crisp set of elements which belong to a fuzzy set $\widetilde{A}$ at least to a degree of γ is called a γ-level set of $\widetilde{A}$ defined by $A_\gamma = \{x \in R | \mu_{\widetilde{A}}(x) \geq \gamma, 0 \leq \gamma \leq 1\}$. If $\widetilde{A}$ is a fuzzy number, then for each γ level, A_γ is a closed interval that can be defined by its *L*ower and *U*pper bounds as $[a_L(\gamma), a_U(\gamma)]$ (Zadeh, 1975).

The definition above paves the way for the transformation of a fuzzy number into a crisp interval by the so-called γ-level cut. The preservation of the whole information of a fuzzy number when this defuzzification process has taken place will be discussed in Sec. 8.3.2.1.

With regard to the defuzzified numbers, Sec. 8.3.2.2 details a transformed mathematical model developed by incorporating the DM's attitude (optimistic or pessimistic) towards the utilities so that decision support can function realistically. In relation to this, the trade-off analysis between the risk of constraint violation and the induced cost is proposed in Sec. 8.3.2.3 to provide information on the tolerance level and reduce the loss caused by uncertainty.

8.3.2.1 Information Reserved in a Defuzzified Fuzzy Number

In the uncertain environment, the mean of the occurrences is always an ideal index for the DMs to base their decision on. Thus, our basis for the defuzzification will be the concept of mean. It should however, be noted that when defuzzifying a fuzzy number by its level set, the information is disaggregated into an interval set. One way to collect the whole information is to integrate all levels. This is in reference to Carlsson and Fuller (2001) who had based their work on the concept of the mean in probability theory, and they used Goetschel and Voxman's (1986) method for ranking fuzzy numbers. To derive a possibilistic mean of a fuzzy number, the arithmetic means of all γ-level sets were used as shown below.

$$\overline{M}(\widetilde{A}) = \int_0^1 \gamma[a_L(\gamma)+a_U(\gamma)]d\gamma = \frac{\int_0^1 \gamma[a_L(\gamma)+a_U(\gamma)]/2d\gamma}{\int_0^1 \gamma d\gamma}$$

$$= \frac{1}{2}\left(\frac{\int_0^1 \gamma a_L(\gamma)d\gamma}{\int_0^1 \gamma d\gamma} + \frac{\int_0^1 \gamma a_U(\gamma)d\gamma}{\int_0^1 \gamma d\gamma}\right) = \frac{1}{2}(M_*(\widetilde{A}) + M^*(\widetilde{A})), \tag{8.34}$$

Therefore, the whole information of a fuzzy number defined by its possibilistic mean can be simply revealed by the average of the lower and upper possibilistic mean values of $\widetilde{A}$. This results in a crisp interval

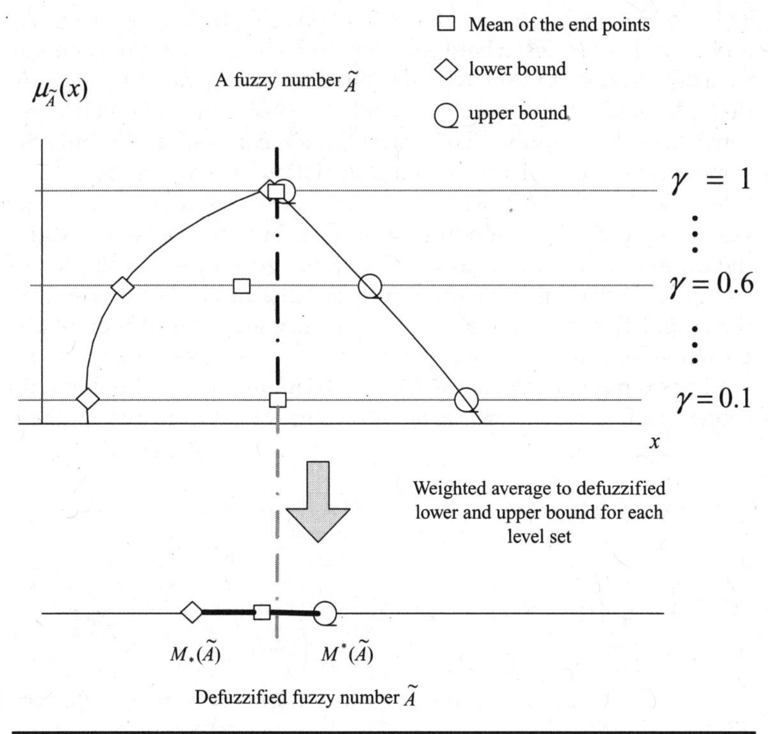

FIGURE 8.4 The defuzzified processes.

denoted by $M(\tilde{A}) \equiv [M_*(\tilde{A}), M^*(\tilde{A})]$ and is closed by the lower and upper possibilistic mean values of $\tilde{A}$. This derivation can be visualized in Fig. 8.4, which shows how a fuzzy number $\tilde{A}$ is transformed into an interval that reserves the whole information weighted by different level sets.

Then, the variance of the fuzzy number was given as below (Carlsson and Fuller, 2001):

$$\mathbf{Var}(\tilde{A}) = \left\{ \int_0^1 \gamma \left[\frac{a_L(\gamma) + a_U(\gamma)}{2} - a_L(\gamma) \right]^2 + \left[\frac{a_L(\gamma) + a_U(\gamma)}{2} - a_U(\gamma) \right]^2 \right\} d\gamma$$

$$= \frac{1}{2} \int_0^1 \gamma [a_U(\gamma) - a_L(\gamma)]^2 d\gamma \tag{8.35}$$

To confirm the transformation, the statistical evidence of the confidence interval is analyzed. That is, if we want to have $1 - \omega$ confidence level with the level of significance, ω, the confidence interval of certain distribution H, $[H_L, H_U]$, should have the property:

$P(H_L \leq H \leq H_U) = 1 - \omega$. Hence it is possible to find numbers H_L and H_U, where H lies in between with probability $1 - \omega$. For example, for a normal distribution, $N(\mu, \sigma)$, the smallest interval of $1 - \omega$ confidence level is between $\mu - z_{\omega/2}\sigma$ and $\mu + z_{\omega/2}\sigma$. For a symmetric and centralized distribution, like Normal, the mean plus and minus the same certain multiplier of standard deviation, $T\sigma$, can get $1 - \omega$ confidence level with shortest interval, $p(\mu - T\sigma \leq x \leq \mu + T\sigma) = 1 - \omega$. For an asymmetric or noncentralized distribution, like Beta, because the mean does not equal to the mode, the mean plus a multiplier of standard deviation, $T_U\sigma$, and minus another multiplier of standard deviation, $T_L\sigma$, can give a $1 - \omega$ confidence level, but it may not be the smallest, $p(\mu - T_L\sigma \leq x \leq \mu + T_U\sigma) = 1 - \omega$ where $T_U\sigma \neq T_L\sigma$.

For example, if $\widetilde{A} = (a, S_L, S_U)$ is a triangular fuzzy number with center a; left spread from a is S_L, and right spread from a is S_U, then

$$\overline{M}(\widetilde{A}) = \int_0^1 \gamma\,[a - (1-\gamma)S_L + a + (1-\gamma)S_U]d\gamma = a + \frac{S_U - S_L}{6},$$

$$Var(\widetilde{A}) = \frac{1}{2}\int_0^1 \gamma(a + S_U(1-\gamma) - (a - S_L(1-\gamma)))^2 d\gamma = \frac{(S_L + S_U)^2}{24}.$$

If it is a symmetric triangular fuzzy number, $S_U = S_L = S$, then $\overline{M}(\widetilde{A}) = a$ and $\mathbf{Var}(\widetilde{A}) = \frac{S^2}{6}$.

Since the fuzzy number of $\widetilde{A}$ is a triangular fuzzy number, it is reasonable to assume that the distribution is a triangular distribution. If it is a symmetric triangular fuzzy number, the shortest confidence intervals with 64.98% and 96.63% confidence levels are

$$\left[a - \frac{S}{\sqrt{6}}, a + \frac{S}{\sqrt{6}}\right] \quad \text{and} \quad \left[a - \frac{2S}{\sqrt{6}}, a + \frac{2S}{\sqrt{6}}\right]$$

respectively. If we want to eliminate the same percentile on both the upper and lower side in an asymmetric triangular distribution, the confidence interval will be $[a - (1-\omega)S_L, a + (1-\omega)S_U]$ with a $1 - \omega$ confidence level. It should be noted that the asymmetric situations have many kinds of ways to obtain the confidence interval from different theories.

Thus, the expected crisp interval can be derived by

$$\begin{aligned} M(\widetilde{A}) &\equiv \left[\frac{\int_0^1 \gamma[a - (1-\gamma)S_L]d\gamma}{\int_0^1 \gamma d\gamma}, \frac{\int_0^1 \gamma[a + (1-\gamma)S_U]d\gamma}{\int_0^1 \gamma d\gamma}\right] \\ &= \left[a - \frac{S_L}{3}, a + \frac{S_U}{3}\right]. \end{aligned}$$

If it's a symmetric triangular fuzzy number,

$$S_U = S_L = S, \quad M(\widetilde{A}) \equiv \left[a - \frac{S}{3}, a + \frac{S}{3}\right].$$

After representing the uncertainty by the expected crisp intervals of the corresponding fuzzy numbers, the general model (GFPM) will be transformed into an interval program as shown below.

General Interval Programming Model (GIPM)

$$\begin{aligned} &\text{Min}\, [\mathbf{c_L}, \mathbf{c_U}]^T \mathbf{x} \\ &\text{s.t.}\, [\mathbf{D_L}, \mathbf{D_U}]\mathbf{x} \geq [\mathbf{F_L}, \mathbf{F_U}] \\ &[\mathbf{E_L}, \mathbf{E_U}]\mathbf{x} \leq [\mathbf{G_L}, \mathbf{G_U}] \\ &\mathbf{x} \geq 0 \end{aligned}$$

Then, solving a fuzzy mathematical model is equivalent to solving an ordinary interval programming model.

In the next section, we shall develop a solution procedure to reach an expected optimal objective of an interval program.

8.3.2.2 The Solution of an Interval Programming Model

Before solving the *interval programming model* (GIPM), the issue of ranking two intervals needs to be dealt with first.

Consider an interval $A = [a_L, a_U]$, then its mid-point and half-width (or simply as "spread") of interval A, can be obtained by

$$m(A) = \frac{1}{2}(a_L + a_U),\ w(A) = \frac{1}{2}(a_U - a_L) \tag{8.36}$$

Sengupta et al. (2001) have defined a ranking function $\phi : I \times I \rightarrow (-\infty, \infty)$ for two intervals of $A = [a_L, a_U]$, $B = [b_L, b_U]$ by

$$\phi(A < B) = \frac{(m(B) - m(A))}{(w(B) + w(A))}, \tag{8.37}$$

where $w(B) + w(A) \neq 0$. Then, $\phi(A < B)$ can be interpreted as the *grade of acceptability* that the "first interval, A is inferior to the second interval, B."

The grade of acceptability of $A < B$ may be classified and interpreted further based on the possible comparative cases of means and spreads of intervals A and B as follow:

$$\phi(A < B) \begin{cases} \leq -1, & \text{if } m(A) > m(B) \text{ and } a_L \geq b_U \\ < 0, > -1 & \text{if } m(A) > m(B) \text{ and } a_L < b_U \\ = 0, & \text{if } m(A) = m(B) \\ > 0, < 1 & \text{if } m(A) < m(B) \text{ and } a_U > b_L \\ \geq 1 & \text{if } m(A) < m(B) \text{ and } a_U \leq b_L \end{cases} \tag{8.38}$$

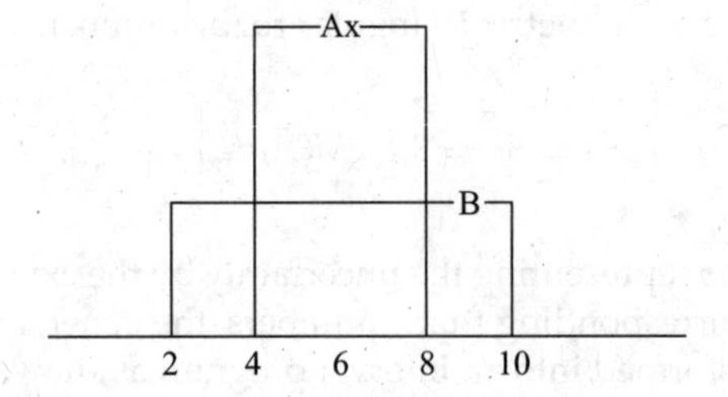

Figure 8.5 *Ax* is nested in *B* (Sengupta et al., 2001).

According to the ϕ-index, the acceptability condition of $Ax \leq B$ for the two intervals, A and B; and a singleton, x, is $\phi(Ax < B) \geq 0$, i.e. $m(Ax) \leq m(B)$. As a result of this, two possible cases follow.

Case I If $a_U x > b_L$ *and* $a_L x > b_L$, then Ax is nested in B as shown in Fig. 8.5 where $[2, 4]x \leq [2, 10]$ with $x = 2$.

Case II If $a_U x > b_U$ *and* $a_L x < b_L$: then B is nested in Ax as shown in Fig. 8.6, where $[0, 6]x \leq [2, 10]$ for $x = 2$.

From the examples above, the following remarks can be given for each of the cases:

1. *For case I:* Since $\phi(Ax < B) \geq 0$, it definitely satisfies the original interval inequality for $x \leq 2$. However, an optimistic DM may not be satisfied because there are possible resources that can be utilized for making more profit. To get higher satisfaction, the DM may increase the value of x by giving a threshold that cannot exceed $\phi(B < Ax)$. On the other hand, the pessimistic DM may want to control the $\phi(Ax < B)$ so that it achieves a higher value.
2. *For case II:* With $\phi(Ax < B) \geq 0$, the original interval inequality condition is not denied even for $x \leq 2$. However, a pessimistic DM may not be satisfied if the right limit of Ax crosses over the right limit of B. To attain his/her required level of satisfaction, the DM may even reduce the value of x so that $a_U x \leq b_U$.

Therefore, based on Moore's concept described in Lemma 8.1, further analysis on incorporating a decision maker's preference into the model can be carried out.

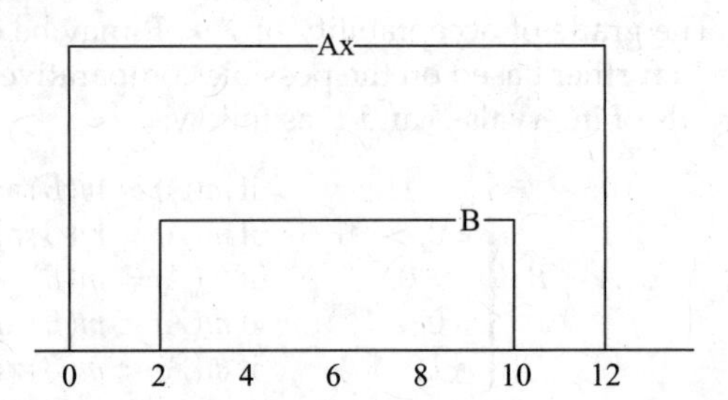

Figure 8.6 *B* is nested in *Ax* (Sengupta et al., 2001).

Lemma 8.1: (Moore, 1979) If an optimal solution does not satisfy the binding condition and if there are alternative optimal solutions, some of them may be the dominated alternatives to the model.

Sengupta et al. (2001) proposed the following relations to simulate one's possible reactions and defined a respective parameter θ in Eqs. (8.39) and (8.40) for DM to reflect his/her preference.

$$Ax \leq B \text{ let } Ax \subset D \quad \text{where } D = [-\infty, b_U].$$

$$Ax \geq B \text{ let } Ax \subset D \quad \text{where } D = [b_L, \infty].$$

$$Ax \leq B \Rightarrow \begin{cases} a_U x \leq b_U, \\ \phi(B < Ax) \leq \theta \in [-1, 1] \end{cases} \tag{8.39}$$

With the limit of $a_U x \leq b_U$, the optimistic DM may set a larger θ to allow the lower bound of Ax to become closer to the upper bound of B. On the contrary, the pessimistic DM may set a smaller θ to make the upper bound of Ax closer to the lower bound of B.

In the same way, for $Ax \geq B$, we have the satisfactory crisp equivalent form in (8.40) below:

$$Ax \geq B \Rightarrow \begin{cases} a_L x \geq b_L, \\ \phi(Ax < B) \leq \theta \in [-1, 1] \end{cases} \tag{8.40}$$

Similarly, the upper bound of Ax is closer to the lower bound of B when θ is increased, but limits it by $a_L x \geq b_L$. In contrast, the smaller the value of θ, the closer is the lower bound of Ax to the upper bound of B. With the result, the lower bound of Ax will never be lower than the lower bound of B even if θ is equal to 1.

Based on Moore's concept, we conclude that $a_U x \leq b_U$ requires that the upper bound of the left-hand side (LHS) cannot exceed the right-hand side (RHS); whereas $a_L x \geq b_L$ requires that the lower bound of the RHS cannot exceed the LHS. Therefore, if a DM is extremely pessimistic, set $\theta = -1$. Otherwise, set $\theta = 1$ if the DM is extremely optimistic towards the decision problems.

Based on the above concepts, if a linear mathematical model contains M uncertain constraints of which m constraints with "$\geq$" and M-m constraints with "$\leq$", we can define θ_1 and θ_2, respectively, based on these two types of uncertain constraints. Then, the interval programming model (GIPM) can be transformed into an ordinary linear form with the two preference parameters, θ_1 and θ_2, as shown below.

Preference Model

$$\text{Minimize } m(Z) = \frac{1}{2} \sum_{j=1}^{n} (c_{Lj} + c_{Uj}) x_j \quad \text{(Mid)}$$

$$\text{subject to } \sum_{j=1}^{n} a_{Lij} x_j \geq b_{Li}, \quad \text{for } i = 1, 2, \ldots, m$$

$$b_{Li} + b_{Ui} - \sum_{j=1}^{n}(a_{Lij} + a_{Uij})x_j \leq \theta_1(b_{Ui} - b_{Li}) + \theta_1 \sum_{j=1}^{n}(a_{Uij} - a_{Lij})x_j$$

$$\sum_{j=1}^{n} a_{Uij}x_j \leq b_{Ui}, \quad \text{for } i = m+1, m+2, \ldots, M$$

$$\sum_{j=1}^{n}(a_{Lij} + a_{Uij})x_j - (b_{Li} + b_{Ui}) \leq \theta_2(b_{Ui} - b_{Li}) + \theta_2 \sum_{j=1}^{n}(a_{Uij} - a_{Lij})x_j$$

$$x_j \geq 0, \quad \forall j$$

In the above processes, the methods can obtain an expected solution with the DM's attribute towards preference levels indicated by θ_1 and θ_2. However, because of uncertainty, the probability exists that some constraints may be violated. If any constraint is violated, then all the effort in planning is rendered useless. Therefore, it is important to know the probability of failure of the program. This is considered a risk when violating the resource constraints and will be discussed in the following section.

8.3.2.3 Risk Level of Interval Programming with Analysis

From the previous description, we have seen that θ is dependent on the DM's preference. In addition, it will influence the solutions. In this section, we discuss the risk that violates the constraints. If we want to compare two intervals, A and B, then three possibilities from each viewpoint of A and B exist. These are:

$$A_1\colon A > B \quad A_2\colon A \sim B \quad A_3\colon A < B$$
$$B_1\colon B < A \quad B_2\colon B \sim A \quad B_3\colon B > A$$

Let us use the example shown in Fig. 8.7 to gain an insight into risk analysis.

First, consider the probability that B is always smaller than A, or A is always larger than B:

From the viewpoint of B, [6, 23] interval B is the range that is always smaller than A, and the probability of range [6, 23] for interval B is

$$\frac{23-6}{24-6} = \frac{17}{18}.$$

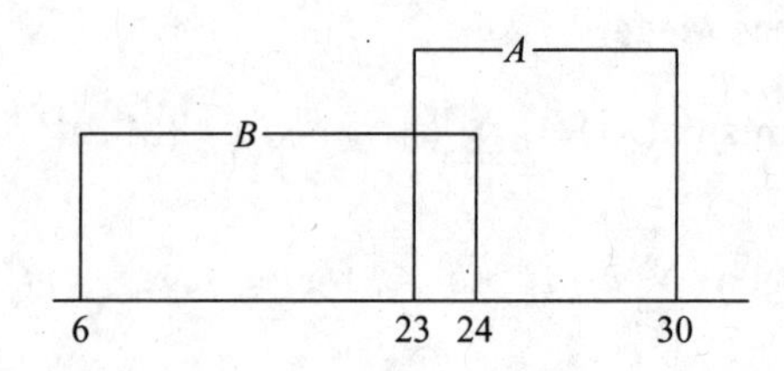

FIGURE 8.7 Comparison between two intervals, A and B.

Alternatively, from the viewpoint of A, [24, 30] interval A is the range that is always larger than B, and the probability of range [24, 30] for interval A is $\frac{6}{7}$.

Since we are certain that $A > B$, if and only if, the range [24, 30] in A or the range [23, 24] in A and [6, 23] is B, thus the probability is

$$P(A > B) = \frac{17}{18} \times \frac{6}{7} + \frac{1}{18} \times \frac{6}{7} + \frac{1}{7} \times \frac{17}{18} = \frac{17}{18} = 0.992063.$$

Second, consider the probability that we cannot render judgment.

In Fig. 8.7, from the viewpoint of B, [23, 24] interval B is the range where it cannot be judged whether $A > B$ or $A < B$. The probability of range [23, 24] for interval B is $\frac{1}{18}$. However, when it falls in the interval [23, 24] of B, A faces two situations: if range [24, 30] falls within interval A, then no matter what the range of B is, we are sure that $A > B$. If range [23, 24] is within interval A, then the probability of range [23, 24] for interval A is $\frac{1}{7}$. However, we are not certain whether $A > B$, if and only if, the range [23, 24] is included in both intervals A and B. Such probability is

$$P(A \sim B) = \frac{1}{7} \times \frac{1}{18} = 0.00794.$$

Third, consider the probability that $A \geq B$ is impossible, that is, $P(A < B)$. In Fig. 8.7, the probability is $P(A < B) = 0$.

The above observation can be summarized from the viewpoint of A as follows: when the occurrence of certain events in A and B is unknown, the probabilities can be estimated by using the information of the *end points* through the following formulas:

$$P(A_1 : A > B) = \frac{a_u - \min(\max(a_L, b_U), a_u)}{a_u - a_L},$$

$$P(A_2 : A \sim B) = \frac{\min(a_u, b_u) - \min(\max(a_L, b_L), b_U)}{a_u - a_L},$$

$$P(A_3 : A < B) = \frac{\min(\max(a_L, b_L), a_U) - a_L}{a_u - a_L}$$

Similarly, from the viewpoint of B, when the occurrence of certain events in B and A is unknown, we can use the information of the *end points* to know the probabilities of $P(B_1)$, $P(B_2)$, and $P(B_3)$ as shown below:

$$P(B_1 : B < A) = \frac{\min(\max(a_L, b_L), b_U) - b_L}{b_u - b_L},$$

$$P(B_2 : B \sim A) = \frac{\min(a_u, b_u) - \min(\max(a_L, b_L), b_U)}{b_u - b_L},$$

$$P(B_3 : B > A) = \frac{b_U - \min(\max(b_L, a_U), b_U)}{b_u - b_L}$$

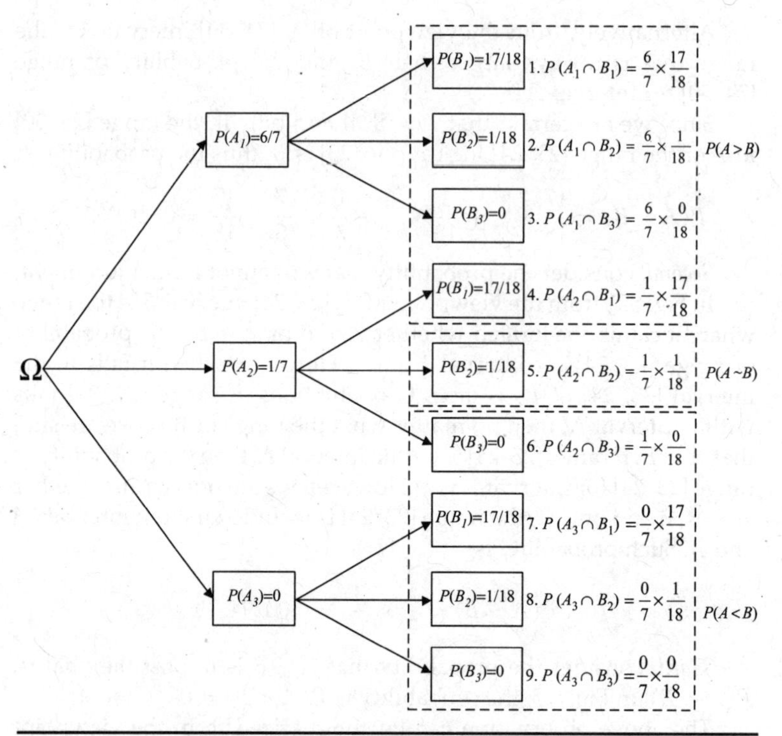

FIGURE 8.8 Tree-based possibility analysis structure for A to B.

It should be noted that because an event in A is actually greater than B, if and only if, any event in B is smaller than the event in A. The situation wherein event A is greater than B and the event B is greater than A all at the same time is an impossibility, that is, $P(A_1 \cap B_3) = P(B_1 \cap A_3) = 0$.

Since each viewpoint has three possibilities, there are $3 \times 3 = 9$ possible situations when comparing two interval numbers. We applied the basic possibility concepts and used the tree-based figure to analyze the situation as shown in Fig. 8.8, which illustrates the case taken from the viewpoint of A in which three three possibilities are shown. For each possibility, three subcases are included. Note that the analysis structure from A to B will get the same result of B to A as stated below.

The Risk Property The probabilities of $P(A > B)$, $P(A < B)$, and $P(A \sim B)$ from both viewpoints of A and B are the same.

From Fig. 8.8, the probability of $A > B$, $P(A > B)$ can be observed from the first four blocks; the uncertain probability that $A > B$ or $A < B$, $P(A \sim B)$, can be observed from the median blocks; and the probability of $A < B$, $P(A < B)$, is shown in the rest of the four blocks.

These three cases can be generalized into the standard forms as follows:

$$P(A > B) = \frac{a_u - \min(\max(a_L, b_U), a_u)}{a_u - a_L} + \frac{\min(a_u, b_u) - \min(\max(a_L, b_L), b_U)}{a_u - a_L} \times \frac{\min(\max(a_L, b_L), b_U) - b_L}{b_u - b_L} \quad (8.41)$$

$$P(A \sim B) = \frac{\min(a_u, b_u) - \min(\max(a_L, b_L), b_U)}{a_u - a_L} \times \frac{\min(a_u, b_u) - \min(\max(a_L, b_L), b_U)}{b_u - b_L} \quad (8.42)$$

$$P(A < B) = \frac{\min(\max(a_L, b_L), a_U) - a_L}{a_u - a_L} + \frac{\min(a_u, b_u) - \min(\max(a_L, b_L), b_U)}{a_u - a_L} \times \frac{b_U - \min(\max(b_L, a_U), b_U)}{b_u - b_L} \quad (8.43)$$

If the DM requires A to be greater than B, then the probability of risk will be $1 - P(A > B) = P(A < B) + P(A \sim B)$. On the contrary, when the DM requires A to be smaller than B, the probability of risk should be $1 - P(A < B) = P(A > B) + P(A \sim B)$.

Summary and Conclusion

Consider an interval program as

$$\begin{aligned} &\textbf{Min} \sum_{i=1}^{n} [c_{iL}, c_{iU}]x_i \\ &\text{s.t} \sum_{i=1}^{n} [a_{ijL}, a_{ijU}]x_i \geq [b_{jL}, b_{jU}], \quad \forall j \\ &x_i \geq 0, \quad \forall i \end{aligned}$$

The program is feasible if and only if all the constraints are satisfied. From the probability's viewpoint, this probability is $p\{\cap_j (A_j > B_j)\} = \Pi_j \{p(A_j > B_j)\}$. On the other hand, any likely unsatisfied constraint would cause the system to fail. Then, the program can be infeasible with probability of $p\{\cup_j (1 - (A_j > B_j))\} = 1 - p\{\cap_j (A_j > B_j)\}$.

Constraint violation or the uncertainty about the situation may contribute to a kind of risk. However, from the efficient management viewpoint, a constraint with the highest risk of violation should be controlled first. In relation to this, there are three risk rules of the system in terms of probabilities. These can be summarized as follows:

1. **Highest risk rule:** with the probability $\max\{1 - p(A_j > B_j), \forall j\}$ that the system might be infeasible from the constraint with the highest risk.
2. **Highest violation rule:** with the probability $\max\{p(A_j < B_j), \forall j\}$ that the system might fail because of the constraint with the highest risk.
3. **No-risk rule:** the probability $\Pi_j\{p(A_j > B_j)\}$ that the system is feasible.

Therefore, the priority rules of risk control can be stated as follows:

1. **Minimizing the probability of highest risk rule.** If more than one constraint has the same risk level, proceed to the second rule.
2. **Minimizing the probability of highest violation rule.** If more than one constraint has the same risk level, proceed to the third rule.
3. **Maximizing the probability of no risk rule.**

8.3.2.4 Trade-Off Analysis Between Risk and Cost

Recall the preference model. Whenever the optimistic level is increased by increasing θ_1 or θ_2, the cost decreases, yet the risk will increase. Therefore, the DMs need to trade off between cost and risk, which is analyzed below.

Without loss of generality, let us consider the incremental changes of the attribute level θ, to the average improved cost (*AIC*) per unit risk deviation.

$$AIC(\theta|\Delta\theta) = \frac{|\Delta TC(\theta|\Delta\theta)|}{|\Delta \text{Risk}(\theta|\Delta\theta)|} \tag{8.44}$$

TC is the total cost, which can be obtained from the object function, and the risk can be measured by the highest risk rule, highest violation rule, and no risk rule in Section 4.3.2. Take note that if the comparison of the value of *AIC* is needed, the same rule must be used to measure the risk. If $\Delta TC(\theta|\Delta\theta) = TC(\theta + \Delta\theta) - TC(\theta)$ is defined and $\Delta\text{Risk}(\theta|\Delta\theta) = \text{Risk}(\theta + \Delta\theta) - \text{Risk}(\theta)$, then the average improved cost will be increased when the cost saving is higher, yet the induced risk is lower. On the other hand, increasing the attribute level (θ) can improve the cost but increase the risk. Therefore, it is suggested that DM should have a more optimistic attitude towards resource management.

8.3.2.5 Numerical Illustration of the Risk Attitude

To illustrate the proposed preference model with risk, a simple example is used.

Example

(GIPM) (Preference Model)

$Min\,[2,\ 4]x_1 - [1,\ 2]x_2$ Transfer $Min\,3x_1 - 1.5x_2$

$s.t\ \ [1,\ 3]x_1 \geq [6,\ 24]$ (i) $\rightarrow$ $s.t\ \ x_1 \geq 6$ (i-a)

$[1,\ 2]x_2 \leq [12,\ 30]$ (ii) $2x_1\theta_1 + 4x_1 \geq 30 - 18\theta_1$ (i-b)

$x_1, x_2 \geq 0$ $2x_2 \leq 30$ (ii-a)

$3x_2 - 1\theta_2 x_2 \leq 18\theta_2 + 42$ (ii-b)

$x_1, x_2 \geq 0$

The number of the constraints corresponds to the classification in the GIPM of which (i) is the constraint with $\geq$, and (ii) is the constraint with $\leq$. Figure 8.9 shows the preferences of different DMs with (a) optimistic, (b) neutral, or (c) pessimistic attitudes.

In the following, the application of the priority rules for trade-off analysis between risk and cost is demonstrated, wherein the higher the values of θ_1 and θ_2, the more optimistic is the DM with the lower objective values and higher risk levels.

To find the risk level of violating constraints that can be taken by a DM, Fig. 8.9 (a) is presented. If $\theta_1 = 1$ this means that $\phi(Ax_1 < B) = 1$.

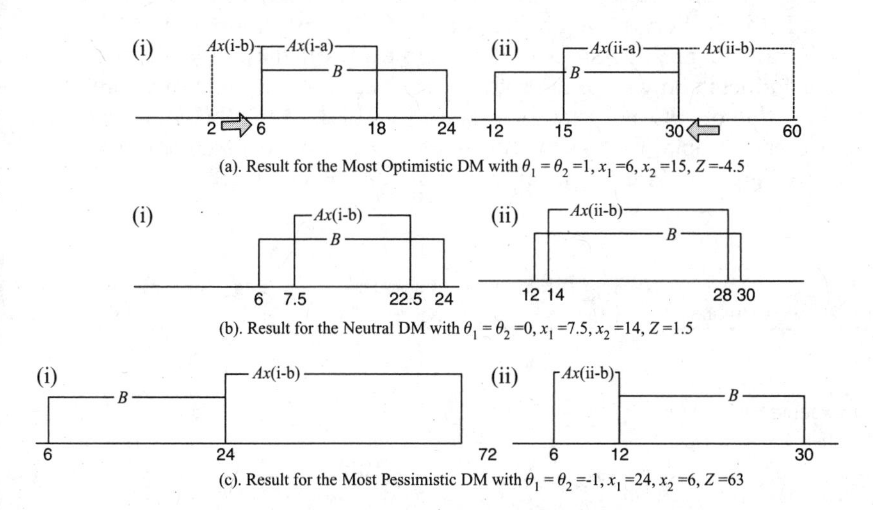

FIGURE 8.9 Example of risk analysis for DMs with different preferences.

Alternatively, if $\theta_2 = 1$, this means $\phi(B < Ax_2) = 1$. The dotted line shows the interval Ax as constrained by inequality (i-b). Since it is also constrained by inequality (i-a), this leads to the feasible range of Ax shown by the bold line when $\theta_1 = \theta_2 = 1$. It can be seen that the lower bound of interval Ax is always equal to the lower bound of interval B in constraint (i), and the upper bound of interval Ax is always equal to the upper bound of interval B in constraint (ii). In this situation, the probability of highest risk rule is found to be $\max\{1 - p(Ax_1 > B) = 1, p(1 - Ax_2 < B) = 1\} = 1$. This means that the possibility of an unsatisfied resource constraint is as high as 1 for the most optimistic DM.

In Fig. 8.9 (b), if $\theta_1 = \theta_2 = 0$, then $\phi(Ax_1 < B) = 0$ and $\phi(B < Ax_2) = 0$. It can be observed that the median of interval Ax is always equal to the interval B. Consequently, the probability of highest risk rule is $\max\{1 - p(Ax_1 > B) = 1 - \frac{1.5}{18} = \frac{16.5}{18}, 1 - p(Ax_2 < B) = 1 - \frac{2}{18} = \frac{16}{18}\} = \frac{16.5}{18}$, which means that the possibility of an unsatisfied constraint from the major resource of risk is $\frac{16.5}{18} = \frac{11}{12}$ for the neutral DM.

In Fig. 8.9 (c), if $\theta_1 = -1$, then $\phi(B < Ax_1) - 1$. Alternatively, if $\theta_2 = -1$, then $\phi(B < Ax_2) = -1$. As such, interval Ax is always larger than B in constraint (i) and interval Ax is always smaller than B in constraint (ii). In this situation, it can be concluded that there is no risk with $\max\{1 - p(Ax_1 > B) = 0, 1 - p(Ax_2 < B) = 0\} = 0$, and it is impossible to violate the constraints for the most pessimistic DM.

In this example with different θ_1 and θ_2, the probability of the highest risk rule is sufficient to judge the risk level. The probabilities are different with different decisions based on the highest risk rule. Furthermore, based on the priority rule, there is no need to consider the second and third rules. The results are summarized in Table 8.6.

In this section, the numerical example is shown by most optimistic, neutral, and most pessimistic. Although trade-off analysis is difficult to conduct in only three cases, the three cases can be used to calculate the value of *AIC*. For example, if $\Delta\theta = 1$ in Eq. (8.47), the values of *AIC* can be obtained as shown in Table 8.6. Since the attribute range is too large, the *AIC* is not sufficient when $\Delta\theta = 1$.

The DM's Attitude	Most Optimistic ($\theta_1 = \theta_2 = 1$)	Neutral ($\theta_1 = \theta_2 = 0$)	Most Pessimistic ($\theta_1 = \theta_2 = -1$)
Probability	100%	$\frac{11}{12} = 91.67\%$	0%
The expected minimum objective value	−4.5	1.5	63
AIC	0.720288	0.670885	Base case

TABLE 8.6 Highest Risk Rule Analysis with Different DM Attitudes

Suppliers	Manufactories	DCs	Customers	Dismantlers	$\widetilde{pc}_l$	$\widetilde{pl}_m$	LC
3	5	3	4	2	(0.1,0.1)	(0.1,.0.1)	20

TABLE 8.7 The Input Scale and Data of the Example

8.3.3 Numerical Illustration

To demonstrate the applicability of the proposed methodology, a numerical example fuzzified from the deterministic case will be presented in the following with the parameter analysis. In addition, discussion will be followed by post-analysis.

8.3.3.1 Step-by-Step Procedure

The procedure of analysis follows the following steps:

Step 1: Specify the input and define the membership functions of the fuzzy data.

In this example, there are three suppliers (i), five manufactories (j), three distributors (k), four customers (l), and two dismantlers (m) with \$20 unit landfilling cost (φ). The input data with the fuzzified parameters $\widetilde{pc}_l$, demand $\widetilde{d}_l$ for each customer l, and $\widetilde{pl}_m$ for dismantler m are listed in Tables 8.7 and 8.8. These fuzzy inputs are assumed to be symmetric triangular and denoted by their respective mode and spread as (mode, spread) as conventional notation.

Step 2: Defuzzify the fuzzy number into a crisp interval form by the possibilistic mean.

Using formula (8.34), the intervals of possibilistic mean value of fuzziness can be obtained as follows:

$$M(\widetilde{pc}_l) = \left[\frac{0.2}{3}, \frac{0.4}{3}\right] \forall l,\ M(\widetilde{pl}_m) = \left[\frac{0.2}{3}, \frac{0.4}{3}\right] \forall m, \text{ and}$$
$$M(\widetilde{d}_1) = [466.667,\ 533.333],\ M(\widetilde{d}_2) = [266.667,\ 333.333]$$
$$M(\widetilde{d}_3) = [366.667,\ 433.333],\ M(\widetilde{d}_4) = [266.667,\ 333.333]$$

Assuming that the intervals of zeros for RHS with respect to constraints (8.2), (8.30), and (8.31) are $M_{8,10,11}(\tilde{0}) = [-5, 5]$, then by calculating the possibilistic mean interval of uncertain parameters, the preference-GCLL model can be applied.

$\widetilde{d}_1$	$\widetilde{d}_2$	$\widetilde{d}_3$	$\widetilde{d}_4$
(500, 100)	(300, 100)	(400, 100)	(300, 100)

TABLE 8.8 The Input Fuzzy Demand $\widetilde{d}_l$ (mode, spread)

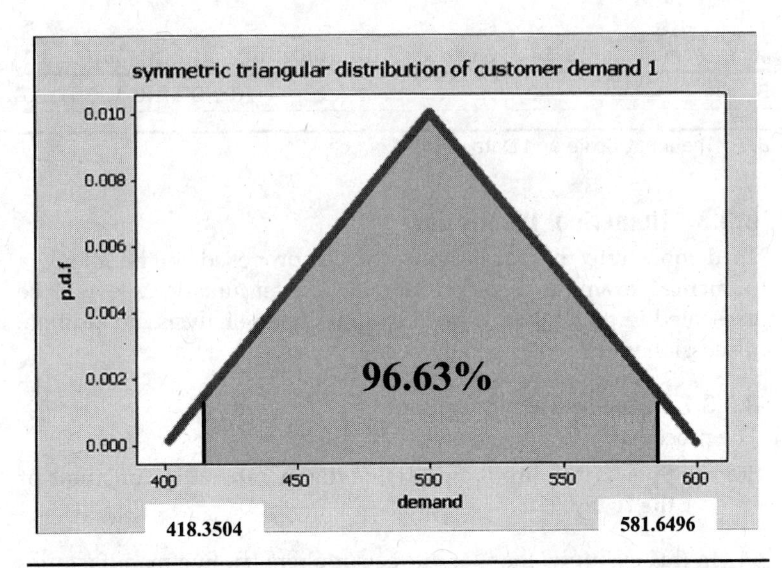

FIGURE 8.10 The symmetric triangular distribution of customer demand 1 with 96.63% C.I.

In addition, the standard deviations of the fuzzy numbers can be calculated from formula (8.35) as follow:

$$\sqrt{\mathbf{Var}(\widetilde{pc}_l)} = \sqrt{\mathbf{Var}(\widetilde{pl}_m)} = \sqrt{\frac{0.1^2}{6}} = 0.04083\ \forall l, m, \text{ and}$$

$$\sqrt{\mathbf{Var}(\tilde{d}_1)} = \sqrt{\mathbf{Var}(\tilde{d}_2)} = \sqrt{\mathbf{Var}(\tilde{d}_3)} = \sqrt{\mathbf{Var}(\tilde{d}_4)} = \sqrt{\frac{100^2}{6}} = 40.8248,$$

Since all the uncertain parameters are described by the symmetric triangular fuzzy numbers, the confidence interval with the mean and standard deviation are relatively easy to obtain from an asymptotic normal probability distribution. If 95.44% confidence level is expected, the confidence interval of a normal probability distribution is between $\mu - 2\sigma$ and $\mu + 2\sigma$. Adapting this to a symmetric triangular possibility distribution, the confidence interval $[\mu - 2\sigma, \mu + 2\sigma]$ is approximated to be 96.63% confidence level. Figure 8.10 shows the range of symmetric triangular distribution of customer demand 1 with 96.63% confidence level.

To observe the utility of this model, three special cases are presented.

1. **The neutral case:** with $\theta_1 = 0$ and $\theta_2 = 0$, the expected objective value of 30896.22 is obtained with the half risk of 50%.

Expected objective value: 30896.22 Interval of the expected objective value: [30776.22, 31016.22] Root mean square imprecision index (RMSII): 146.988 96.63% confidence interval: [30602.25, 31190.20]					
x_{ij}	$x_{13} = 100$	$x_{15} = 400$	$x_{22} = 444$	$x_{33} = 390$	
y_{jk}	$y_{22} = 550$	$y_{31} = 490$	$y_{51} = 318.33$	$y_{52} = 141.67$	
z_{kl}	$z_{12} = 108.33$	$z_{13} = 400$	$z_{14} = 300$	$z_{21} = 500$	$z_{22} = 191.67$
o_{km}	$o_{12} = 61.67$	$o_{21} = 118.33$			
Rd_{mj}	$Rd_{12} = 106$	$Rd_{15} = 1.56$	$Rd_{25} = 58.44$		
Rz_{lk}	$Rz_{11} = 61.67$	$Rz_{22} = 35$	$Rz_{32} = 48.33$	$Rz_{42} = 35$	
α_j	$\alpha_2 = 1$	$\alpha_3 = 1$	$\alpha_5 = 1$		
β_k	$\beta_1 = 1$	$\beta_2 = 1$			
δ_m	$\delta_1 = 1$	$\delta_2 = 1$			

Table 8.9 Optimal Solution for a Neutral DM of the Example

2. **The most optimistic case:** with $\theta_1 = 1$ and $\theta_2 = 1$, the best objective value of 28525.51 is obtained with the highest risk of 100%.
3. **The most pessimistic case:** with $\theta_1 = -1$ and $\theta_2 = -1$, the worst objective value of 34082.31 is obtained with the lowest risk of 0%.

The total objective mean of the cost for the neutral case, TC, with intervals $[M_*(\widetilde{TC}), M^*(\widetilde{TC})]$, 2σ, with 96.63% confidence level are listed in Table 8.9 as an example. The visual graph of the solution is shown in Fig. 8.11 as a logistic pattern of the case.

From Table 8.9, we can observe how the proposed model functioned from the respective crisp solutions. Apart from the expected objective value and the interval value, the standard deviation and confidence interval can help the DM to understand the scale of the uncertainty. In this example where 2σ is adopted with 3.37% significance level, the DM has 96.63% confidence level to control the solution between $[30896.22 - 2\sigma, 30896.22 + 2\sigma]$ from the current solution.

Since θ_1 and θ_2 reflect the optimistic or pessimistic level of DM's attribute and control the probability of the supply of the resources, further analysis on their characteristics is carried out in the following section.

Step 3: Solve the preference–GCLL model with different optimistic or pessimistic levels for the DM.

In the preference–GCLL model, θ_1 and θ_2 represent the DM's attribute (optimistic or pessimistic level) for "$\geq$" and "$\leq$" constraints, respectively. If θ_1 and θ_2 are higher, a better expected objective value is present, but this is coupled with a higher risk of insufficiency in

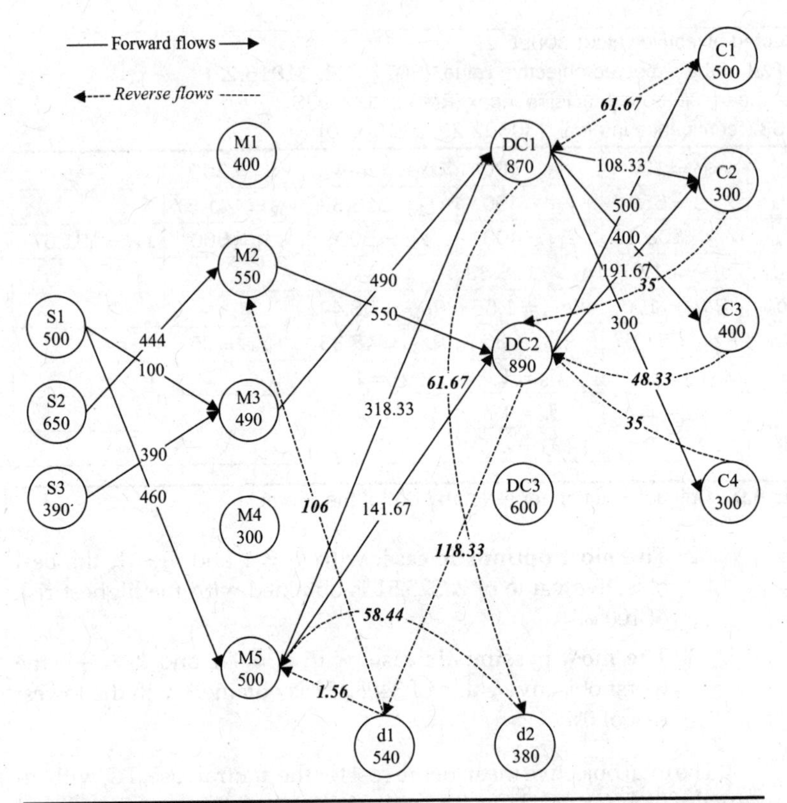

FIGURE 8.11 The expected logistic pattern for a neutral DM of the numerical example.

the resource supply. Trade-off analysis between the expected cost and risk needs to be carried out.

The influence between θ_1 and θ_2 by their individual behavior is observed first. As shown in Table 8.10, if $\theta_2 = 0$ changing the value of θ_1 will result in different expected objective values significantly; whereas changing the value of θ_2 with any of the values of θ_1 will not affect the objective values. This is shown in Appendix. With respect to the preference–GCLL model, the parameter θ_2 in the constraint (10b) implies that whether the DM overestimates or underestimates the landfilling rates, the uncertainty in the estimated ranges will not affect the decisions significantly. Therefore, in the problem of this study, the key factor is θ_1 to which more attention should be given. Further discussion is given below.

Step 4: Conduct the risk analysis for the preference-GCLL model.

The preference–GCLL model fails if any of its constraints does not hold. As seen in Table 8.10, the higher the values of θ_1, the better the

θ_1	θ_2	**Expected Interval (mid ± width)**	**96.63% Confidence Interval (mid ± 2 × *RMSII*)**	**Expected Risk**	$\frac{\lvert\Delta TC(\theta_1)\rvert}{\lvert\Delta Risk(\theta_1)\rvert}$
−1	0	[33923.79, 34240.83]	34082.31±388.34	0%	–
−0.9		[33671.02, 33975.82]	33823.42±373.35	5%	51.778
−0.8		[33420.91, 33713.59]	33567.25±358.51	10%	51.234
−0.7		[33180.80, 33664.95]	33321.14±343.81	15%	49.222
−0.6		[32940.99, 33209.79]	33075.39±329.26	**23%**	**30.71875**
−0.5		[32705.62, 32964.14]	32834.88±316.66	**32.1%**	**26.42967**
−0.4		[31938.19, 32189.62]	32063.91±307.99	**41.1%**	**85.66333**
−0.3		[31522.26, 31769.37]	31645.81±302.69	**48.4%**	**57.27397**
−0.2		[31273.58, 31518.32]	31395.95±299.78	**49%**	**416.4333**
−0.1		[31024.90, 31267.67]	31146.09±296.88	**50%**	**499.73**
0		[30776.22, 31016.22]	30896.22±293.98	50%	
0.1		[30538.94, 30776.57]	30657.75±291.07	55%	47.694
0.2		[30303.20, 30538.46]	30420.83±288.17	60%	47.384
0.3		[30067.47, 30300.36]	30183.92±285.27	65%	47.382
0.4		[29831.74, 30062.26]	29947.00±282.36	70%	47.384
0.5		[29596.01, 29824.16]	29710.09±279.46	75%	47.382
0.6		[29360.28, 29586.06]	29473.17±276.56	80%	47.384
0.7		[29124.55, 29347.96]	29236.25±273.66	85%	47.384
0.8		[28888.82, 29109.86]	28999.34±270.75	90%	47.382
0.9		[28653.09, 28871.75]	28762.42±267.85	95%	47.384
1		[28417.36, 28633.65]	28525.51±264.95	100%	47.382

TABLE 8.10 Comparison of θ_1 with $\theta_2 = 0$

objective value is, yet the higher is the risk defined by the probability of violating the resource constraints. Therefore, trade-offs between the expected cost and the risk of insufficient resources are further analyzed.

From formulas (8.41)–(8.43), we can calculate the probability of unsatisfied constraint j, and Sec. 8.3.2.4 shows that the expected risk of violating any constraint in the preference–GCLL Model is max{1− the probability of satisfied constraint j, $\forall j$}. Therefore, if $\theta_1 = 1$ and $\theta_2 = 0$, max{1 − the probability of satisfied constraint j, $\forall j$} = 1 and implies the probability of 100% that constraints will be violated, which refers to constraint (8.29) on customer demands. Since the preference–GCLL model has a certain chance to fail, the DM should not adopt the solution despite the fact that the expected solution may be better than the other θ_1 values.

In Table 8.10, the risks mainly resulted from constraints (8.28) and (8.29), recovery rate and uncertain customer demand, respectively.

The risks in *italic* and **boldface** in Table 8.10 are influenced by constraint (8.28), uncertain recovery rate, while other impacts that come from uncertain customer demand showed a very regular pattern. In Table 8.10, we can observe the uncertain recovery rate increasing the risk at a very fast rate with θ_1. In the beginning, the impact of recovery rate for the risk is not sufficient with customer demand, but after $\theta_1 = -0.7$, it apparently dominates the impact of customer demand. After $\theta_1 = -0.1$, constraint (8.28) is limited by Moore's concept, and the risk is only influenced by customer demand.

In conclusion, the prediction of customer demand should be a very careful process. In addition, the recovery rate is the main factor when a neutral decision is made with the *trade-off* processes. Uncertain demands with uncertain recovery rates cause uncertain recovery amounts. Uncertain recovery amounts with uncertain landfilling rates cause uncertain landfilling amounts, as well as uncertain reuse amounts. This becomes a chain reaction in this closed-loop logistic system.

All expected objects come into existence only when the resources are satisfied. The possibility of violated-constraint is called *expected risk* and is calculated from the expected interval to examine the possible inaccessibility to the expected goal. The DM should consider the risk before deciding on the suitable optimistic or pessimistic level. Since, from the analysis, θ_2 has no sufficient influence in our problem, the constraints with "$\geq$" are more important than those with "$\leq$" in GSC logistics system. The DM can then decide on θ_1 from Table 8.10 for the trade-off analysis as detailed below.

Step 5: Analyze the trade-off between expected cost and risk.

Using Eq. (8.44) and considering the incremental increase of the attribute level, θ_1 by 0.1, the *average improved cost* (AIC) per 1% risk can be estimated as follows.

$$\mathbf{AIC}(\theta_1|\Delta\theta_1 = 0.1) = \frac{|\Delta TC(\theta_1|\Delta\theta_1 = 0.1)|}{|\Delta Risk(\theta_1|\Delta\theta_1 = 0.1)|}$$

The AIC is used to observe the improvement of the total cost per 1% risk when $\Delta\theta_1$ is given. As shown in the last column in Table 8.10, the average improvement of the expected cost between two θ_1 levels can be foreseen when an increase of 1% in the risk results. If the improvement is large enough for the DM to take the risk, the DM may increase his/her attribute level. Figure 8.12 shows the DM's attribute, θ_1, versus the AIC from where we can observe a sufficient improvement between $\theta_1 = -0.3$ and $\theta_1 = 0$.

From Table 8.10 and Fig. 8.12, it is known that the values of AIC between $\theta_1 = -0.3$ and $\theta_1 = 0$ are relatively larger than others. This

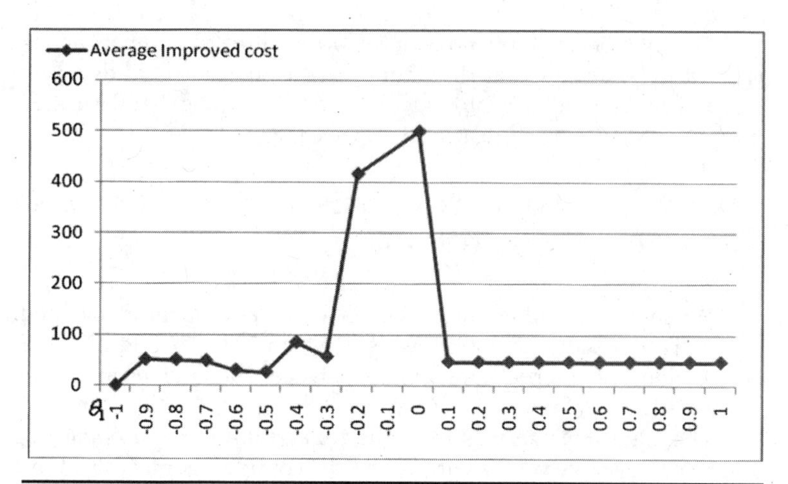

FIGURE 8.12 The results of the DM's attribute, θ_1, versus the AIC.

means that if the attribute level is as small as −0.3, the risk is 48.4%. Based on this, the DM may consider setting his/her attribute level directly to $\theta_1 = 0$, because the average improvement is sufficiently large that increasing the risk level is worth doing to save the expected cost as

$$AIC(\theta_1|\Delta\theta_1 = 0.3) = \frac{|\Delta TC(\theta_1|\Delta\theta_1 = 0.3)|}{|\Delta Risk(\theta_1|\Delta\theta_1 = 0.3)|} = \frac{745.59}{1.6\ \%} = 468.494.$$

On the other hand, if the attribute level is as large or near 0, the DM should consider setting the value to 0, because the saved cost can induce too much risk. In the case of the closed-loop logistics, we may conclude the neutral case ($\theta_1 = 0$) as containing 50% risk for the most suitable decision, because the risk is neither too large nor the most effective to save the cost.

Therefore, this trade-off process is used to compare the cost saving containing the same probability of risk with the measure *AIC*.

Step 6: Estimate the confidence interval from the confidence level to realize the uncertain environment.

This can be seen from column 3 of Table 8.10. For example, if $\theta_1 = 1$, the objective value of the mean is 28525.51, and the standard deviation of objective is

$$\sqrt{\mathrm{Var}(\widetilde{T}C)} = \sqrt{\mathrm{Var}\left(\varphi \sum_m \widetilde{p}l_m \sum_k o_{km}\right)}$$
$$= \varphi \sum_k o_{km} \sqrt{\mathrm{Var}\left(\sum_m \widetilde{p}l_m\right)} = 132.4708.$$

The membership functions of the original fuzzy numbers are assumed to be a triangular distribution. Thus, the standard deviation used to get a confidence interval is utilized. Assuming that we use 2σ, by using the triangular distribution, we can get

$$P(\mathbf{28525.51 - 2 \times 132.4708 \leq x \leq 28525.51 + 2 \times 132.4708}) = 0.9663$$
$$\Rightarrow P(28260.5684 \leq x \leq 28790.4516) = 96.63\%$$

When $\theta_1 = 1$ and $\theta_2 = 0$, a 96.63% confidence level is obtained, which is to say that the objective is between 28260.57 and 28790.45. The confidence interval comes into existence when the resources can be satisfied.

The expected intervals from the fuzzy numbers can obtain the expected objective values with θ, and the confidence intervals with a certain confidence level can help us realize the uncertain environment. If a DM makes a decision for a certain θ, we can use the confidence interval with a confidence level to describe the variation of the expected goal from the original uncertain pattern described by the fuzzy membership function.

8.3.3.2 Post-Analysis and Discussion

The above processes can provide useful information for both the optimistic and pessimistic DMs. Through the information, they can consider the expected goal and risk to do a "trade-off" and find out the most suitable decision for their companies. In this section, we show the post-analysis of the solution range in Table 8.11 by recording all solutions the parameter analysis has undertaken. The solution range means that no matter what attitude (optimistic or pessimistic) the DM has, the optimal solution must fall in the range specified in Table 8.11. This information can help eliminate some impossible feasible solution when the attitude levels of the DMs, as well as the risks, are changed. In addition, the decision in a "not bad" situation can also be controlled even if no information about the DM's attitude level is available. This not only provides a rough structure of the solution, but also provides the idea of what variables are important and must be focused on.

The variables can be divided into two parts: basic variables and nonbasic variables. The basic variables are of constant concern. In Table 8.11, if the lower bound is not zero, the variable is the basic variable. There are two kinds of basic variables: constant value and interval value. We can observe that although different kinds of DMs may make decisions, some of the decision variables remain with the same constant values. This kind of decision variables will never change their values, thus prior decisions can be made on these decision variables no matter what the attitudes of the DMs are. From the management viewpoint, this kind of constant value is not critical.

Objective value: [28525.51, 34082.31]					
x_{ij}	$x_{11} = [0, 0]$ $x_{21} = [0, 0]$ $x_{31} = [0, 0]$	$x_{12} = [0, 0]$ $x_{22} = [326.08, 444]$ $x_{32} = [0, 0]$	$x_{13} = [98.44, 100]$ $x_{23} = [0, 0]$ $x_{33} = [390, 390]$	$x_{14} = [0, 0]$ $x_{24} = [0, 83.93]$ $x_{34} = [0, 0]$	$x_{15} = [400, 401.56]$ $x_{25} = [0, 0]$ $x_{35} = [0, 0]$
y_{jk}	$y_{11} = [0, 0]$ $y_{23} = [0, 0]$ $y_{42} = [0, 0]$	$y_{12} = [0, 0]$ $y_{31} = [490, 490]$ $y_{43} = [0, 291.15]$	$y_{13} = [0, 0]$ $y_{32} = [0, 0]$ $y_{51} = [0, 322.78]$	$y_{21} = [0, 0]$ $y_{33} = [0, 433.33]$ $y_{52} = [90, 141.67]$	$y_{22} = [422.08, 550]$ $y_{41} = [0, 300]$ $y_{53} = [0, 0]$
z_{kl}	$z_{11} = [0, 0]$ $z_{22} = [42.22, 191.67]$ $z_{33} = [0, 433.33]$	$z_{12} = [108.33, 333.33]$ $z_{23} = [0, 0]$ $z_{34} = [0, 0]$	$z_{13} = [366.67, 400]$ $z_{24} = [0, 0]$	$z_{14} = [266.67, 333.33]$ $z_{31} = [0, 0]$	$z_{21} = [466.67, 533.33]$ $z_{32} = [0, 0]$
o_{km}	$o_{11} = [0, 0]$ $o_{32} = [0, 0]$	$o_{12} = [57.22, 87.78]$	$o_{21} = [105, 150]$	$o_{22} = [0, 0]$	$o_{31} = [0, 0]$
Rd_{mj}	$Rd_{11} = [0, 0]$ $Rd_{21} = [0, 0]$	$Rd_{12} = [0, 106]$ $Rd_{22} = [0, 0]$	$Rd_{13} = [0, 1.56]$ $Rd_{23} = [0, 0]$	$Rd_{14} = [0, 135]$ $Rd_{24} = [0, 81.07]$	$Rd_{15} = [0, 0]$ $Rd_{25} = [0, 58.44]$
Rz_{lk}	$Rz_{11} = [57.22, 76.11]$ $Rz_{23} = [0, 0]$ $Rz_{42} = [30.56, 37.78]$	$Rz_{12} = [0, 0]$ $Rz_{31} = [0, 0]$ $Rz_{43} = [0, 0]$	$Rz_{13} = [0, 0]$ $Rz_{32} = [43.89, 62.78]$	$Rz_{21} = [0, 0]$ $Rz_{33} = [0, 0]$	$Rz_{22} = [30.56, 49.44]$ $Rz_{41} = [0, 11.67]$
α_j	$\alpha_1 = [0, 0]$	$\alpha_2 = [1, 1]$	$\alpha_3 = [1, 1]$	$\alpha_4 = [0, 1]$	$\alpha_5 = [1, 1]$
β_k	$\beta_1 = [1, 1]$	$\beta_2 = [1, 1]$	$\beta_3 = [0, 1]$		
δ_m	$\delta_1 = [1, 1]$	$\delta_2 = [1, 1]$			

TABLE 8.11 The Valid Intervals for All Variables in All Situations

Some of the decision variables are not constant and their lower bounds are not equal to zero. From the mathematical viewpoint, this kind of variables is always basic and more important than other variables. Based on this information, it can be known what decision variables must be paid more attention to in terms of control to avoid any mistake.

8.4 Conclusions

Based on green issues, closed-loop logistics have become more and more important in recent years, and their resolution technologies have been critical for production companies. The reduction of primary resource use, pollution prevention, waste management, and policies governing sustainable products have thus become the focuses of modern industrial societies and environmental policies. Closed-loop logistics is one of the most essential keys in relation to the cost incurred by companies.

Since every part of the mentioned issues is related to the overall logistics in a product life cycle system, closed-loop logistics—with its overall cost—has become an urgent concern for companies in their supply-demand chain management. Closed-loop supply chain management was then studied to integrate conventional forward logistics with reversed logistics. In this chapter, we introduced both deterministic and uncertain situation with the corresponding analytical models and solutions.

For the deterministic situation, a mathematical programming model for general applications has been introduced. To retain the integral properties, the model was formulated with Gussian symbols that are not algebraically computable. Therefore, a transformation procedure was proposed to convert the model into an integer linear programming model with additional $2M$ decision variables and M constraints. A numerical example was provided for illustration.

Based on the deterministic model, a more realistic model to cope with uncertain situation is introduced by adopting the fuzzy approach. To avoid the shortcoming of using level cuts in the conventional solutions, therefore, in this chapter the interval programming approach from the concept of possibilistic mean is introduced so that all information of level cuts can be integrated towards providing an effective solution.

For an uncertain decision environment, developing a decision support procedure is especially important for actual management practice. It is likewise an effective decision support system that should be able to adapt to the DM's preference. Therefore, in this chapter, when transforming from the fuzzy program to an interval program, the DM's preference was taken into account in the model formulation. Consequently, a trade-off mechanism was developed to allow the DMs to evaluate between the expected cost of logistics and the expected risk of resource utilization through the defined parameter θ. The higher the value of θ, the better is the objective to be reached, and the higher the risk that should be taken. In order to provide more information for managerial purposes, the analysis of the confidence interval on the objective values was provided.

From the results of both solution and risk analysis, it was realized that customer demand and recovery rate are the main factors in an uncertain GSC environment. Therefore, accurate forecasting on the demand and recovery rate should be carried out in the GSC logistics planning activities.

Through the observation of the whole solution range, some decision variables are always constant regardless of the risk or the attribute of the DM. This provides guidance for managers as to what critical variables should be controlled when facing an uncertain environment.

References

Bellman, R. E., and L. A. Zadeh. (1970). Decision-making in a fuzzy environment. *Management Science,* 17, 141-164.

Baumgarten H., B. Christian, F. Annerous, and S.-D.Thomas. (2003). Supply chain management and reverse logistics-integration of reverse logistics processes into supply chain management approaches. *International Symposium on Electronics and the Environment,* 79-83.

Bass, B., G. Huang, and J. Russo. (1997). Incorporating climate change into risk assessment using grey mathematical programming. *Journal of Environmental Management,* 49(1), 107-123.

Biehl, M., E. Prater, and J. R. Matthew. (2007). Assessing performance and uncertainly in developing carpet reverse logistics systems. *Computers & Operations Research,* 34(2), 443-463.

Campos, L. and J. L. Verdegay. (1989). Linear programming problems and ranking of fuzzy numbers. *Fuzzy Sets and Systems,* 32(11), 1-11.

Carlsson, C., and R. Fuller. (2001). On possibilistic mean value and variance of fuzzy numbers. *Fuzzy Sets and Systems,* 122(2), 315-326.

Chiang, J. (2001). Fuzzy linear programming based on statistical confidence interval and interval-valued fuzzy set. *European Journal of Operational Research,* 129(1), 65-86.

Davila, E., and N. Chang. (2005). Sustainable pattern analysis of a publicly owned material recovery facility in a first-growing urban setting under uncertainty. *Journal of Environmental Management,* 75(4), 337-351.

Dippon, J. (1998). Globally convergent stochastic optimization with optimal asymptotic distribution. *Journal of Applied Probability,* 353(2), 395-402.

Dubois, D., and H. Prade. (1987). The mean value of a fuzzy number. *Fuzzy Sets and Systems,* 24(3), 279-300.

Fleischmann, M., H. R. Krikke, R. Dekker, and S. D. P. Flapper. (2000). A classification of logistics networks for product recovery. *Omega*, 28(6), 653-666.

Goetschel, R., and W. Voxman. (1986). Elementary fuzzy calculus. *Fuzzy Sets and Systems*, 18(1), 31-43.

Hsu, H-W., and H-F. Wang. (2009). Modeling of green supply logistics, in: H-F. Wang (Ed), *Web-Based Green Products Life Cycle Management Systems: Reverse Supply Chain Utilization*, IGI Global Publication, pp. 268-282 Inuiguchi M., and J. Ramik. (2000). Possibilistic linear programming: a brief review of fuzzy mathematical programming and a comparison with stochastic programming in portfolio selection problem. *Fuzzy Sets and Systems*, 111(1), 3-28.

Kongar, E. (2004). Performance measurement for supply chain management and evaluation criteria determination for reverse supply chain management. *Environmentally Conscious Manufacturing*, 6, 106-117.

Laan, E. van de, M. Salomon, R. Dekker, and L. Van Wassenhove. (1999). Inventory control in hybrid systems with remanufacturing. *Management Science*; 45 (5): 733-747.

Lu, Q., C. Vivi, A. S. Julie, and R. Taylor. (2000). A practical framework for the reverse supply chain. *International Symposium on Electronics and the Environment*, 266-271.

Mahadevan, M., D. F. Pyke, and F. Morit. (2003). Periodic review, push inventory policies for remanufacturing. *European Journal of Operational Research*, 151(3), 536-551.

Moore, R. E. (1979). *Methods and Applications of Interval Analysis*, Philadelphia: SIAM Publications.

Ovidiu L., and R. Dekker. (2005). A stochastic approach to a case study for product recovery network design. *European Journal of Operational Research*, 160(1), 268-287.

Ovidiu, L. (2007). A genetic stochastic model for supply-and-return network design. *Computers & Operations Research*, 34(2), 417-442.

Rommelfanger, H. (1996). Fuzzy linear programming and application. *European Journal of Operational Research*, 92(3), 512-527.

Salema, M. I. G., A. P. Barbosa-Povoa, and A. Q. Novais. (2007). An optimization model for the design of a capacitated multiproduct reverse logistics network with uncertainty. *European Journal of Operational Research*, 179(3), 1063-1077.

Schultmann, F., Z. Moritz, and R. Otto. (2006). Modeling reverse logistic tasks within closed-loop supply chains: An example from the automotive industry. *European Journal of Operational Research*, 171(3), 1033-1050.

Sengupta A., K. P. Tapan, and D. Chakraborty. (2001). Interpretation of inequality constraints involving interval coefficients and a solution to interval linear programming. *Fuzzy Sets and Systems*, 119(1), 129-138.

Tanaka, H., T. Okuda, and K. Asai. (1974). On fuzzy mathematical programming. *Journal of Cybernetics*, 3(4), 37-46.

Hsu, H-W, and H.-F. Wang. (2009). Modeling of green supply logistics. *Web-Based Green Products Life Cycle Management Systems: Reverse Supply Chain Utilization*, H. F. Wang (ed.), IGI Global Publication, 2009.

Wang, H. F., and H. W. Hsu. (2009.) A closed-loop logistic model with a spanning-tree-based genetic algorithm. *Computers & Operations Research*. doi:10.1016/j.cor.2009.06.001.

Zadeh, L. (1975). The concept of a linguistic variable and its application to approximate reasoning. *Information Sciences*, 8, 199-249.
Zikopoulos, C., and G. Tagaras. (2007). Impact of uncertainty in the quality of returns on the profitability of a single-period refurbishing operation. *European Journal of Operational Research*, 182(1), 205-225.
Zimmermann, H.-J. (1976). Description and optimization of fuzzy systems. *Internal Journal of General Systems*, 2(1), 209-215.

Appendix: Comparison of the Expected Objective Value Between θ_1 and θ_2

See Table 8A.1 on p. 168.

θ_1 \ θ_2	**−1**	**−0.8**	**−0.6**	**−0.4**	**−0.2**	**0**	**0.2**	**0.4**	**0.6**	**0.8**	**1**
−1	34082.31	34082.31	34082.31	34082.31	34082.31	**34082.31**	34082.31	34082.31	34082.31	34082.31	34082.31
−0.8	33567.25	33567.25	33567.25	33567.25	33567.25	**33567.25**	33567.25	33567.25	33567.25	33567.25	33567.25
−0.6	33075.39	33075.39	33075.39	33075.39	33075.39	**33075.39**	33075.39	33075.39	33075.39	33075.39	33075.39
−0.4	32063.91	32063.91	32063.91	32063.91	32063.91	**32063.91**	32063.91	32063.91	32063.91	32063.91	32063.91
−0.2	31395.95	31395.95	31395.95	31395.95	31395.95	**31395.95**	31395.95	31395.95	31395.95	31395.95	31395.95
0	30896.22	30896.22	30896.22	30896.22	30896.22	**30896.22**	30896.22	30896.22	30896.22	30896.22	30896.22
0.2	30420.83	30420.83	30420.83	30420.83	30420.83	**30420.83**	30420.83	30420.83	30420.83	30420.83	30420.83
0.4	29947	29947	29947	29947	29947	**29947**	29947	29947	29947	29947	29947
0.6	29473.17	29473.17	29473.17	29473.17	29473.17	**29473.17**	29473.17	29473.17	29473.17	29473.17	29473.17
0.8	28999.34	28999.34	28999.34	28999.34	28999.34	**28999.34**	28999.34	28999.34	28999.34	28999.34	28999.34
1	28525.51	28525.51	28525.51	28525.51	28525.51	**28525.51**	28525.51	28525.51	28525.51	28525.51	28525.51

Table 8A.1 Comparison of the Expected Objective Value Between θ_1 and θ_2

CHAPTER 9

Green Customers

Features and Identification

9.1 Introduction

In recent years, as consciousness about the worldwide need to protect the earth's environment has increasingly gained ground, tasks involving environmental protection have been conducted globally. Governments have been promoting energy saving and carbon reduction. But there is a dilemma. On the one hand, consumers are not always aware of green products, nor are they always conscientious protecting the environment. On the other hand, protection of the environment through the use of green products has not been totally accepted by manufacturers. Therefore, unless there are enough customers identified by manufacturers, the latter usually will not voluntarily produce green products that carry higher production costs than common goods. This leads to an important issue of identifyin target customers by cluster analysis.

Since cluster analysis is based on factors that show the characteristics of the groups, this chapter begins in Sec. 9.2 with observations of consumption behavior and identification of factors that may affect green-consumption behavior. Based on these features, a sample questionnaire is provided in Sec. 9.3 to illustrate the possible questions to be asked to obtain the information needed. Then, basic statistical methods for sampling and cluster analysis are introduced in Sec. 9.4. Finally, in Sec. 9.5, a numerical example is given to show how to use the sample data to identify target customers.

9.2 Features of the Green Consumers

We cannot avoid purchasing goods to satisfy daily needs. People who only purchase goods that have the minimum adverse impact on the environment are classified as green consumers. Peattie (1993) talks about the green consumer by saying that when a consumer realizes that environment-related problems are adversely affecting quality of life, he would start purchasing goods that are environmentally friendly. This reduces degradation of the environmental protection of which is actually the responsibility of a community.

Roberts (1996) defines "ecologically conscious consumer behavior (ECCB)" as "the behavior that prompts a consumer to buy products that are friendly to the environment." Therefore, a green consumer considers the ecological impact of products and services in making his daily purchases. He even determines whether or not raw materials and processes used in the manufacture of products are harmful to the environment.

Now that the definition and behavior of green consumers have been stated, the question to ask is, "What are the factors influencing green consumption behavior?" Many studies point out that demographics of consumers, which include gender, income, age, and so on, affect green-consumption behavior (Green et al., 1993). Moreover, environment-related psychological characteristics of consumers also affect their green-consumption behavior. Understanding how these factors affect consumption behavior is necessary to effectively promote green goods. Such understanding allows a manufacturer to generate higher profit and encourage people to adopt green-consumption behavior, which are necessary to reduce injury to the environment and to attain sustainable development (Bohlen et al., 1993; Schlegelmilch, 1996; Diamantopoulos et al., 2003).

Existing literature says that there are two types of personal characteristics.

9.2.1 Demographic Characteristics

The demographic characteristics that affect green consumption behavior are discussed below.

Gender This is one of the major demographic factors that affect green-consumption behavior. There are differences between males and females in terms of attitude and behavior concerning green consumption. Straughan and Roberts (1999) point out that females are more inclined than males to engage in green consumption. Many different studies also support this view.

Income Income is another factor that is said to determine inclination of a person to be a green consumer. When a person has higher income, he is willing to spend more money on purchasing green products. Straughan and Roberts (1999) and Zimmer (1994) point out that income is a predictive operator of green-consumption behavior, noting that income and green consumption are positively related.

Age Many studies point out that younger consumers are more sensitive than older people to the environment. Diamantopoulos and

Schlegelmilch (1993) say younger consumers consider the environment more than older ones in making purchases, and thus buy more green products. On the contrary, many older consumers take more care of the environment they live in by cleaning their surroundings and practicing waste segregation. Other studies have reported different findings. Roberts (1996) says older consumers are more inclined than younger ones to manifest green-consumption behavior because of their social position, which makes society expect them to be more responsible, and the standards they have adopted over the years.

Education Level Generally speaking, someone who has a high level of education should have knowledge of the importance of protecting the environment and of green products. Those who have higher education, therefore, are expected to be more likely to become green consumers. Zimmer (1994) and Roberts (1996) say education level is highly positively correlated with green-consumption behavior.

Other Factors Other factors including social position, occupation, location, and so on, have been mentioned by research to be affecting green-consumption behavior.

In addition to demographic factors, psychological characteristics of consumers also affect their green-consumption behavior. Although previous research had not thoroughly discussed these characteristics in terms of their impact on green-consumption behavior, their significance is discussed in this chapter.

9.2.2 Psychographic Characteristics

Environmental Concern (EC) Environmental concern refers to knowledge, such as cognition of green products, related to environment protection. Roberts and Bacon (1997) use regression analysis to point out that concern for the environment is relevant to green-consumption behavior.

Perceived Consumer Effectiveness (PCE) Many studies point out that awareness of the need to protect the environment makes one assess and understand whether his/her behavior positively or negatively affects the environment. Roberts (1996) calls this understanding *perceived consumer effectiveness* (PCE). In his research, PCE is said to be an important factor, in fact, even more important than demographic and psychological characteristic variables that affect green-consumption behavior. Therefore, if consumers understand that what they do affects the environment, then they will be more willing to manifest behavior that is protective of the environment.

Environment-protection Behavior Consumers' environmental protection behavior, such as resource recovery, reuse of materials, and so on, shows their likelihood of becoming green consumers. Schlegelmilch (1996) points out that environmental protection behavior and green-consumption behavior are highly correlated.

Other Factors Many studies say there are other factors that affect green-consumption behavior. These include political views. Roberts (1996) says liberalism, a left-leaning political stand, espouses democracy and openness. Results of his study show that a consumer's degree of liberalism dictates his/her inclination to practice green-consumption behavior. Meanwhile, Stern (1993) introduces the concept of altruism as affecting one's concern of the environment. Stern says that if a person is altruistic, which means he/she has concern for others, he/she pays more attention to environment-related concerns. Therefore, altruism and green-consumption behavior are correlated.

9.3 Questionnaire Design

Taking into consideration the above-mentioned characteristics of green consumers, the questionnaire for this study was designed in a manner that differentiates consciousness of green consumption and actual green product-purchase behavior. Table 9.1 shows the design of the questionnaire concerning consumption of green electronic products, which are some of the major products required to meet 4R directives. The questionnaire has five parts:

1. General preference towards commodity.
2. Environmental knowledge and consciousness, including environmental knowledge, attitude towards environmental protection, and ecofriendly behavior.
3. Purchasing behavior—in general, and for electronic and electric products.
4. Cost consideration vs. eco-level.
5. Sociodemographics.

9.4 Analytical Methods

Before conducting the survey, two issues need to be considered: sampling method and size. After collecting the data, two tasks also need to be proceeded: data cleaning and clustering analysis. The corresponding methods are introduced in the following sections.

9.4.1 Sample Size Determination

To determine the sample size, some concepts are first defined (Degroot and Schervish, 2002).

Dear Participant:

This questionnaire is a survey on how and why you purchase green electronic products. The purpose is to understand the purchasing behavior of customers so that better services can be developed for your benefit. The information gathered here will be used only for academic research and will never be made public.

Thank you for your cooperation.

I. The following asks about your personal concepts and views on product purchase. Please put "x" in the box corresponding to the degree of your approval.	**Totally disagree**	**Quite disagree**	**Moderate**	**Somehow agree**	**Very much agree**
1. Products should be valued in terms of their utilization, thus I normally choose those with good quality.	☐	☐	☐	☐	☐
2. I seldom believe advertisements because they usually exaggerate information.	☐	☐	☐	☐	☐
3. At the market, products offering similar functions are essentially the same, thus the purchase is okay as long as the product is cheap.	☐	☐	☐	☐	☐
4. I am usually curious about new products and normally want to try them out.	☐	☐	☐	☐	☐

TABLE 9.1 Sample Questionnaire for Target Customers

5. I like to buy fancy products and those that are different from others.	☐	☐	☐	☐	☐
6. From what one buys, we can see the buyer's taste.	☐	☐	☐	☐	☐
7. When choosing a product, I am usually influenced by advertisements.	☐	☐	☐	☐	☐
8. I like to buy popular products that are on sale.	☐	☐	☐	☐	☐
9. I am attracted to and tend to buy products that have elegant and eye-catching packaging.	☐	☐	☐	☐	☐
10. It is okay if a product is a bit expensive as long as it is made in good quality.	☐	☐	☐	☐	☐
11. When choosing a product, I first check on its price.	☐	☐	☐	☐	☐
12. No matter how beautiful the packaging, this is useless since it will be thrown out anyway after the product has been unpacked or used up.	☐	☐	☐	☐	☐
• **Where do you usually get information on different kinds of products? Please choose according to importance: 1—most important; 2—second important; 3—third important.** ☐ 1. TV ☐ 2. Newspapers ☐ 3. Magazines ☐ 4. Internet ☐ 5. As recommended by friends and relatives ☐ 6. Store displays ☐ 7. Bus and outdoor billboards ☐ 8. Radio ☐ 9. Others (specify) ________					

Table 9.1 Sample Questionnaire for Target Customers (*Cont.*)

II. The following asks about your personal concepts and views on green products. Please put a check mark in the boxes that correspond to your understanding, approval, and inclination on the items below.					
(1) Understanding Environmental Issues. Please put "x" in the box correspond-ing to the degree of understanding.	**Totally do not under-stand**	**Under-stand a little**	**Moderate**	**Under-stand well**	**Under-stand perfectly**
1. Do you under-stand the term "global warming"?	☐	☐	☐	☐	☐
2. Do you under-stand the term "greenhouse effect"?	☐	☐	☐	☐	☐
3. Do you under-stand what is meant by the term "green products"?	☐	☐	☐	☐	☐
4. From the perspective of environmental contamination, do you know the difference between "general commodity" and "green products"?	☐	☐	☐	☐	☐
(2) Attitude towards environmental protection. Please put "x" in the box corresponding to the degree of approval.	**Absolutely do not agree**	**Do not quite agree**	**Moderate**	**Quite agree**	**Absolutely agree**
1. Because environmental protection is an important issue, we should pay a certain amount of money to protect the environment.	☐	☐	☐	☐	☐

TABLE 9.1 Sample Questionnaire for Target Customers (*Cont.*)

2. I think carefully about my actions because environmental protection is of great importance.	☐	☐	☐	☐	☐
3. When buying electronic products, I consider whether using them could jeopardize the environment or not (e.g., ease of recycling or energy savinga).	☐	☐	☐	☐	☐
4. Buying green electronic products has positive contribution to environmental protection.	☐	☐	☐	☐	☐
5. I actively look for information on green electronic products.	☐	☐	☐	☐	☐
(3) Eco-friendly behavior. Please put "x" in the box corresponding to your degree of approval.	**Absolutely not**	**Probably not**	**Moderate**	**Probably**	**Absolutely**
1. When at work, in school, or going out, I throw paper/glass/plastic/aluminum into their respective recycling bins.	☐	☐	☐	☐	☐
2. At home, I classify garbage and dispose of it in its designated place.	☐	☐	☐	☐	☐

Table 9.1 Sample Questionnaire for Target Customers (*Cont.*)

3. I use green appliances (e.g., my own tableware for a meal).	☐	☐	☐	☐	☐
4. If stores/shops provide disposable tableware, I refuse to use this and instead ask for ecofriendly tableware.	☐	☐	☐	☐	☐
III. The following asks about your personal purchasing behavior toward green products, in particular green electronic products.					
(1) In General Please put "x" in the box corresponding to the degree of approval.	**Absolutely do not agree**	**Do not quite agree**	**Moderate**	**Quite agree**	**Absolutely agree**
1. When there is only a slight difference in price and quality, I buy electronic products with green labels.	☐	☐	☐	☐	☐
2. If I am aware that some products are hazardous to the environment, I do not buy those products.	☐	☐	☐	☐	☐
3. In accordance with environmental protection, I will replace my current products that are not ecofriendly.	☐	☐	☐	☐	☐
4. When I see advertisements about green products, I am willing to buy them.	☐	☐	☐	☐	☐

Table 9.1 Sample Questionnaire for Target Customers (*Cont.*)

5. To protect the environment, I choose ecofriendly products regardless of price.	☐	☐	☐	☐	☐
6. When I buy certain green products, I recommend them to my friends and relatives.	☐	☐	☐	☐	☐

(2) The (10) items below list the evaluation indexes of a product's ecofriendliness. Please score their importance in terms of your purchase guide for electronic products. 1—lowest; 10—highest.
1. The design of the product must reduce the usage of environmentally hazardous materials (e.g., it does not contain lead, mercury, or cadmium). ___________
2. The design of the product must consider the choice of raw materials (e.g., product weight or recyclability). ___________
3. The design of the product must consider the choice of raw materials (e.g., product weight or recyclability). ___________
4. The design of the product must consider the choice of raw materials (e.g., product weight or recyclability). ___________
5. The final design of the product must be environmentally commendable (e.g., it is recyclable or easy to disassemble). ___________
6. The design of the product must extend the life cycle of a product (e.g., it is upgradable or the components can be substituted). ___________
7. The design of the product must consider aspects of energy saving (e.g., it conforms to regulations on energy conservation). ___________
8. The eventual management of discarded products must be considered (e.g., buyers are provided services for recycling, such as for discarded batteries). ___________
9. The product manufacturer has a commendable record on environmental achievements (e.g., the company conforms with ISO14001). ___________
10. The packing materials for the product must be considered (e.g., hazardous substances are reduced, and materials are easy to disassemble or easy to recycle). ___________

Table 9.1 Sample Questionnaire for Target Customers (*Cont.*)

IV. Cost vs. Ecofactors: A total of 28 subcategories further specify the ten (10) ecofriendly indexes mentioned above. The box below illustrates the various marks corresponding to the electronic products environmental assessment tool (EPEAT), according to which electronic products must conform. All of the products with bronze, silver, or gold marks are categorized as green products. Products with gold marks are the most ecofriendly, followed by silver-marked products, then by bronze-marked products.

Electronic Products Environmental Assessment Tool (EPEAT)	
Marks	**Standards**
Bronze	Conforms to 23 compulsory standards in the subcategories.
Silver	Conforms to 23 compulsory standards in the subcategories and at least 50% of elective standards.
Gold	Conforms to 23 compulsory standards in the subcategories and at least 75% of elective standards.

Consider the following scenarios based on the information above.

1. Currently, there are two electronic products—laptops for NTD 36000—that match your demand. There is no big difference in terms of their performance and specifications. However, one of them is a green product conforming to the **bronze mark standards** mentioned above; the other is a product with no green marks. What price is acceptable to you?
   The green product should be more expensive by ☐ 0–5% ☐ 5–10% ☐ 10–15% ☐ 15–20% ☐ 20%
2. Under the same conditions, when the green product conforms to the **silver mark standards,** what then is the price acceptable for you?
   The green product should be more expensive by ☐ 0–5% ☐ 5–10% ☐ 10–15% ☐ 15–20% ☐ 20%
3. Also under the same conditions, but this time with the green product conforming to **gold mark standards,** what price is acceptable for you?
   The green product should be more expensive by ☐ 0–5% ☐ 5–10% ☐ 10–15% ☐ 15–20% ☐ 20%

V. The next segment asks a few more questions on your background. These are for reseach purposes only and will never be made public.

Sex: ☐ M ☐ F

Age: ☐ Under 18 ☐ 18–24 ☐ 25–31 ☐ 32–37 ☐ 38–50 ☐ 51–65 ☐ 65 or above

Occupation: ☐ Student ☐ IT (software) ☐ IT (hardware) ☐ Finance/Insurance ☐ Education ☐ Architecture ☐ Medical service ☐ Manufacturing ☐ Communications ☐ Entertainment ☐ Publishing ☐ Tourism ☐ Catering ☐ General merchandising ☐ Brokerage ☐ Government ☐ Retired

TABLE 9.1 Sample Questionnaire for Taget Customers (*Cont.*)

Education: ☐ Primary school ☐ Junior high school ☐ Senior high school or equivalent ☐ University ☐ Master degree or above

Marital Status: ☐ Single ☐ Married

Monthly Income (NT$): ☐ Unemployed ☐ Under $20,000
☐ $20,000–$30,000 ☐ $30,000–$40,000 ☐ $40,000–$50,000
☐ $50,000–$60,000 ☐ $60,000–$70,000 ☐ $70,000 or above

Number of Families residing **in the same house :** ☐ 1 ☐ 2 ☐ 3 ☐ 4
☐ 5 ☐ 6 ☐ 7 or above

Place of residence ☐ Northern ☐ Central ☐ Southern ☐ Eastern

And it is located in? ☐ Downtown ☐ Suburb area

END OF QUESTIONNAIRE.
Thank you very much.

TABLE 9.1 Sample Questionnaire for Taget Customers (*Cont.*)

Definition 9.1: Target Population The totality of elements that are under discussion and about which information is desired will be called the *target population.*

Definition 9.2: Sample Population Let $X_1, X_2, \ldots, X_n$ be a random sample from a population with density $f(.)$. This population is called the *sample population.*

Definition 9.3: Sample Mean and Variance Let $X_1, X_2, \ldots, X_n$ be a random sample from a population with density $f(.)$; the sample mean, $\overline{X}$, and the sample variance, s^2, are defined in Eq. (9.1)

$$\overline{X} = \frac{\sum_{i=1}^{n} X_i}{n} \qquad s^2 = \frac{\sum_{i=1}^{n} (\overline{X} - X_i)^2}{n-1} \tag{9.1}$$

The numerator of the sample variance measures the *sum of squares* (SS) of deviations from the mean, and the denominator is the *degree of freedom* (df). The sample mean and variance are used to estimate the expected mean and variance as below.

Theorem 9.1 Let $X_1, X_2, \ldots, X_n$ be a random sample from a population with density $f(.)$, which has mean $\mu$and variance σ^2. Then,

$$E[\overline{X}] = \mu \quad \text{and} \quad \text{var}[\overline{X}] = \frac{1}{n}\sigma^2 \tag{9.2}$$

For a large sample size, the values of sample mean tend to be more concentrated around its population mean; the spread of the sample mean around the population mean will be smaller. This leads to another issue of how many samples are needed to make the pattern of the sample data sufficient to represent the total population. The issue about the required number of samples is addressed by the Chebyshev inequality, which is shown below.

Theorem 9.2: Chebyshev Inequality (Moody et al., 1976) Let $f(.)$ be a density with mean (μ) and finite variance (σ^2), and let $\overline{X}$ be the sample mean of a random sample of size (n) from $f(.)$. Let $\varepsilon > 0$ and $0 < \delta < 1$ be any two specified numbers. If n is any integer and $n > \sigma^2/(\varepsilon^2\delta)$, then

$$P[-\varepsilon < \overline{X} - \mu < \varepsilon] \geq 1 - \delta \tag{9.3}$$

Therefore, if a probability of 0.95 that the sample mean will lie within 0.5 of the population mean is to be ensured, then, according Theorem 9.2, the following holds:

$$\sigma^2 = 1,\ \varepsilon = 0.5, \text{ and } \delta = 0.05;$$
$$n > \frac{\sigma^2}{\delta\varepsilon^2} = \frac{1}{0.05(0.5)^2} = 80.$$

That is, the sample size is required to be at least 80.

If for any reason, one cannot collect a sufficient sample, there are two methods that may be employed to resolve this problem: one is to group the data to create a large sample and the other is to set up a higher significance level.

9.4.2 Data Analysis

Given the sample size, the methods to collect the required data are numerous. The most commonly used methods are through the Internet, post mail, and personal interview. Among these methods, personal interview is the most effective in ensuring good quality of data, but is also the most labor-intensive. After the survey has been conducted, statistic analysis of the significance of the chosen criteria and the validity of the identified groups should be carried out.

The common method to ensure the significance of the result is the one-factor or two-factor analysis of variance, also referred to as the ANOVA test. Basically, the purpose of the ANOVA test is to investigate which independent variables will cause significant differences to the dependent variables to support the decisions. Therefore, the ANOVA test is a statistical tool to test whether the means of several samples are equal based on the variance as defined in Eq. (9.1).

When carrying out ANOVA test, three measures of SS are calculated when independent sample groups are considered: between groups (SS_b), within group (SS_w), and total variance (SS_t). The ratio of the corresponding MSEs contributes to the F ratio for statistic test as specified below.

If there are $k \geq 2$ mutually independent sample groups, then:

1st group with n_1 samples: $X_{11}, X_{11}, \ldots, X_{1n}$;
2nd group with n_2 samples: $X_{21}, X_{22}, \ldots, X_{2n}$;
...
kth group with n_k samples: $X_{k1}, X_{k2}, \ldots, X_{kn}$;

$N = n_1 + n_2 + \cdots + n_k$ is the total sample number. Each group has different sample distributions with different sample means, X_1, $X_2, \ldots, X_k$ and variance, $\sigma_1^2, \sigma_2^2, \ldots, \sigma_k^2$. Define

$$SS_b = \sum n_k(\overline{X}_k - \overline{X}_G)^2,\ df_b = k - 1;$$

$$SS_w = \sum\sum(X_{ik} - \overline{X}_k)^2,\ df_w = N - k, \text{ and}$$

$$SS_t = \sum_{i=1}^{N}(X_i - \overline{X}_G),\ df_t = N - 1, \text{ then } SS_t = SS_b + SS_w \text{ and}$$

$$F = \frac{MS_b}{MS_w} = \frac{\dfrac{SS_b}{df_b}}{\dfrac{SS_w}{df_w}} \tag{9.4}$$

The single-factor ANOVA test is aimed at determining which of the following hypotheses holds:

$$H_0 : \mu_1 = \mu_2 = \cdots = \mu_k$$
$$H_1 : \text{not all are equal}$$

where μ represents the mean of the total population of each group.

ANOVA is a test that compares the F value with the p-value using a preset threshold for acceptance of a hypothesis. If the p-value is smaller, then H_o is rejected and the test is considered *significant*. Then the mean values of these k groups are different. The details and the statistical tables can be seen from any statistic textbooks, such as that by Degroot and Schervish (2002).

9.4.3 Cluster Analysis

Cluster analysis is to find *homogeneous* and *well-separated* subsets in a universe of discourse. It is one of the most important activities in the human learning process.

Given a sample of n observations, the problem of cluster analysis is to devise a clustering scheme for grouping the observations into c clusters with $n > c \geq 2$ so that while the dimension of the observed data is reduced, the characteristics of the sample can be revealed.

1. **Feature Selection:** Realizing the pattern of the data by finding the properties of the data items.
2. **Clustering Analysis:** Based on the selected features, the structure in data sets or samples is searched.
3. **Classification:** To identify "new" data so that the pattern of total populations can be recognized.

Two major categories for cluster analysis can be applied and they are briefly introduced below.

Hierarchical Clustering Method

A *hierarchical clustering method* is a common approach to conduct cluster analysis based on a sequential search for different patterns of clusters according to the different degrees of relations among the objects. The procedure is as follows:

1. To establish a *"relation,"* R, between each pair of objects denoted by r_{ij} according to the selected features. For instance:

$$R \triangleq \begin{array}{c} \\ x_1 \\ x_2 \\ x_3 \\ x_4 \\ x_5 \\ x_6 \end{array} \begin{array}{c} \begin{array}{cccccc} x_1 & x_2 & x_3 & x_4 & x_5 & x_6 \end{array} \\ \begin{bmatrix} 1 & 0.2 & 1 & 0.6 & 0.2 & 0.6 \\ 0.2 & 1 & 0.2 & 0.2 & 0.8 & 0.2 \\ 1 & 0.2 & 1 & 0.6 & 0.2 & 0.6 \\ 0.6 & 0.2 & 0.6 & 1 & 0.2 & 0.8 \\ 0.2 & 0.8 & 0.2 & 0.2 & 1 & 0.2 \\ 0.6 & 0.2 & 0.6 & 0.8 & 0.2 & 1 \end{bmatrix} \end{array}$$

How to establish this relation matrix? The commonly used methods are introduced in the following. It can be noted that each method will produce a *similar matrix,* which satisfies *symmetric* properties (i.e., $r_{ij} = r_{ji}$, for all i, j) and *reflective* properties (i.e. $r_{ii} = 1$, for all i).

a. **By Numerical Product**

$$r_{ij} = \begin{cases} 1, & \text{if } i = j \\ \sum_{k=1}^{m} x_{i_k} x_{j_k} / M, & \textbf{otherwise} \end{cases} \tag{9.5}$$

where $M > 0$ is selected such that

$$M \geq \max_{ij} \left(\sum_{k=1}^{m} x_{i_k} x_{j_k} \right).$$

b. **By Correlation Coefficient**

$$r_{ij} = \frac{\sum_{k=1}^{m} |x_{i_k} - \overline{x}_i| \cdot |x_{j_k} - \overline{x}_j|}{\sqrt{\sum_{k=1}^{m} (x_{i_k} - \overline{x}_i)^2}\sqrt{\sum_{k=1}^{m} (x_{j_k} - \overline{x}_j)^2}} \tag{9.6}$$

where

$$\overline{x}_i = \frac{1}{m}\sum_{k=1}^{m} x_{i_k},\ \overline{x}_j = \frac{1}{m}\sum_{k=1}^{m} x_{j_k}.$$

c. **By Max/Min Method**

$$r_{ij} = \frac{\sum_{k=1}^{m} \min(x_{i_k}, x_{j_k})}{\sum_{k=1}^{m} \max(x_{i_k}, x_{j_k})}. \tag{9.7}$$

d. **By Arithmetic Mean/Min Method**

$$r_{ij} = \frac{\frac{1}{2}\sum_{k=1}^{m} (x_{i_k} + x_{j_k})}{\sum_{k=1}^{m} \min(x_{i_k}, x_{j_k})}. \tag{9.8}$$

e. **Geometric Mean/Min Method**

$$r_{ij} = \frac{\sum_{k=1}^{m} \sqrt{x_{i_k} \cdot x_{j_k}}}{\sum_{k=1}^{m} \min(x_{i_k}, x_{j_k})}. \tag{9.9}$$

f. **Absolute Exponential Method**

$$r_{ij} = e^{-\sum_{k=1}^{m} |x_{i_k} - x_{j_k}|}. \tag{9.10}$$

g. **Absolute Subtraction Method**

$$r_{ij} = \begin{cases} 1, & \textit{if } i = j; \\ 1 - c\sum_{k=1}^{m} |x_{i_k} - x_{j_k}|, & \textbf{otherwise.} \end{cases} \tag{9.11}$$

where c is selected such that $0 \leq r_{ij} \leq 1$.

To conduct cluster analysis, an *equivalent matrix,* which possesses the third property of *transtivity* (i.e., $r_{ij}r_{jk} = r_{ik}$, for all i, j, k), is necessary. To obtain an equivalent matrix based on a similar matrix, Theorem 9.3 gives a clue that will be illustrated by the following example.

Theorem 9.3 For an $n \times n$ similar matrix $A = [a_{ij}]$, by max–min composite operation defined by $[a_{ik}] = \max(\min_j(r_{ij}r_{jk}))$ and $A^{p+q} = A^p A^q$, then $I \leq A \leq \cdots \leq A^{m-1} = A^m = A^{m+1} = \cdots$ holds, the matrix, A^m, is an equivalent matrix, i.e., $R^* = A^m$.

Taking the example above, we may notice that R is a similar matrix as well as an equivalent matrix. So, $R^* = R$.

Then, based on this equivalent matrix, cluster analysis can be done with respect to different degree of homogeniety defined by Definition 9.4.

Definition 9.4: α-degree of Homogeniety An α-degree of homogeniety of a set is defined in the equivalent relation defined by

$$\tilde{R}_\alpha = \{r_{ij} | \text{ where } r_{ij} = 0, \text{ if } r_{ij} < \alpha; \text{otherwise}, r_{ij} = 1, \text{ for all } i,\ j\}$$

Following the example above with equivalent matrix R^*, we have different clusters with respect to different degrees of homogeneity as shown in Fig. 9.1.

Although the sequence of clustering, one can recognize which one is closer to which. Being *not iterative* during the procedure is a disadvantage. Thus, they cannot change the assignment of objectives to clusters that have been made on proceeding levels. An alternative approach based on the concept of optimization has been developed as so-called *c-means method,* as introduced in the next section.

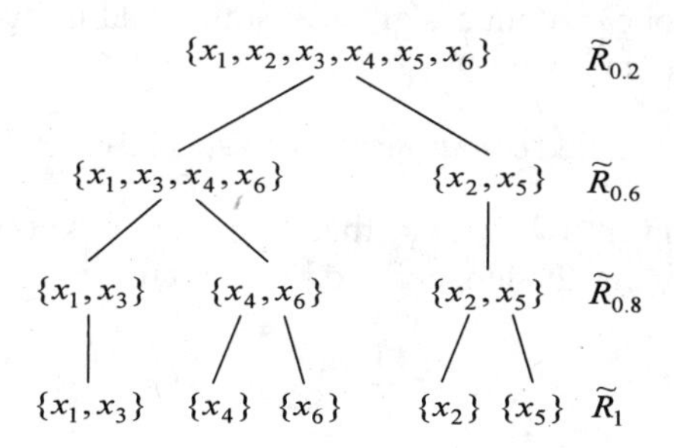

FIGURE 9.1 Hierachical structure of clustering.

c-Mean Method

Of the e-mean method measures, the "desirability" of clustering candidates for each cluster c by an objective function, and allows the most precise formulation of the clustering criterion. One method that is frequently used is the so-called c-mean algorithm, which defines "center of clusters" and minimizing the total "spread" around those centers.

Definition 9.5: Fuzzy c-Mean Method Let the data set $\mathbf{X} \triangleq \{x_1, \ldots, x_n\} \in R^p$ be a subset of the real p-dimensional vector space R^p. $\tilde{S}_i$ for I = 1, 2, ..., c are the fuzzy clusters with membership functions defined by

$$\mu_{\tilde{S}_i} : \mathbf{X} \rightarrow [0, 1]$$

then we would assign each data to the appropriate clusters with

$$x_k \to \mu_{ik} \triangleq \mu_{S_i}(x_k)$$

Definition 9.6 For a given integer $2 \leq c < n$, let

$$V_{cn} \triangleq \{\text{all real matrix with dimensionality } c \times n\}.$$

The matrix $U = [u_{ik}] \in V_{cn}$ is a "c-partitioning," if

(a). $\mu_{ik} \in \{0, 1\}$, for $1 \leq i \leq c$, $1 \leq k \leq n$.
(b). $\sum_{i=1}^{c} u_{ik} = 1$, for $1 \leq k \leq n$.
(c). $0 < \sum_{i=1}^{c} u_{ik} < n$, for $1 \leq i \leq c$.
then $M_c \triangleq = \{\text{all } c - \text{partitioning of } \mathbf{X}\}$.

Note 1. Condition (a) says that for each data x_k, one can either assign it to one cluster S_i or not.
Note 2. Condition (b) ensures that each data is assigned to one cluster.
Note 3. Condition (c) describes the basic clustering pattern that one can not assign *none* of the data in interests nor assign all data in one cluster.

This is called *c-mean method* as a special case of *fuzzy c-mean method*.

Definition 9.7 Given that $1 < c < n$ is known, $U \in M_c$ and S_1, $S_2, \ldots, S_c$ are clusters defined by U, then

$$v_i \triangleq \frac{1}{|S_i|} \sum_{x_k \in S_i} x_k, \quad i = 1, \ldots, c \tag{9.12}$$

are called *cluster centers*. A cluster center is measured by the weighted average of the data and the *dissimilarity* between the points in a cluster, and its cluster center is measured by Euclidean distance as:

$$d_{ik} \triangleq d(x_k, v_i) = \left[\sum_{j=1}^{p} (x_{kj} - v_{ij})^2 \right]^{1/2} \tag{9.13}$$

thus, the *variance of a cluster* can be obtained by

$$\sum_{x^k \in S_i} d_{ik}^2. \tag{9.14}$$

Based on the definitions above, the c-mean method takes the minimum variance as objective function and considers the following problem.

c-Mean Model

$$\textbf{Min } z(U) = \sum_{i=1}^{c} \sum_{k=1}^{n} ||x_k - v_i||^2$$

$$s.t.\ v_i = \frac{1}{|S_i|} \sum_{k=1}^{n} x_k,\ \forall i$$

$$U \in M_c.$$

Then, n sample data described by p features can be clustered into c clusters with minimum variance within each cluster to show the homogeneity of the clusters.

In summary, to identify the target customers is a process of pattern recognition which is based on the significant features to group customers into different homogeneous groups. Since the features of the customers are collected from survey, in this section, the questionnaire design and the basic statistical methods for sampling were introduced with two major types of clustering methods. Figure 9.2 summarizes the overall procedure.

The next section demonstrates numerically the analysis based on the survey conducted using the sample questionnaires.

9.5 Target Consumer Identification: Numerical Illustration

Past studies have pointed out that different results were obtained from different studies focusing on sociodemographic characteristics, that is, it is society-specific. In this section, the methods illustrated in Sec. 9.4 are used to identify the target consumer with a sample data collected from Taiwan using the questionnaires listed in Sec. 9.2. Based on the characteristics of the customers' clusters, the target customers can be found, and this will be the foundation of marketing.

9.5.1 Sample Size Determination

First, the number of samples to be collected for effective analysis should be decided when a simple sampling method is adopted. Based on formula (9.3) of the Chebyshev inequality (Mood et al., 1976), the probability 0.99 that the sample mean will lie within 0.5 of the population mean is to be ensured, then, the required data is at least 400 as

$$\sigma^2 = 0.25,\ \varepsilon = 0.5,\ and\ \delta = 0.01.;$$

$$n > \frac{\sigma^2}{\delta \varepsilon^2} = \frac{0.25}{0.01(0.5)^2} = 400.$$

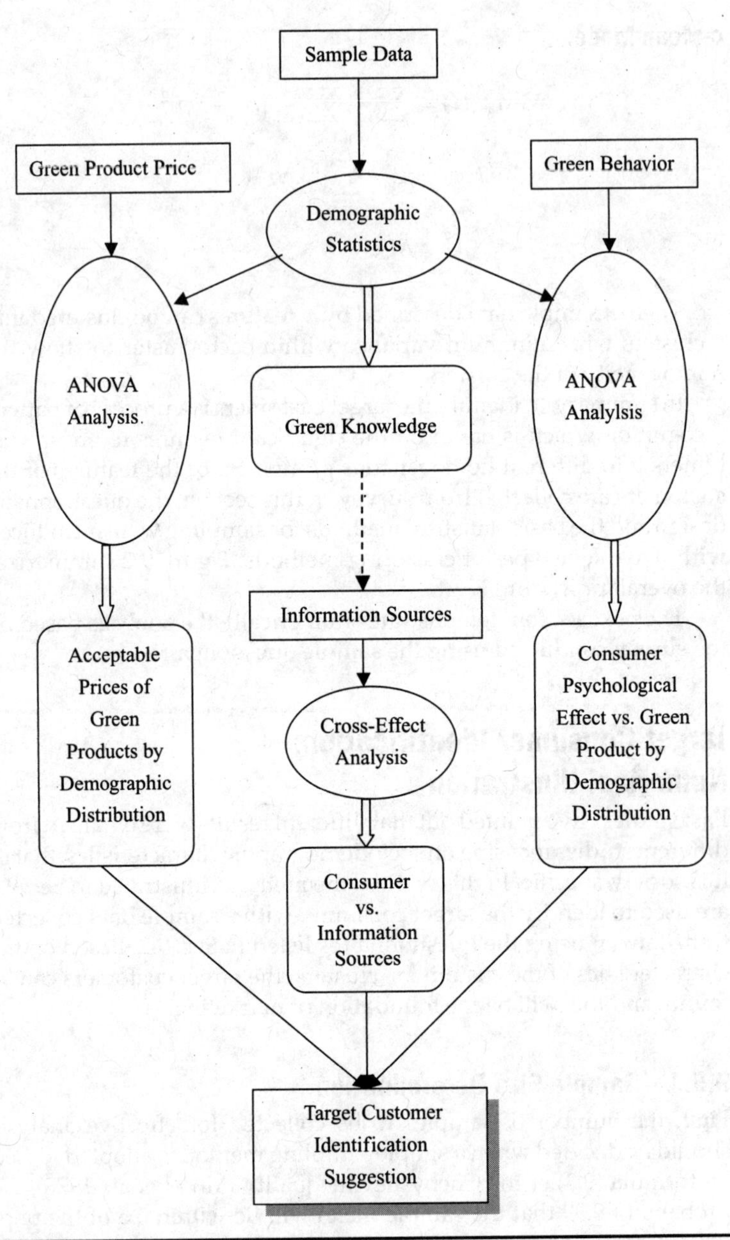

FIGURE 9.2 Framework of data analysis.

I	Totally disagree	Quite disagree	Moderate	Certain agree	Very much agree
II (1)	Totally do not understand	Not quite understand	Moderate	Understand well	Understand perfectly
II (2) III (1)	Absolutely not agree	Not quite agree	Moderate	Quite agree	Absolutely agree
II (3)	Absolutely not	Probably not	Moderate	Probably	Absolutely
Score	1	2	3	4	5

TABLE 9.2 Quantification of the Linguistic Terms

Based on the demographic record, 450 people from all over Taiwan were selected for interview. Records of 414 of them were used for analysis after the data-cleaning process was undertaken.

The following illustrates how the target customers were identified through statistical analysis.

9.5.2 Quantification of the Survey Data

In this section, the target customers are identified based on their features analyzed from the questionnaires. To perform the statistic, the different levels articulated from the interview are quantified according to the scores given in Table 9.2.

The analysis is conducted based on the collected information from each part of the questionnaires. The obtained effective sample of this research is 414. The top information source is TV network followed by newspapers and magazines.

9.5.3 Distributions of Sociodemographic Variables

Figure 9.3 shows the sample distribution related to each sociodemographic factor. It can be seen that with equivalent gender distribution, the unmarried no-income college students resident in North Taiwan are the majority in our sample set.

9.5.4 Factor Analysis

To find the significant variables of demographic factors to the environment variables categorized from our questionnaire as in Table 9.3, factor analysis was conducted and the result was shown in Table 9.4 where the significant variables with respect to the environmental categories are marked with "*".

Based on Table 9.4, the selected demographic variables are valid and based on regression analysis. Environmental attitude, purchasing behavior, environmental knowledge, age, and occupation were selected to be the factors for clustering.

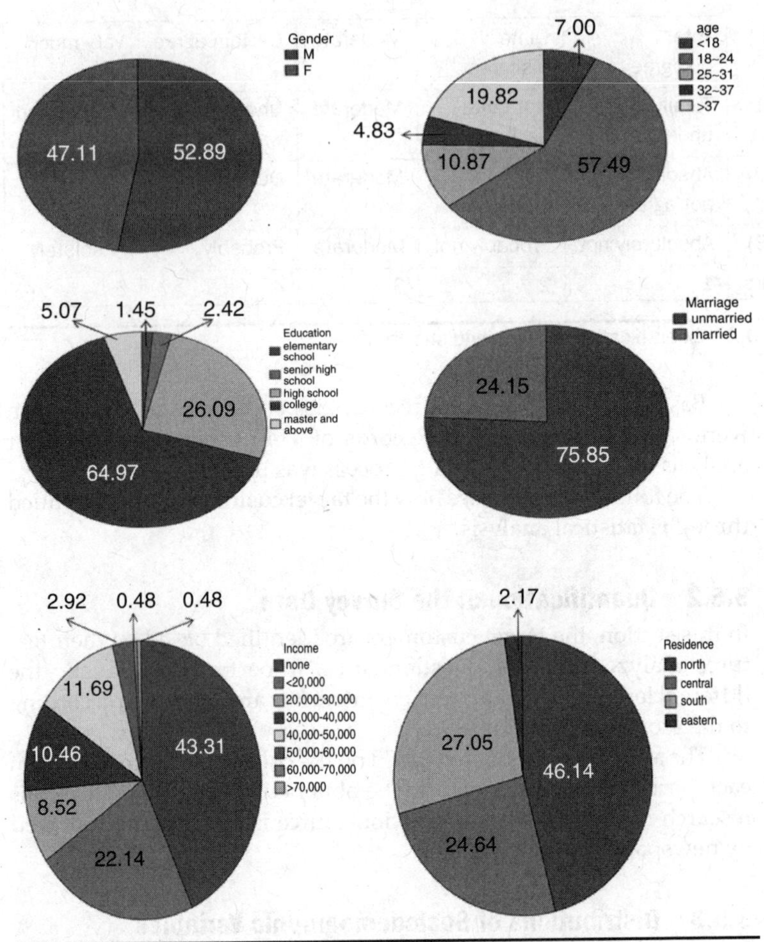

Figure 9.3 Framework of data analysis. Statistics of demographic distribution w.r.t each factor.

9.5.5 Target Customer Identification

Cluster Analysis

Based on *C-mean method*, Table 9.5 shows the number of sample people with the chariacters described.

Cluster I includes:

- university, college students, higher education, younger (under age 24), and lower income than those of Cluster III;
- higher environmental knowledge, attitude and purchasing behavior than in Cluster II;

Category I	General preference
Category II	Environmental knowledge and consciousness
Category III	Purchasing behavior
Category IV	Cost consideration vs. ecofriendliness

Table 9.3 Major Categories for Analysis

- when choosing a product, the degree to which they are influenced by advertisements is higher than in Cluster III;
- when choosing a product, price is not the first concern.

Cluster II includes:

- senior high school, university students, older (25–37 years old), and lower education than those in Cluster I;
- passive in terms of participation in nonconsumption environmental behavior;
- products offering similar functions are essentially the same; cheaper is better; they like to buy popular products that are on sale.

Cluster III includes:

- older (38 years old and above), higher income but lower education than others;
- highest level of environmental knowledge and attitude;
- when choosing a product, they seldom are influenced by advertisements because they believe advertisements usually exaggerate information;
- beautiful packaging is not the first concern;

Category Demographic Factor	(I) General Preference	(II) Environmental Knowledge and Consciousness	(III) Purchasing Behavior	(IV) Cost Consideration vs Eco-Level
Gender	*	—	—	—
Age	—	*	*	*
Marital status	*	*	*	*
Educational	*	*	*	*
Location	—	*	*	*
Monthly income	*	*	*	*

Table 9.4 Factor Analysis of Demographic Variable

Clusters of Consumption Types	Number of People	Percentage
Cluster I	222	53.6
Cluster II	58	14
Cluster III	134	32.4
Total	414	100.0

TABLE 9.5 Distribution of Number of People in Each Cluster

- the acceptable price is the highest of three-level-mark products of EPEAT, which can be more expensive than traditional products.

ANOVA Test Among Clusters

To evaluate the significance of the difference between clusters, the ANOVA test was carried out and the result is shown in Table 9.6.

Then according to the score, we may classify cluster III as "dark green," cluster II as "light green," and cluster I as "blue green." In other words, cluster III is our target customers.

9.5.6 Conclusion of the Numerical Example

Some observation of this numerical example can be drawn to reflect the purchasing behavior of Taiwan as follows:

1. **The resource of the green electronic product information:** The proportion of respondents who have access to green product information is high. As respondents get older, their use of newspapers, magazines, and television advertisements as sources of information increases.

Weight Score Factors	3	2	1
Preferences	Cluster III	Cluster I	Cluster II
Environmental knowledge and attitude	Cluster III	Cluster I	Cluster II
Purchasing behavior	Cluster III	Cluster I	Cluster II
Total Score	9	6	3
Age	Cluster III	Cluster II	Cluster I
Monthly income	Cluster III	Cluster II	Cluster I
Educational	Cluster I	Cluster II	Cluster III

TABLE 9.6 ANOVA Test

2. **The one-way ANOVA of the acceptable green electronic product price and the demographic variable of the consumer:** There is no remarkable difference among the acceptable prices for various groups. Therefore, manufacturers do not need to use different pricing strategies for different consumer groups in selling green products.
3. **The dependency sample, one-way ANOVA of the acceptable green electronic product price and the green degree of the product for the customer:** The acceptable price of green products vary significantly with the different degrees of greenness of said products. As the degree of greenness of a product increases, a consumer becomes willing to spend more for it.
4. **The ANOVA of the population demographic variables with respect to consumer behavior, environmental consciousness, environmental attitude, and ecofriendly behavior of customers can be outlined from such study as below:**
   a. *The concept of environmental protection:* Results of this study show that environment protection only varies remarkably with occupation, indicating that there generally is no difference in knowledge of the environment and recognition of green products. It is noteworthy that results show that environmental protection by consumers engaged in the science and technology industries (occupations) is worse than consumers in other industries.

   b. *The environmental protection attitude:* Respondents with high incomes (above NT$50,000 a month) got higher scores in the "environmental protection attitude" category, indicating that environment awareness increases with income.

   c. *Environmental protection behavior:* In this research, gender, civil status, occupation, income, and age remarkably influence environmental protection behavior. Females, married individuals, those working for the government, nongovernmental organizations, those who earn higher incomes, and those who are older were all more inclined to engage in environmental protection.

   d. *Green consumer behavior:* In this research, income, age, and occupation remarkably affect green-consumption behavior. Income has a positive correlation with green-consumption behavior, indicating that consumers with larger incomes are more inclined to purchase green electronic products. Older consumers who have been working for quite some time are more inclined than university and postgraduate students are to purchase green products.

Consumers working for the government and nongovernmental organizations are more inclined than other working individuals to purchase green electronic products.

5. **The two-way ANOVA of the environmental protection attitude and green-consumption behavior:** The analysis shows that:
   a. Consumers working for the government and nongovernmental organizations (NGOs) are more inclined than other working individuals to purchase green products.
   b. Consumers with higher incomes (above NT$30,000 per month) are more inclined than lower-income earners to purchase green products.
   c. Older consumers (above 32 years old) are more inclined than younger ones are to purchase green products.

   Moreover, among consumers aged 25 to 31 years old, and who have just entered the labor force, those with incomes of above NT$30,000 per month are the better targets for marketing green electronic products.
6. **Analysis of the relationship between psychological characteristics of consumers and their green-consumption behavior:**
   Environmental protection concept, attitude, and behavior are positively correlated with green-consumption behavior. Among the three, environmental protection attitude has the highest correlation with green-consumption behavior.
7. **Analysis of the green-degree appraisal target for the consumer:**
   When green consumers purchase green products, the most important appraisal has to do with this requirement: "the product design must reduce the material harmful for the environment." Therefore, when a manufacturer designs a product, it needs to consider reduction of the use of materials harmful to the environment to be able to sell more. Moreover, whether a product is energy efficient is also one of the most common aspects of appraisal. Meanwhile, environment-related achievements and performance of a company that produces a product are usually not taken into account by consumers in appraising a product.

9.6 Summary and Conclusion

This chapter introduces a method to identify target green customers for manufactures to promote green products. Because the perception toward environmental issues differs from people with different sociodemographic backgrounds, this chapter introduces the procedure through statistic analysis based on a questionnaire given by a sampled population. Then, the views of consumers about the green concept, green products, and green-consumption behavior are summarized and the consumers are clustered based on the significant factors determined by factor analysis.

Taking into account the results of the various data analyses conducted in Taiwan, the following suggestions are made as a reference of anlaysis.

Younger consumers mostly use their networks to obtain green information, and the older group relies on newspapers, magazines, and television advertisements. Therefore, different media should be used in promoting green products among different groups of people.

Consumers from various groups find the same acceptable price for a green product falling under a specific degree of greenness. When the degree of greenness of a product is higher, consumers from all groups are willing to spend more to purchase this product. Therefore, manufacturers can raise the prices of products as the degrees of greenness increases without turning off consumers.

This research also discusses the relation of the various psychological characteristics of consumers and their green-consumption behavior. Results show that environmental protection concept, attitude, and behavior all are positively correlated with green-consumption behavior. It is worth noting that the degree of understanding of environment-related knowledge and concepts (such as global warming, green products, and so on) assumes a low correlation with green-consumption behavior.

Moreover, individual purchasing behavior assumes moderate and high correlation with the environmental protection attitude and green-consumption behavior. Therefore, *merely educating the public about the energy crisis, the worsening condition of the global environment, and green products only increases understanding of the issues, but does not necessarily prompt people to purchase green products.* However, doing so can enhance the environmental protection attitude of consumers, who may be made to understand that their actions have a bearing on the environment. If consumers believe so, they will be willing to exert effort to help protect the environment and eventually be inclined to become green consumers.

If manufacturers aim to promote green products and if the government wants to encourage people to help protect the environment, they should utilize various kinds of marketing methods or advertisement slogans. For example, they may explain that purchasing green products would lead to reduced energy consumption and fewer trees cut. They may also say that replacing motorcycles with bicycles would reduce carbon emissions and air pollution. Doing so would make the public understand the importance of adopting environmental protection behavior and be aware that green consumers positively contributes to sustainability of the environment.

Analysis of the results shows three main groups of consumers of green electronic products, namely, (1) those who work for the government or NGOs, (2) those with higher incomes (above NT$30,000 per month), and (3) older ones (above 32 years old). If manufacturers want to promote green electronic products, they may be more efficient in doing so if they aim at targeting these three groups. If they want to expand the market, however, they need to aim at consumers beyons those belonging to these three groups.

During the appraisal of products by consumers, the aspect they are keenest about is whether materials harmful to the environment are used in the products' production and whether energy efficiency is observed in their manufacture. Environment-related achievements of manufacturers are not taken into account by consumers. In fact, the greenest electronic products nowadays are explained only in terms of which environmental protection standard the manufacturer has conformed with and which prize category the product belongs (whether bronze, silver, or gold). Unless manufacturers provide specifications of the markers, consumers would never know. Therefore, when promoting green electronic products, manufacturers should explain which marker the product conforms to and what its positive contribution is to the environment. If consumers have more information on these points, their chances of buying green products will increase.

References

Bohlen, G. M., Diamantopoulos, A., Schlegelmilch, B. B., 1993. *Marketing Intelligence and Planning*, 11(1), 37-48.

DeGroot, M. H., Schervish, M. J., 2002. *Probability and Statistics*, 3rd ed., New York: Addison Wesley.

Mood, A. M., Graybill F. A., Boes, D. C., 1976. *Introduction to the Theory of Statistics*, 3rd ed., Taiwan: McGraw-Hill.

Roberts, J. A., 1996. Green consumers in the 1990s: profile and implications for advertising. *Journal of Business Research*, vol. 36, no. 3, pp. 217-231.

Roberts, J. A., Bacon, D. R., 1997, Exploring the subtle relationships between environmental concern and ecologically conscious consumer behavior. *Journal of Business Research*, 40, 1, pp. 79-89.

Schlegelmilch, B. B., Bohlen, G. M., and Diamantopoulos, A., 1996, The link between green purchasing decisions and measures of environmental consciousness, *European Journal of Marketing*, 30, 5, pp, 35-55.

Stern, P. C., Dietz, T., and Kalof, L., 1993, Value orientations, gender, and environmental concern, *Environment and Behavior*, 25, 3, 32-48.

Straughan Robert D., Roberts, James A., 1999, Environmental segmentation alternatives: a look at green consumer behavior in the new millennium. *The Journal of* 160 *Consumer Marketing*, Vol. 16, Issue. 6, pp. 558-575.

Zimmer, M. R., 1994, Green issues: dimensions of environmental concern, *Journal of Business Research*. 30, 1, pp. 63-74.

CHAPTER 10

End-of-Life Management

Disassembly and Reuse

10.1 Introduction

Many countries are aggressively working on waste minimization and resource recovery because of severe environmental problems. One of the most well-known environmental regulations, the Waste Electrical and Electronic Equipment (WEEE) Directive, founded by the European Union (EU) in 2003, is a good example of environmental management policy implementation. EU members were asked to establish a product return stream and execution of recovery process to reduce incineration or landfill of electronic and electrical equipment waste.

However, product recovery cannot only be considered as an environmental issue or as regulations followed by businesses; it is also an economical operations strategy through appropriate management. When products are returned, they can be reprocessed to bring quality back to a reusable level, and to obtain recoverable components and materials. Sometimes, it is much cheaper to use returned components and recycled materials to remanufacture than using new parts by procurement.

One of the most important techniques in product recovery is disassembly. It can be defined as systematic separation of an assembly into its components or subassemblies and allows for the selective separation of desired parts or materials. The basic concept of disassembly planning is to determine the optimal disassembly degree or depth of *end-of-life* (EOL) of products, and the subsequent recovery options of disassembled components. From the perspective of a business that sells new products to customers and simultaneously recycles used products from customers, returned products can be remanufactured into the same (or other) products to fulfill the same market demand.

This situation addresses the concept of *demand-driven disassembly planning* problem.

If all remanufactured products in the market follow the return stream provided by the original manufacturer and become recoverable products, these products will be retained in the supply chain unless there is no benefit in recovering them. Such a supply chain framework is called a *closed-loop supply chain*. Although there are many studies on the mathematical programming of a closed-loop supply chain, there are few studies on the effect a closed-loop mechanism on return, disassembly, and recovery strategies. Thus, it is necessary to provide a decision support model that includes market mechanisms in the analysis.

10.2 Current Development

In recent years, increasing concern arising from environmental consciousness has resulted in a great deal of research and developments related to green management. Experts and scholars have completed numerous studies to investigate, analyze, and provide decision support models or systems for companies facing restrictions of regulations and growing environmental consciousness among customers. In this section, we will introduce related issues. We begin with a review of the closed-loop supply chain mechanism and life-cycle effects. The second section addresses the strategies of EOL alternative selection and the effects of the quality of returns. The third section reviews product structure representation, disassembly modeling, and planning. Finally, we conclude this chapter as a base to develop an analytical model.

10.2.1 Issues with a Closed-Loop Supply Chain

Firms and manufacturers have faced growing environmental concerns from consumers and government regulations. They are also burdened with the responsibility of sustainable management. The closed-loop manufacturing system illustrated in Fig. 10.1 shows that the concept of a recoverable manufacturing system is widely promoted. This system aims to return used products and then repair, remanufacture, or recycle them to extend the product life cycle and reduce waste. These systems offer potential advantages, including increased profitability through reduced requirements for materials by procurement and improved market share based on environmental image.

From a process flow and business perspective, there are three major activities in the closed-loop supply chains as shown in Fig. 10.1.

1. **Product returns management:** The timing, quantity, and quality of returns significantly influence recovery and remanufacturing strategies. Consequently, it is important to manage product acquisition from the front end of a reverse supply

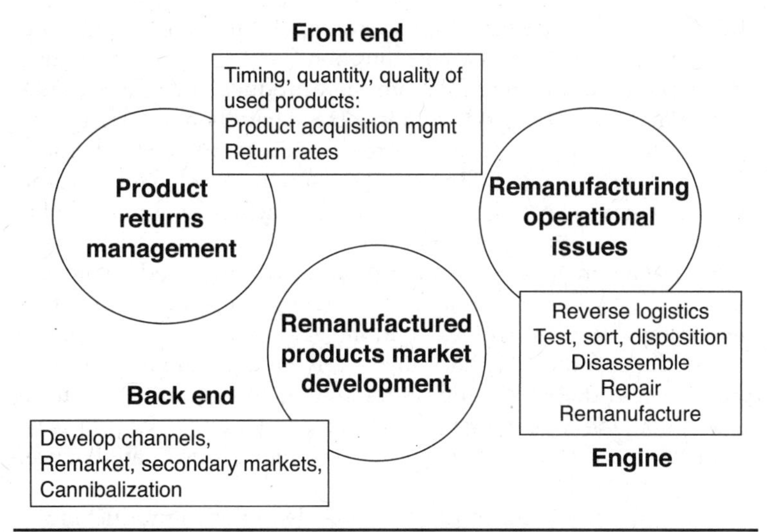

FIGURE 10.1 Activities in the closed-loop supply chain.

chain. Acquisition price and returns quality/quantity have a positive relationship due to economic incentives perceived by consumers.

2. **Remanufacturing operational issues:** Numerous research aims to plan and control remanufacturing-related activities for recovery process costs to be at reasonable levels, and to maximize profit. Major tasks are reverse logistics, testing, sorting, disassembly, remanufacturing, and disposition.
3. **Remanufactured product market development:** The problem of consumer behavior and product valuation occurs at the back end of a closed-loop supply chain. Cannibalization and diffusion rate/path are important but not widely studied marketing strategies of remanufactured products.

Each of these three major activities plays an indispensible role. However, the bulk of research on closed-loop supply chains focuses on technical and operational issues. There are few studies on the market of remanufactured products and the acquisition of used products. To make closed-loop supply chains more profitable, Guide and Wassenhove (2009) suggested that all managerial bottlenecks must be removed; therefore, all subprocesses should be integrated to release their hidden value.

10.2.2 Life-Cycle Effects on the Quantity and Quality of Returned Products

Although "product life cycle" has been studied for a long time, there are some conflicting definitions from the literature. Kumar et al. (2007)

divided the life cycle into three stages: (1) manufacturing (value creation), (2) usage (value consumption), and (3) recovery (value reclamation). Raw materials become new products and value is added during manufacturing. When products are bought and used by consumers, the remaining value decreases due to the deterioration of quality or performance. With time, the residual value, quality, or product performance falls to the minimum satisfaction level perceived by consumers, who then decide to discontinue using the products. Post-use decisions include maintaining and upgrading products to continue use, or selling them to EOL product processors to recover the remaining value. The difference in the value perceived by each consumer leads to the difference in the remaining value of end-of-use products and the uncertainty in the timing and quantity of returns. In addition, the remaining value perceived by EOL processors is not identical to the value perceived by consumers; thus, there is uncertainty in the quality of returns.

An earlier—and the most common—definition of product life cycle was given by Cox (1967) as the evolution in sales volume of a product over time. When a product comes into the market, demand will go through four phases during its life cycle: (1) introduction, (2) growth, (3) maturity, and (4) decline. When the products sold to the market end their lives, consumers might take them back to the EOL processor for recycling. Potential remanufacturing volumes are limited to the volumes of returns/supply of EOL products in a closed-loop supply chain system due to the importance of matching supply and demand for remanufacturing strategies. Thus, Östlin et al. (2009) proposed an improved plan in which remanufactured products are upgraded to the latest specification to add a new function or increase performance, which is a viable option for extending the life cycle of products, as shown in Fig. 10.2. Demand for upgraded remanufactured products is therefore regenerated; potential remanufacturing volumes increase; and supply and demand are matched.

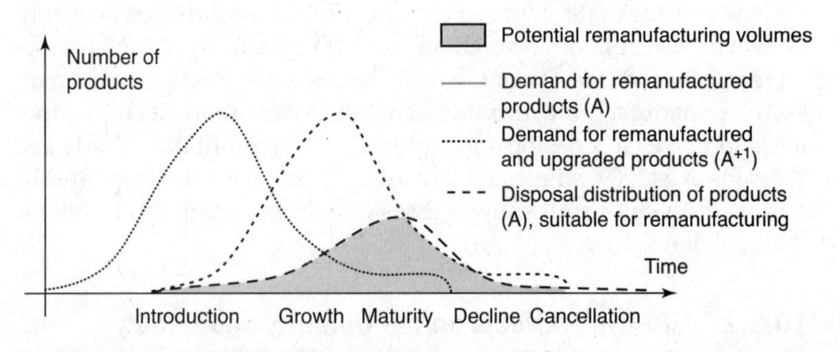

FIGURE 10.2 Potential remanufacturing volumes when upgrading to the latest technology is a viable option.

10.2.3 EOL and End-of-Use Recovery Selection

When an EOL or end-of-use product returns, it can be reused, repaired, remanufactured, recycled, incinerated, or dumped in a landfill. All options are suitable for products and components. The EOL/end-of-use recovery options are defined as follows (Gungor and Gupta, 1999; Ilgin and Gupta, 2010).

1. **Reuse:** Aims to return used products for use in the same application. Examples include reusable plastic bottles and containers.
2. **Repair:** Involves fixing or replacing broken parts in a returned product to improve its quality/performance standards to a good condition. However, the quality of repaired products is generally less than the quality of new products.
3. **Remanufacture:** In the remanufacturing process, used products are disassembled and inspected. Reusable parts and modules are retained, and repairable parts and modules are fixed. All available components are reassembled into remanufactured products with new parts that replace worn-out ones.
4. **Recycle:** Recycling aims to obtain reusable materials from used products without conserving the original product structure. Recycled materials can be used as resources in producing original and other parts. Kongar and Gupta (2006) mentioned that destructive disassembly is often performed for recycling.
5. **Incineration and landfilling:** Waste products, modules, and parts with no feasible or advantageous recovery options are incinerated or dumped in landfills. These are poor choices for the environment.

Table 10.1 summarizes the differences in characteristics of product recovery options, which include the required level of disassembly, quality, and resulting outcomes.

After a comprehensive review of definitions of recovery options, the next step is to determine the appropriate option in treating EOL products. This depends on quantitative or qualitative criteria. This problem is concerned with different types of users such as original producers, third-party recyclers, or environmentalists, who have their own objectives and preferences. In other words, even if different decision makers consider the same EOL products and the same criteria, it is possible that different appropriate approaches exist for each EOL component for recovery purposes.

Many studies focus on developing the decision support model or system for EOL/end-of-use recovery strategies. Mathematical models, especially *multicriteria decision making* (MCDM) methodologies have been extensively presented. Bufardi et al. (2004) proposed

Recovery Options	Level of Disassembly	Quality Requirements	Resulting Outcomes
Reuse	Not required	To be as good as new	As in the same application
Repair	To product or module level	Restore to be in good condition	Fixed or repaired products/modules
Remanufacturing	To part level	Inspect all components and replace/fix to be as good as new	Remanufactured product containing used and new parts
Recycling	To material level	Less or not required	Reusable materials
Incineration or landfilling	To module, part, or material level	Not required	Wastes and garbage

TABLE 10.1 Comparison of Product Recovery Options

MCDM to support the decision maker in selecting the best compromise EOL alternatives based on their preferences and the performance of EOL products with respect to the relevant environmental, social, and economic criteria. Jun et al. (2007) presented a multi-objective evolutionary algorithm to select the best EOL recovery option for a turbocharger case that maximizes its recovery value and deals with the trade-off between recovery cost and quality after the recovery process.

Disassembly is an important process to separate desired parts and materials for possible recovery options; thus, EOL selection problems are simultaneously investigated with product disassembly in some studies. Lee et al. (2001a) determined the optimal EOL options of components and the level of disassembly using a multi-objective methodology that simultaneously consider environmental impact and economic value. Xanthopoulos and Iakovou (2009) presented a remanufacturing-driven reverse supply chain framework and developed a two-phased algorithm. In the first phase, goal programming was used to determine the most attractive subassemblies and components to be disassembled for recovery. In the second phase, a mixed-integer programming model was proposed to find optimal remanufacturing/disassembly planning for multiple products with common components, as well as to satisfy demand for remanufactured products during each period. In the next section, we will discuss disassembly modeling and planning.

10.3 Concept of Disassembly

As mentioned previously, remanufacturing is the transformation of used products into products that satisfy the as-new quality and other

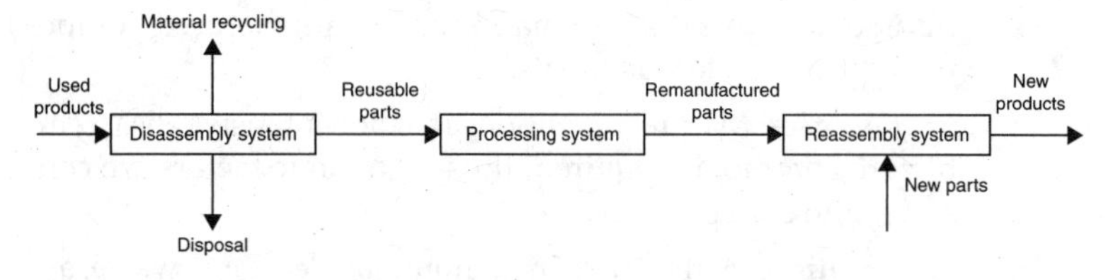

FIGURE 10.3 General description of remanufacturing systems (Guide, Jr., 1997).

standards through a series of industrial processes. Generally, these processes are partial/complete disassembly; reprocessing, which contains testing, sorting, evaluation, and the operations required to bring the disassembled parts and modules back to an as-new condition; and reassembly of remanufactured and new parts by procurement into remanufactured products as Fig. 10.3 illustrates.

One of most important techniques in product remanufacturing is disassembly, since it allows for selective separation of desired parts, modules, and materials (Gungor et al., 1999; Ilgin et al., 2010). The three main functions of disassembly are to recover pure materials, isolate hazardous substances, and separate reusable parts and modules (Lee et al., 2001a). From the perspective of degree of disassembly, it can be divided into two types: selective/partial disassembly and complete disassembly. In the former, one or more parts are detached from used products for possible recovery options and to remove hazardous materials. In the latter, all parts and modules in used products are simultaneously separated (O'Shea et al., 1998). Related research on disassembly has been widely studied. Ilgin and Gupta (2010) classified them into several categories, as follows.

1. **Sequencing:** Disassembly sequencing is concerned with determining the best order/sequence in the disassembly process of products to obtain the desired parts and modules.
2. **Scheduling:** Disassembly scheduling is defined as setting an order and planning the disassembly of EOL products to fulfill demand for the parts and components over time (Gupta and Taleb, 1994).
3. **Disassembly line balancing:** This determines the assignment of a set of disassembly tasks to different work stations while satisfying the disassembly precedence constraints, minimizing the number of stations needed, and balancing the process times among all stations (McGovern and Gupta, 2011).
4. **Disassembly-to-order (DTO) system:** DTO is a demand-driven disassembly planning and recovery selection problem. It determines optimal lot sizes of multiple EOL products to

disassemble to satisfy demand for remanufactured components and recycled materials.

5. **Ergonomics:** Manual operations in labor-intensive disassembly are commonly required; thus ergonomic factors are considered in some studies.
6. **Automation:** With increasing amount of electronic waste, automation of disassembly process is sure to decrease labor costs and enhance productivity.

The research problems summarized above manage and control the disassembly process from different points of view. However, there is strong interdependence among these different problems. For example, the output of disassembly sequencing, such as required times and ways to disassemble the specific parts or modules, can be used as input data for disassembly scheduling. To develop a suitable model for disassembly analysis and planning, it is necessary to build a clear and precise mathematical representation of product structure or generate all possible disassembly sequences, especially in the disassembly sequencing field. Such good representation is conducive to enhancing the modeling process.

10.3.1 Design for Disassembly Representation

It is important to provide a basis representation of possible disassembly sequences. One of the earliest and well-known methods is the AND/OR graph introduced by Homem de Mello and Sanderson (1990). Generally, AND/OR graph can be defined as a directed graph, in which each node represents its possible subassemblies/components, and each arc represents every feasible disassembly operations. Edge is in "AND relation" if its parent node (assembly or product) can only be disassembled by a single operation into two corresponding subassemblies/parts simultaneously. A set of edges with the same parent node are in the "OR relation" if it is possible to obtain more than two kinds of subassemblies/parts through different methods of disassembly from their parent node. Figure 10.4 shows a simple example for AND/OR graph.

The AND/OR graph representation of product structure is an enumeration method that generates all possible disassembly operations and subassemblies. Thus, extremely time-consuming computation is needed to solve a disassembly sequencing problem in the case of a complicated product structure. Lambert and Gupta (2002; 2005) pointed out that an AND/OR graph can only deal with disassembly sequencing problems, which are applied restrictedly to a situation in which a given returned product has to be disassembled to obtain the desired parts/subassemblies for possible recovery, but without any restriction of demand on market.

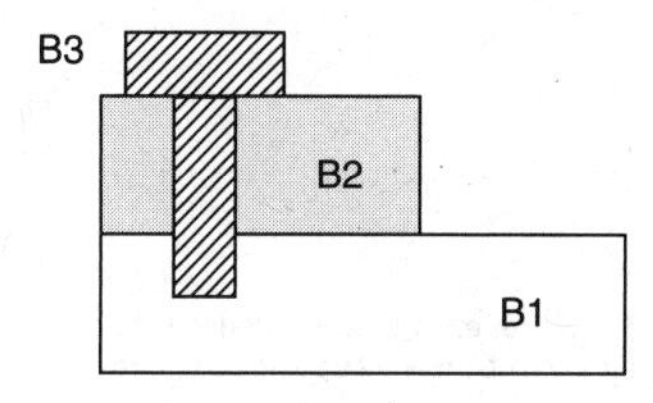

(a) Product structure

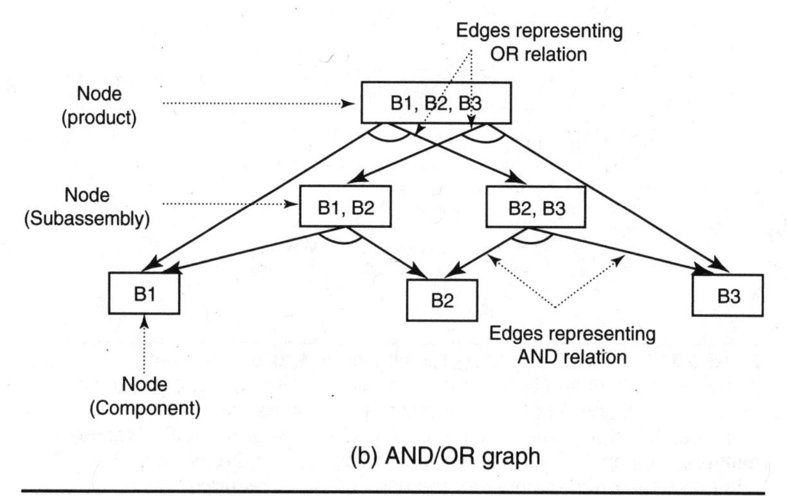

(b) AND/OR graph

FIGURE 10.4 (a) A product structure showing that part B3 is a screw used to combine the two boards, B1 and B2. (b) AND/OR graph for the product in part (a).

From the perspective of *material requirements planning* (MRP), the configuration of a product is usually represented by a tree structure, which is based on the information derived from the *bill of materials* (BOM). Regular MRP transforms demand for products into a requirement of its multiple components. That is, MRP is used for an assembly system, in which parts/components converge to a single demand source of product, as shown in Fig. 10.5 (a). However, in a disassembly system, each returned product diverges to its own types of parts/components to meet multiple demands. This kind of difference between the assembly and disassembly systems was called *convergence* and *divergence* properties by Brennan et al. (1994) and Kim et al. (2007). Consequently, regular MRP cannot be used to solve disassembly scheduling problems. Fortunately, the difficulty did not last long. Gupta and Taleb (1994) and Taleb and Gupta (1997) suggested "reverse" MRP, which is a basic form of disassembly scheduling considering multiple returned products, components

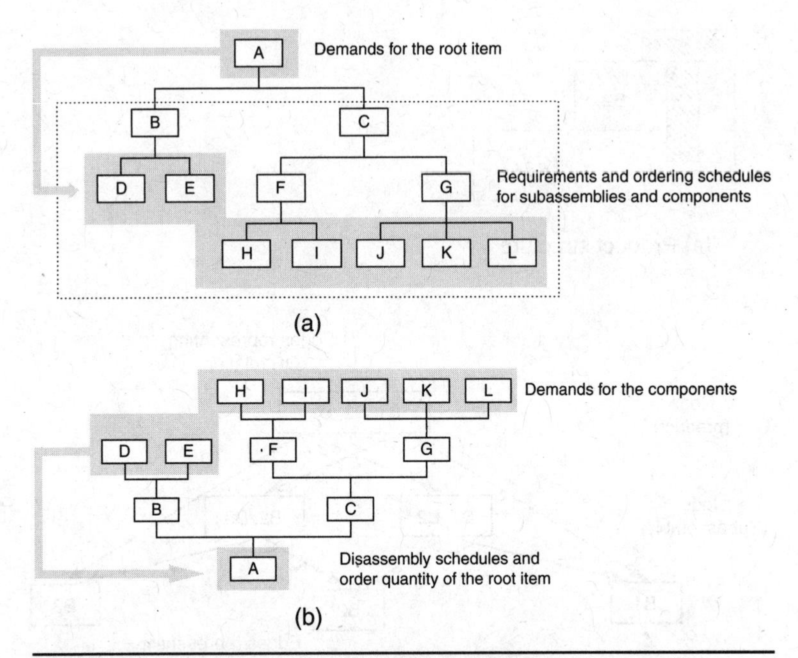

FIGURE 10.5 (a) Tree structure of a product in regular MRP. Demand for a root item A is decomposed to all its descendants, intermediate, and leaf items (B, C, D, E, . . . , etc.) to generate ordering schedules for parts and subassemblies. (b) Tree structure of a product in reverse MRP. Demands for multiple components (leaf items) must trace back to the root item to determine the order quantity for the used products required.

commonality, and demands for parts and components, as illustrated in Fig. 10.5 (b).

The tree structure in Fig. 10.5 (b) is also called a *disassembly product structure*, and it can be obtained by generalizing all possible parts/subassemblies and the necessary disassembly operations, such as AND/OR graph. This representation of product structure reduces complexity, and takes important steps in the direction of research on the disassembly scheduling problem.

Lambert and Gupta (2005) extended the concept of reverse MRP to a tree network model. Returned products were described by a reverse BOM with component multiplicity and commonality. Product data were further formulated into two matrices: tree structure matrix and yield matrix. The former showed that the disassembly sequence of each node (subassemblies) must abide by some precedence relationship, while the latter provided the multiplicity of each leaf (part). The objective of the model was to determine the optimal lot-size of used products and disassembly depth, minimize disassembly costs and disposal cost of redundant components, maximize sale revenue of reusable parts, and meet a given demand for different types of

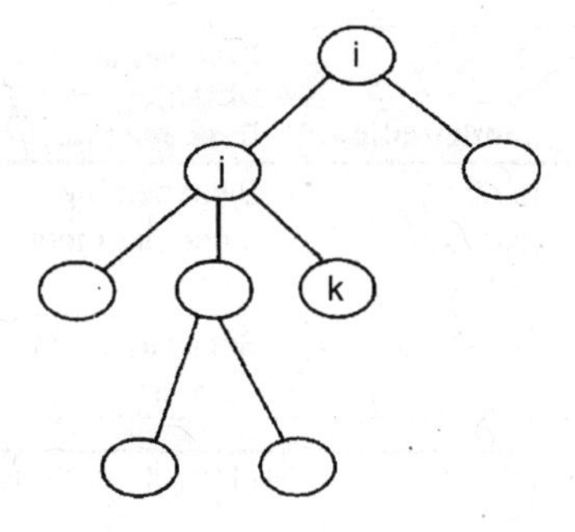

FIGURE 10.6 Example for a product structure representation.

parts over the planning horizon. This became the first line of research on demand-driven lot-sizing problems for disassembly planning or so-called DTO problems.

Meacham et al. (1999) developed an algorithm to solve the disassembly scheduling problem considering common components among multiple products and limited recoverable returned products. The product structure representation they used is similar to the tree network model proposed by Lambert and Gupta (2005), as shown in Fig. 10.6. A hierarchical structure was also used, thus the precedence of disassembly must be followed. The only difference between them is the permission to partially disassemble modules/subassemblies in the "OR" relation. However, in Lambert and Gupta's opinions, all nodes in the tree network model were considered "virtual" modules, and they abided by the following disassembly rule.

To obtain a desired part (or subassembly), the parent node of this part needs to be disassembled by a single operation, and all descendants of the node are separated simultaneously.

In Meacham's observation, modules that follow this rule can be considered the "AND" type of node. A common example of this type of node is a subassembly that represents a joining technique combining all its descendant parts together. After the joining technique is removed, all parts or subassemblies belonging to this AND node can be obtained. If a module can be defined as an "OR" type of node, which was extensively considered in Meacham's study, it abides by the following disassembly rule.

To obtain a desired descendant part (or subassembly) of this node, it is not necessary to obtain all descendants. That is, selective/partial separation for desired part(s) from the node is permitted.

An example for this type of node is a circuit board, in which different kinds of components are welded. We can remove them one by one, depending on the requirements or their recovery costs and profits. Table 10.2 summarizes the two types of modules.

Module	Configuration	Process of Obtaining Descendants	Partial Disassembly
AND module	Joining technique (virtual)	Disassembly (once per module)	Not permitted
OR module	Substantial	Separation (once per part)	Permitted

Table 10.2 Comparison Between AND Module and OR Module

10.3.2 Demand-Driven Disassembly Planning

In the preceding section, we introduced a demand-driven disassembly planning problem in which the objective is to determine the return timing or quantity of different types of EOL products to disassemble for possible recovery process to satisfy all demands for remanufactured products, reused parts, repaired modules, and recycled materials. Kongar and Gupta (2002) defined this as the DTO system. In the system, a collector pays a certain acquisition cost to take back EOL products from end users. Then, these EOL products are cleaned, inspected, and sorted for disassembly operations. There are different types of demand; thus, EOL products are either nondestructively disassembled into the parts/modules level to remanufacture/reuse/repair/storage/dispose, or destructively disassembled into the material level to recycle/dispose. In addition, the disassembly operation can be selective or complete. The former represents a partial section in an EOL product being disassembled, and the latter represents all items included in the product being disassembled.

Kongar and Gupta (2002) also proposed single-period goal programming for a DTO system to determine the best combination of the number of each type of EOL product to be returned and selectively/completely disassembled to meet the demand for parts and materials under a variety of physical, financial, and environmental constraints.

In the following, we will formally state the problem of disassembly to demand and introduce a mathematical model with a numerical example.

10.4 Disassembly to Demand (D2D): Modeling and Analysis

Due to economic and environmental issues, a manufacturer establishes a closed-loop supply chain system to pursue a compromise between sustainability and profitability. The closed-loop supply chain system consists of all activities involving product return and recovery, as shown in Fig. 10.1. It is assumed that the manufacturer does not adopt a subcontractor (such as third-party logistics) to perform product return and recovery operations. The manufacturer sells multiple products to consumers in a product market. With time, when the value or quality remaining in the products decreases to a certain level, consumers will decide whether to discard the products or return them to the collection center to gain a certain subsidy provided by the original manufacturer. Then, the manufacturer can determine the type and amount of EOL products acquired for storage or disassembly operations. Parts or modules completely/partially disassembled from the returned products can be stored or recovered. Remanufactured products and their inventory, which are accomplished by the remanufacturing process with both disassembled parts and new parts from procurement, fulfill market demand. Parts and modules can also be repaired to sell to the secondary component market, recycled to sell to the material market, or disposed to landfills. Products or its components will be retained in the closed-loop supply chain over the entire planning horizon until the following situations happen.

1. Consumers discard the used products or return to other recycling streams.
2. The manufacturer rejects returned products because of poor quality or overaged returned products.
3. Parts and modules are sold to secondary markets, recycled, and then sold to material markets and disposed of in landfills.

Thus, we mark a system boundary according to situations in which products can be retained in a closed-loop relationship (see Fig. 10.7).

10.4.1 Representation of Product Structure

Given detailed information on the BOM of products and the hierarchical relationship among parts and modules, we can represent it as a tree network such as the one shown in Fig. 10.8 (Lambert and Gupta, 2005).

The structure can be divided into three levels: product, module, and part. The basic level of a product is constructed by part-level items expressed as leaf nodes of the tree network. The module-level items consist of at least one or more descendant nodes that are part-level items. The highest-level represents the product itself.

From the representation, the *yield parameters* can be written in mathematical form and defined as follows.

$A^{l}_{i,j}$: The number of part item i that is/are the descendant(s) of module node j for product type l

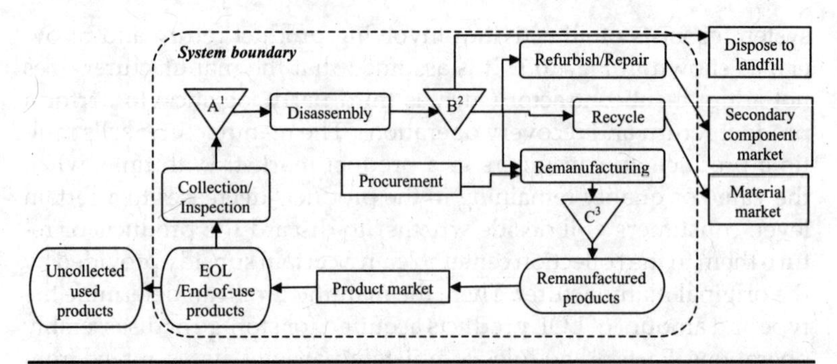

FIGURE 10.7 Illustration of problem framework and system boundary. KEY: 1 represents the inventory of returned EOL/end-of-use products; 2 represents the inventory of completely disassembled parts or partially disassembled modules; and 3 represents the inventory of remanufactured products.

For example, if $A^l_{i,j} = 3$, then we can obtain three part item i simultaneously through the disassembly operations from module node j of a product type l. The yield parameter can also be transformed into a matrix form. Here is an example:

For a specific product type l', the *yield matrix* $\mathbf{A}^{l=l'} = \begin{bmatrix} 1 & 0 & 3 \\ 0 & 1 & 0 \\ 1 & 1 & 0 \\ 2 & 0 & 0 \\ 0 & 0 & 1 \end{bmatrix}_{(i\times j)}$

Its tree network representation is shown in Fig. 10.8. The rows correspond to the leaves in the tree network, i.e., the part level items 1, 2, 3, 4, and 5 in Fig. 10.8, while the columns correspond to the intermediate nodes, i.e., the module level items M1, M2, and M3. The nonzero elements in the matrix represent the number of parts that exist in the corresponding (i, j) position. Furthermore, to keep the consistency in matrix size for different types of products with common parts, the number of rows in all yield matrices of product is fixed to the total number of types of parts in our system. Also, the number of columns in all yield matrices is determined by the "maximum" total number of module nodes in all product types. In the next section, we will further discuss the different types of module configurations and the corresponding disassembly operations.

10.4.2 Disassembly Configurations for Modules

To obtain a desired part for recovery, its parent node-module item should be disassembled to remove the part from it. However, the compatible disassembly configuration depends on the types of module

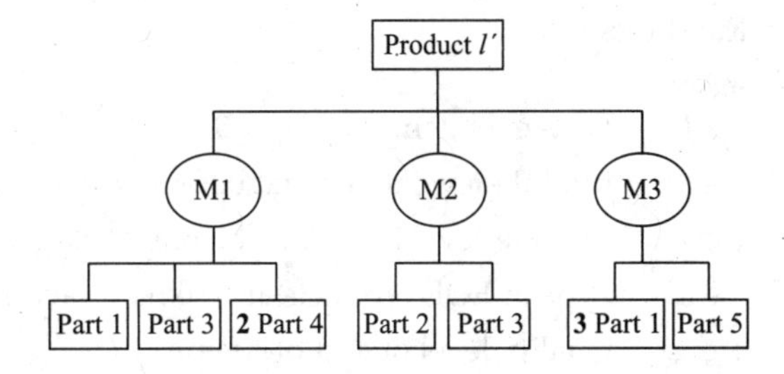

Figure 10.8 Graphical illustration for specific product type l'.

structure. As Meacham et al. (2003) pointed out, module nodes in a product tree network model can be classified into two types: "AND node" and "OR node." We summarize the properties and corresponding disassembly configurations of these two modules in Table 10.2. The mathematical representation for two types of modules can be easily done by separating the yield parameter into two distinct forms, $A^l_{i,j}$ and $O^l_{i,k}$, where the index k denotes the number of OR node modules in a specific product type.

10.4.3 Restrictions of Recovery Options

Figure 10.1 shows that after EOL products are disassembled, they can be recovered for various purposes including refurbish/repair, recycle, remanufacturing, and disposal. We quote the restrictions of different recovery options from Thierry et al. (1995) in Table 10.3.

10.4.4 Mathematical Model

In this section, we shall introduce a disassembly model based on the functions we discussed in the previous section. Notations are first defined, then the model is formulated.

Recovery Options	Required Degree of Disassembly	Required Quality	Resulting Outcomes
Refurbish/Repair	To module (which keep intact) or part level	Good condition	Repaired modules/parts
Remanufacturing	To part level	Good condition	Remanufactured product containing used and new parts
Recycling	To module or part level	Not required	Reusable materials
Disposal	To module or part level	Not required	Wastes

Table 10.3 Restrictions of Recovery Options

Notations

Indices

$i \in I$ Part-level items

$j \in J$ Module-level items that belong to the "AND" node

$k \in K$ Module-level items that belong to the "OR" node

$l \in L$ Type of EOL products and remanufactured products

$q \in Q$ Quality level of end-of-life products

$m \in M$ Recycled material items

$t \in T$ Time period

Sets

$I = \{1, 2, \ldots, N\}$ The set of part-level items

$J = \{1, 2, \ldots, V\}$ The set of module-level items that belong to the "AND" node

$K = \{1, 2, \ldots, W\}$ The set of module-level items that belong to the "OR" node

$L = \{1, 2, \ldots, H\}$ The set of type of EOL products and remanufactured products

$Q = \{0, 1\}$ The set of quality level of EOL products
$q = 0$ means the quality is in "good condition"
$q = 1$ means the quality is "malfunctioning"

$M = \{1, 2, \ldots, G\}$ The set of recycled material items

$T = \{1, 2, \ldots, Z\}$ The set of time period

Parameters

(1) Product structure

$A^l_{i,j}$ Yield matrix of AND module:
The number of part i that are descendants of module node j in product l

$O^l_{i,k}$ Yield matrix of OR module:
The number of part i that are descendants of module node k in product l

$\overline{O}^l_{i,k}$ $= 1$, when $O^l_{i,k} > 0$; $= 0$, otherwise

(2) Quality related

$W^l(t')$ The proportion represents that the length of usage time for product l is t' periods

$P^l(q/t')$ The probability that the quality of returned EOL product l belong to level q given that the length of usage time is t' periods

(3) Related cost and revenue

(a) Product level

CC^l Unit collection cost of EOL product l

CU^l Unit penalty/opportunity cost due to uncollected EOL product l

HR^l Unit holding cost for EOL product l

HM^l Unit holding cost for remanufactured product l

VR^l Unit market value of remanufactured product l

(b) Module level

CD_j^l Unit disassembly cost of module level item j in product l

$CFA_{j,q}^l$ Unit refurbish/repair cost of module level item j under quality level q in product l

$CFO_{k,q}^l$ Unit refurbish or repair cost of module level item k under quality level q in product l

CRA_j^l Unit recycling cost of module level item j of product l

CRO_k^l Unit recycling cost of intact module item k of product l

CWA_j^l Unit disposal cost of module level item j of product l

CWO_k^l Unit disposal cost of intact module item k

HA_j^l Unit holding cost for module level item j of product l

HO_k^l Unit holding cost for intact module item k of product l

SD_j^l Setup cost of disassembly process of module j of product l

SFA_j^l Setup cost of refurbish/repairing process of module j of product l

SFO_k^l Setup cost of refurbish/repairing process of module k of product l

SRA_j^l Setup cost of recycling process of module j of product l

SRO_k^l Setup cost of recycling process of intact module k of product l

VA_j^l Unit market value of reusable module j of product l

VO_k^l Unit market value of reusable module k of product l

(c) Part level

$CS_{i,k}^l$ Unit separation cost of part i from module k of product l

CP_i Unit procurement cost and remanufacturing cost for part i

$CM_{i,q}^l$ Unit remanufacturing (reassembly) cost of part-level item i of product l under quality level q

$CFP_{i,q}$ Unit refurbish or repair cost of part level item i under quality level q

CRP_i Unit recycling cost of part i

$CRO^l_{i,k}$ Additional recycling cost for module k of product l due to part(s) i which has (have) not been separated from module k

$SS^l_{i,k}$ Setup cost of separation process of module k of product l with its descendant part i

$SRO^l_{i,k}$ Setup cost of recycling process of partially disassembled module k of product l with its descendant part i

CWP_i Unit disposal cost of part-level item i

HP_i Unit holding cost for part-level item i

SM^l_i Setup cost of remanufacturing process of part i for product l

SFP_i Setup cost of refurbishing process of part i

SRP_i Setup cost of recycling process of part i

VP_i Unit market value of reusable part i

(d) Material level

VM_m Unit market value of recycle material m

(4) Capacities and others

$Y_{m,i}$ Amount of reusable material m yielded from one unit of part level item i by recycling process

UM^l_i Remanufacturing capacity for part i of product l

UD^l_j Disassembly capacity of module j of product l

UFA^l_j Refurbish/repairing capacity of module j of product l

UFO^l_k Refurbish/repairing capacity of module k of product l

UFP_i Refurbish/repairing capacity of part i

URA^l_j Recycling capacity of module j of product l

URO^l_k Recycling capacity of intact module k of product l

URP_i Recycling capacity of part i

$US^l_{i,k}$ Separation capacity of module k of product l with its descendant part i

$URO^l_{i,k}$ Recycling capacity of partially disassembled module k of product l with its descendant part i

$D^{l,t}$ Amount of demand of remanufactured products l from product market at time t

M A sufficiently large positive number

Decision Variables

(1) Collection/inventory for EOL products and disassembly/separation processes

(a) Product level

$a_q^{l,t}$ The amount of EOL products l in quality level q collected for possible recovery process at period t

$u_q^{l,t}$ The amount of uncollected EOL products l in quality level q at period t

$rm^{l,t}$ The amount of products l to be remanufactured at period t

$ir_q^{l,t}$ The inventory of EOL product l in quality level q at the end of period t

(b) Module level

$d_{j,q}^{l,t}$ The number of module j of EOL product l in quality level q that is to be disassembled at time t

$s_{k,q}^{l,t}$ The number of intact module k of EOL product l in quality level q, which is used to perform the separation process at time t

$sp_{k,q,i}^{l,t}$ The number of module k of EOL product l in quality level q whose part i is to be separated at time t

$zd_j^{l,t}$ $= 1$, when the disassembly process of module j of product l take place in period t
$= 0$, otherwise

(c) Part level

$zs_{i,k}^{l,t}$ $= 1$, when the separation process of module k of product l with its descendant part i take place in period t
$= 0$, otherwise

(2) Inventory of completely (partially) disassembled parts (modules)

(a) Module level

$ia_{j,q}^{l,t}$ The inventory of module j of EOL product l in quality level q at the end of period t

$io_{k,q}^{l,t}$ The inventory of intact module item k of EOL product l in quality level q at the end of period t

$io_{k,q,i}^{l,t}$ The inventory of disassembled module level item k of EOL product l in quality level q with its descendant part(s) i which has (have) not been separated at the end of period t

(b) Part level

$ip_{i,q}^{t}$ The inventory of disassembled part i in quality level q at the end of period t

(3) Recovery options and Inventory for remanufactured products

(a) Product level

$im^{l,t}$ The inventory of remanufactured product l at the end of period t

(b) Module level

$fa_{j,q}^{l,t}$ The quantity of module level item j of EOL product l in quality level q to be refurbished/repaired at time t

$fo_{k,q}^{l,t}$ The quantity of module level item k of EOL product l in quality level q to be refurbished/repaired at time t

$ra_{j,q}^{l,t}$ The quantity of module level item j of EOL product l in quality level q to be recycled at time t

$ro_{k,q}^{l,t}$ The quantity of intact module item k of EOL product l in quality level q to be recycled at time t

$ro_{k,q,j}^{l,t}$ The quantity of disassembled module level item k of EOL product l in quality level q with its descendant part(s) i which has(have) not been separated to be recycled at time t

$wa_{j,q}^{l,t}$ The quantity of module j of EOL product l in quality level q to be disposed at time t

$wo_{k,q}^{l,t}$ The quantity of intact module item k of product l in quality level q to be disposed at time t

$wo_{k,q,i}^{l,t}$ The quantity of disassembled module level item k of EOL product l in quality level q with its descendant part(s) i which has(have) not been separated to be disposed at time t

$zfa_{j}^{l,t}$ = 1, when refurbish/repairing process of module j of product l take place in period t
= 0, otherwise

$zfo_{k}^{l,t}$ = 1, when refurbish/repairing process of module k of product l take place in period t
= 0, otherwise

$zra_{j}^{l,t}$ = 1, when recycling process of module j of product l take place in period t
= 0, otherwise

$zro_{k}^{l,t}$ = 1, when recycling process of intact module k of product l take place in period t
= 0, otherwise

(c) Part level

$m_{i,q}^{l,t}$ The quantity of disassembled part level item i of product l in quality level q to be remanufactured at time t

$fp_{i,q}^{t}$ The quantity of disassembled part level item i in quality level q to be refurbished or repaired at time t

$rp_{i,q}^{t}$ The quantity of disassembled part level item i in quality level q to be recycled at time t

$wp_{i,q}^{t}$ The quantity of disassembled part level item i in quality level q to be disposed at time t

$p_{i}^{l,t}$ The quantity of procurement of part i for remanufactured product l at time t

$zm_{i}^{l,t}$ = 1, when remanufacturing process of part i for product l take place in period t
= 0, otherwise

zfp_{i}^{t} = 1, when refurbishing process of part i take place in period t
= 0, otherwise

zrp_{i}^{t} = 1, when recycling process of part i take place in period t
= 0, otherwise

$zro_{k,i}^{l,t}$ = 1, when recycling process of partially disassembled module k of product l with its descendant part i take place in period t
= 0, otherwise

The Mixed-Integer Programming Optimization Model

The disassembly for demand (D*f*D) model to maximize the total profit is formulated below:

D*f*D Model

Maximize

$$RMV + RFB + RCY - DPS - INV - DSM - COL \tag{10.1}$$

where

Remanufacturing profit = RMV

$$= \sum_{t} \left[\begin{array}{l} \sum_{l} VR^{l} \cdot rm^{l,t} \\ - \left(\sum_{i,q,l} CM_{i,q}^{l} \cdot m_{i,q}^{l,t} + \sum_{i,l} (CP_{i} + CM_{i,q=0}^{l}) \cdot p_{i}^{l,t} + \sum_{i,l} SM_{i}^{l} \cdot zm_{i}^{l,t} \right) \end{array} \right]$$

Refurbish/repair profit = RFB

$$= \sum_{t} \left[\sum_{i} VP_{i} \cdot \left(\sum_{q} fp_{i,q}^{t} \right) + \sum_{j,l} VA_{j}^{l} \cdot \left(\sum_{q} fa_{j,q}^{l,t} \right) + \sum_{k,l} VO_{k}^{l} \cdot \left(\sum_{q} fo_{k,q}^{l,t} \right) \right]$$

$$- \sum_{t} \left[\begin{array}{l} \sum_{i,q} CFP_{i,q} \cdot fp_{i,q}^{t} + \sum_{j,q,l} CFA_{j,q}^{l} \cdot fa_{j,q}^{l,t} + \sum_{k,q,l} CFP_{k,q}^{l} \cdot fo_{k,q}^{l,t} \\ + \sum_{i} SFP_{i} \cdot zfp_{i}^{t} + \sum_{j,l} SFA_{j}^{l} \cdot zfa_{j}^{l,t} + \sum_{k,l} SFO_{k}^{l} \cdot zfo_{k}^{l,t} \end{array} \right]$$

Recycle profit = RCY

$$= \sum_t \left[\sum_m \left(\begin{array}{l} \sum_i \left[Y_{m,i} \cdot \sum_q rp_{i,q}^t \right] + \sum_i \left(Y_{m,i} \cdot \sum_{j,l} \left[A_{i,j}^l \cdot \sum_q ra_{j,q}^{l,t} \right] \right) \\ + \sum_i \left(Y_{m,i} \cdot \sum_{k,l} \left[O_{i,k}^l \cdot \sum_q \left(ro_{k,q,i}^{l,t} + ro_{k,q}^{l,t} \right) \right] \right) \end{array} \right) \right]$$

$$- \sum_t \left[\begin{array}{l} \sum_i \left(CRP_i \cdot \sum_q rp_{i,q}^t \right) + \sum_{j,l} \left(CRA_j^l \cdot \sum_q ra_{j,q}^{l,t} \right) \\ + \sum_{i,k,l} \left(CRO_{i,k}^l \cdot \sum_q ro_{k,q,i}^{l,t} \right) + \sum_{k,l} \left(CRO_k^l \cdot \sum_q ro_{k,q}^{l,t} \right) \\ + \sum_i SRP_i \cdot zrp_i^t + \sum_{j,l} SRA_j^l \cdot zra_j^{l,t} + \sum_{k,l} SRO_k^l \cdot zro_k^{l,t} \\ + \sum_{i,k,l} SRO_{i,k}^l \cdot zro_{i,k}^{l,t} \end{array} \right]$$

Disposal cost = DPS

$$= \sum_t \left[\begin{array}{l} \sum_i \left(CWP_i \cdot \sum_q wp_{i,q}^t \right) + \sum_{j,l} \left(CWA_j^l \cdot \sum_q wa_{j,q}^{l,t} \right) \\ + \sum_{k,l} \left(CWO_k^l \cdot \sum_q wo_{k,q}^{l,t} \right) + \sum_i \left(CWP_i \cdot \sum_{k,q,l} wo_{k,q,i}^{l,t} \right) \end{array} \right]$$

Inventory cost = INV

$$= \sum_t \left[\begin{array}{l} \sum_i \left(HP_i \cdot \sum_q ip_{i,q}^t \right) + \sum_{j,l} \left(HA_j^l \cdot \sum_q ia_{j,q}^{l,t} \right) + \sum_{k,l} \left(HO_k^l \cdot \sum_q io_{k,q}^{l,t} \right) \\ + \sum_i \left(HP_i \cdot \sum_{k,q,l} io_{k,q,i}^{l,t} \right) + \sum_l \left(HR^l \cdot \sum_q ir_q^{l,t} \right) + \sum_l HM^l \cdot im^{l,t} \end{array} \right]$$

Disassembly/Separation cost = DSM

$$= \sum_t \left[\begin{array}{l} \sum_{j,l} \left(CD_j^l \cdot \left[\sum_q d_{j,q}^{l,t} \right] \right) + \sum_{i,k,l} \left(CS_{i,k}^l \cdot \left[\sum_q sp_{i,q,k}^{l,t} \right] \right) \\ + \sum_{j,l} SD_j^l \cdot zd_j^{l,t} + \sum_{i,k,l} SS_{i,k}^l \cdot zs_{i,k}^{l,t} \end{array} \right]$$

Collection/Uncollection cost = COL

$$= \sum_{l,t} \left[CC^l \cdot a^{l,t} + CU^l \cdot u^{l,t} \right]$$

Constraints

Remanufactured product demand constraints

$$\left(\sum_j NA^l_{i,j} + \sum_k NO^l_{i,k}\right) \cdot rm^{l,t} \leq \sum_q m^{l,t}_{i,q} + p^{l,t}_i \qquad (10.2)$$
$$\forall i \in I, \quad \forall l \in L, \quad \forall t \in T$$

$$D^{l,t} + im^{l,t} = rm^{l,t} + im^{l,t-1} \qquad \forall l \in L, \forall t \in T \qquad (10.3)$$

Quality and quantity properties of returned products and balance constraints:

$$a^{l,t}_q + u^{l,t}_q + ir^{l,t}_q = \left\lfloor \sum_{t'}^{t-1} \left[P^l\left(q \,|t'\right) \cdot W^l\left(t'\right) \cdot D^{l,t-t'} \right] \right\rfloor + ir^{l,t-1}_q$$
$$\forall l \in L, \quad \forall q \in Q, \quad \forall t \geq 2 \qquad (10.4)$$

$$a^{l,t}_q + ia^{l,t-1}_{j,q} = d^{l,t}_{j,q} + fa^{l,t}_{j,q} + ra^{l,t}_{j,q} + wa^{l,t}_{j,q} + ia^{l,t}_{j,q}$$
$$\forall j \in J, \quad \forall q \in Q, \quad \forall l \in L, \quad \forall t \in T \qquad (10.5)$$

$$a^{l,t}_q + io^{l,t-1}_{k,q} = s^{l,t}_{k,q} + fo^{l,t}_{k,q} + ro^{l,t}_{k,q} + wo^{l,t}_{k,q} + io^{l,t}_{k,q}$$
$$\forall k \in K, \quad \forall q \in Q, \quad \forall l \in L, \quad \forall t \in T \qquad (10.6)$$

Disassembly and separation process constraints:

$$\sum_{j,l}\left(A^l_{i,j} \cdot d^{l,t}_{j,q}\right) + \sum_{k,l}\left(O^l_{i,k} \cdot sp^{l,t}_{k,q,i}\right) + ip^{t-1}_{i,q}$$
$$= \sum_l m^{l,t}_{i,q} + fp^t_{i,q} + rp^t_{i,q} + wp^t_{i,q} + ip^t_{i,q} \quad \forall i \in I, \forall q \in Q, \quad \forall t \in T \quad (10.7)$$

$$sp^{l,t}_{k,q,i} + ro^{l,t}_{k,q,i} + wo^{l,t}_{k,q,i} + io^{l,t}_{k,q,i} = \overline{O}^l_{i,k} \cdot s^{l,t}_{k,q} + io^{l,t-1}_{k,q,i},$$
$$\forall i \in I, \quad \forall q \in Q, \quad \forall k \in K, \quad \forall l \in L, \quad \forall t \in T \qquad (10.8)$$

$$\sum_{q,t} sp^{l,t}_{k,q,i} \leq M \cdot O^l_{i,k} \quad \forall i \in I, \quad \forall k \in K, \quad \forall l \in L \qquad (10.9)$$

$$\sum_{q,t} ro^{l,t}_{k,q,i} \leq M \cdot O^l_{i,k} \quad \forall i \in I, \quad \forall k \in K, \quad \forall l \in L \qquad (10.10)$$

$$\sum_{q,t} wo^{l,t}_{k,q,i} \leq M \cdot O^l_{i,k} \quad \forall i \in I, \quad \forall k \in K, \quad \forall l \in L \qquad (10.11)$$

$$\sum_{q,t} io^{l,t}_{k,q,i} \leq M \cdot O^l_{i,k} \quad \forall i \in I, \quad \forall k \in K, \quad \forall l \in L \qquad (10.12)$$

Capacity constraints:

$$p_i^{l,t} + \sum_q m_{i,q}^{l,t} \le UM_i^l \cdot zm_i^{l,t} \quad \forall i \in I, \quad \forall l \in L \quad \forall t \in T \qquad (10.13)$$

$$\sum_q fa_{j,q}^{l,t} \le UFA_j^l \cdot zfa_j^{l,t} \quad \forall j \in J, \quad \forall l \in L, \quad \forall t \in T \qquad (10.14)$$

$$\sum_q fo_{k,q}^{l,t} \le UFO_k^l \cdot zfo_k^{l,t} \quad \forall k \in K, \quad \forall l \in L, \quad \forall t \in T \qquad (10.15)$$

$$\sum_q ra_{j,q}^{l,t} \le URA_j^l \cdot zra_j^{l,t} \quad \forall j \in J, \quad \forall l \in L, \quad \forall t \in T \qquad (10.16)$$

$$\sum_q ro_{k,q}^{l,t} \le URO_k^l \cdot zro_k^{l,t} \quad \forall k \in K, \quad \forall l \in L, \quad \forall t \in T \qquad (10.17)$$

$$\sum_q ro_{k,q,i}^{l,t} \le URO_{i,k}^l \cdot zro_{i,k}^{l,t} \quad \forall i \in I, \quad \forall k \in K, \quad \forall l \in L, \quad \forall t \in T \qquad (10.18)$$

$$\sum_q d_{j,q}^{l,t} \le UD_j^l \cdot zd_j^{l,t} \quad \forall j \in J, \quad \forall l \in L, \quad \forall t \in T \qquad (10.19)$$

$$\sum_q sp_{i,q,k}^{l,t} \le US_{i,k}^l \cdot zs_{i,k}^{l,t} \quad \forall i \in I, \quad \forall k \in K, \quad \forall l \in L, \quad \forall t \in T \qquad (10.20)$$

Binary and non-negativity constraints:

$$zm_i^t, zfp_i, zfa_j^{l,t}, zfo_k^{l,t}, zrp_i, zra_j^{l,t}, zro_k^{l,t}, zro_{i,k}^{l,t}, zd_j^{l,t}, zs_{i,k}^{l,t} \in \{0, 1\}$$
$$\forall i \in I, \quad \forall j \in J, \quad \forall k \in k, \quad \forall l \in L, \quad \forall t \in T \qquad (10.21)$$

$$a_q^{l,t}, u_q^{l,t}, rm^{l,t}, ir_q^{l,t}, d_{j,q}^{l,t}, s_{k,q}^{l,t}, sp_{k,q,i}^{l,t}, ia_{j,q}^{l,t}, io_{k,q}^{l,t}, io_{k,q,i}^{l,t}, ip_{i,q}^t, im^{l,t}1,$$
$$fa_{j,q}^{l,t}, fo_{k,q}^{l,t}, ra_{j,q}^{l,t}, ro_{k,q}^{l,t}, ro_{k,q,i}^{l,t}, wa_{j,q}^{l,t}, wo_{k,q}^{l,t}, wo_{k,q,i}^{l,t}, m_{i,q}^{l,t}, fp_{i,q}^t,$$
$$rp_{i,q}^t, wp_{i,q}^t, p_i^{l,t} \ge 0 \quad \text{and} \quad \text{integer},$$
$$\forall i \in I, \quad \forall j \in J, \quad \forall k \in k, \quad \forall l \in L, \quad \forall q \in Q, \quad \forall t \in T \qquad (10.22)$$

10.5 An Illustrative Example

We employ a numerical example to illustrate the use of our model. Consider two different products that are returned, disassembled, recovered and resold (see Fig. 10.9 for their structures). The objective of the model is to determine the best combination of the number of each

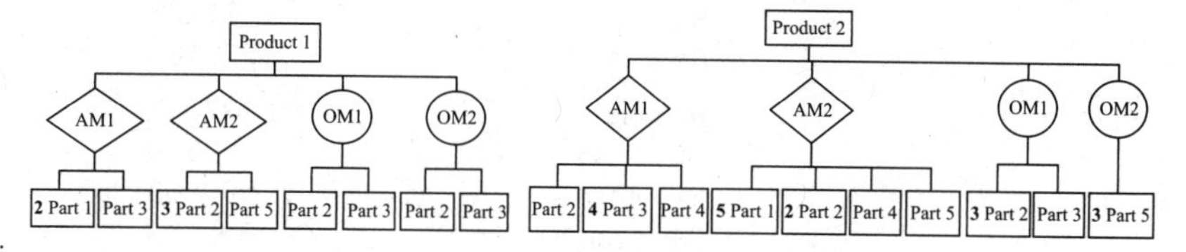

FIGURE 10.9 Product structure for the case example.

product type to be returned, disassembled, and recovered to meet the period demands for remanufactured products (as shown in Table 10.4). We used the ILOG/CPLEX solving tool to obtain the optimal decision. The output is summarized in Table 10.5. The number of constraints was 2020, and the number of variables was 4880. The optimal solution with the objective value is 6,261,067. Also, Fig. 10.10 and Fig. 10.11 show the partial optimal solution.

10.6 Conclusion and Future Research Direction

In this chapter, we present a mathematical model, namely (DfD), for a demand-driven disassembly planning under a closed-loop supply chain system. The model contributes to the sustainable operations for a business through profitable product recovery strategies and waste deduction. In particular, we take into account the market mechanism under a closed-loop environment, including timing, quality, and quantity

Period	Product I	Product II
1	355	100
2	475	100
3	515	150
4	520	150
5	470	250
6	350	280
7	280	390
8	250	480
9	190	550
10	190	550
11	150	540
	150	435

TABLE 10.4 Period demands for the case example

Variable		Value
Profit or revenue	Remanufactured	8377678
	Repair	71129
	Recycle	253105
Cost	Recycle	127063
	Disposal	2850
	Inventory	580863
	Disassembly	888109
	Collection	841960

TABLE 10.5 Detailed total profit, revenue, and cost

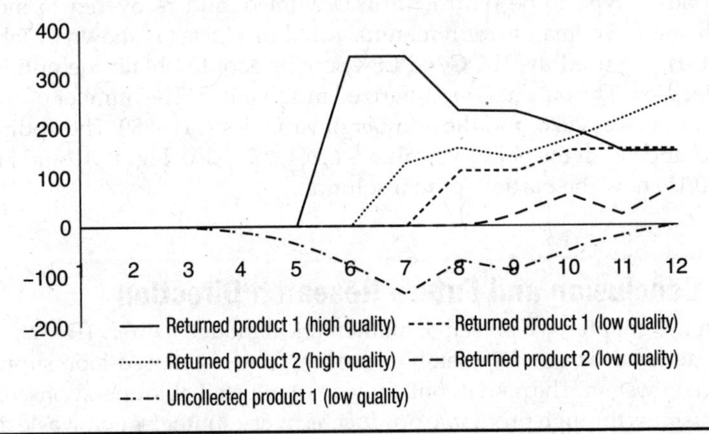

FIGURE 10.10 Comparison between return products acquired (represented by positive value) and uncollected products (represented by negative value).

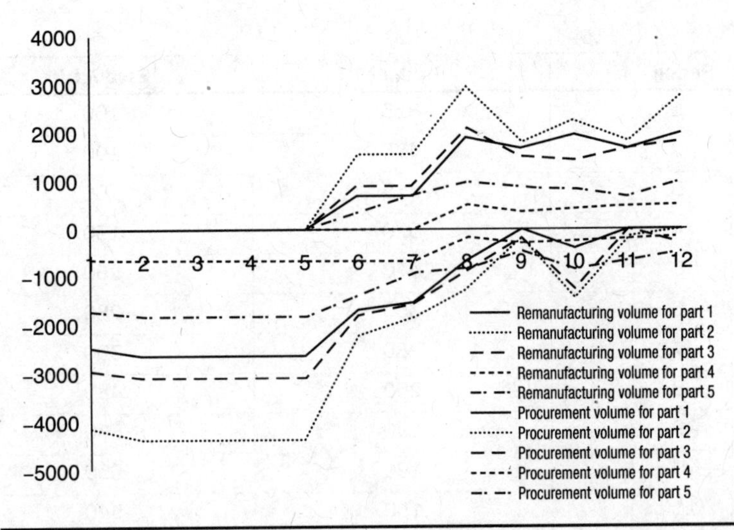

FIGURE 10.11 Comparison between remanufacturing volumes (represented by positive value) and procurement volumes (represented by negative value) for each type of parts.

issues of returned EOL products. Sensitivity analysis for these key factors can be easily done.

In fact, closing the loop also helps to mitigate the undesirable environmental footprint of supply chains. How to measure the positive and negative environmental impacts in a product life cycle is a crucial problem; moreover, the trade-off between environmental cost and business profits is also a necessary consideration for future research.

References

Barba-Gutierrez, Y., Adenso-Diaz, B., Gupta, S.M., 2008, Lot sizing in reverse MRP for scheduling disassembly, *International Journal of Production Economics*, 111, 741-751.

Brennan, L., Gupta, S.M., Taleb, K.N., 1994, Operations planning issues in an assembly/disassembly environment, *International Journal of Operations and Production Management*, 14, 57-67.

Bufardi, A., Gheorche, R., Kiritsis, D., Xirouchakis, P., 2004, Multicriteria decision-aid approach for product end-of-life alternative selection, *International Journal of Production Research*, 15, 16, 3139-3157.

Cox, Jr., W.E., Product life cycles as marketing models, 1967, *Journal of Business*, 40, 4, 375-384.

Fleischmann, M., Bloemhof-Ruwaard, J.M., Dekker, R., Van der Laan, E., van Nunen, J., Van Wassenhove, L.N., 1997, Quantitative models for reverse logistics: A review, *European Journal of Operational Research*, 103, 1-17.

Guide, Jr., V.D.R., 1997, Scheduling with priority dispatching rules and drum-buffer-rope in a recoverable manufacturing system, *International Journal of Production Economics*, 53, 101-116.

Guide, Jr., V.D.R., Jayaraman, V., Srivastava, R., Benton, W.C., 2000, Supply-chain management for recoverable manufacturing systems, *Interfaces*, 30, 3, 125-142.

Guide. Jr., V.D R., Van Wassenhove, L., 2009, The evolution of closed-loop supply chain research. *Operations Research*, 57, 1, 10-18.

Guide, Jr., V.D.R., Van Wassenhove, L., 2006, Closed-loop supply chains: An introduction to the feature issue (Part 1), *Production and Operations Management*, 15, 3, 345-350.

Guide, Jr., V.D.R., Teunter, R.H., Van Wassenhove, L., 2003, Matching demand and supply to maximize profits from remanufacturing, *Manufacturing & Service Operations Management*, 5, 4, 303-316.

Gungor, A., Gupta, S.M., 1999, Issues in environmentally conscious manufacturing and product recovery: A survey, *Computers & Industrial Engineering*, 36, 811-853.

Gupta, S.M., Taleb, K.N., 1994, Disassembly scheduling, *International Journal of Production Research*, 32, 1857-1866.

Homem de Mello, L.S., Sanderson, A.C., 1990, AND/OR graph representation of assembly plans, *IEEE Transactions on Robotics and Automation*, 6, 188-199.

Ilgin, M.A., Gupta, S.M., 2010, Environmentally conscious manufacturing and product recovery (ECMPRO): A review of the state of the art, *Journal of Environmental Management*, 91, 563-591.

Inderfurth, K., Langella, I.M., 2006, Heuristics for solving disassemble-to-order problems with stochastic yields, *OR Spectrum*, 28, 73-99.

Jayaraman, V., Guide, V.D.R., Srivastava, R., 1999, A closed-loop logistics model for remanufacturing, *Journal of Operational Research Society*, 50, 497-508.

Jorjani, S., Leu, J., Scott, C., 2004, Model for the allocation of electronics components to reuse options, *International Journal of Production Research*, 42, 1131-1145.

Jun, H.B., Cusin, M., Kiritsis, D., Xirouchakis, P., 2007, A multi-objective evolutionary algorithm for EOL product recovery optimization: Turbocharger case study, *International Journal of Production Research*, 45, 18-19, 4573-4594.

Kim, H.J., Lee, D.H., Xirouchakis, P., Zust, R., 2003, Disassembly scheduling with multiple product types, *CIRP Annals—Manufacturing Technology*, 52, 403-406.

Kim, K., Song, I., Kim, J., Jeong, B., 2006, Supply planning model for remanufacturing system in reverse logistics environment, *Computers & Industrial Engineering*, 51, 279-287.

Kim, H.J., Lee, D.H., Xirouchakis, P., 2007, Disassembly scheduling: Literature review and future research directions, *International Journal of Production Research*, 45, 18-19, 4465-4484.

Kongar, E., Gupta, S.M., 2002, A multi-criteria decision-making approach for disassembly-to-order systems, *Journal of Electronics Manufacturing*, 11, 2, 171-183.

Kongar, E., Gupta, S.M., 2006, Disassembly-to-order system under uncertainty, *Omega*, 34, 550-561.

Krikke, H.R., van Harten, A., Schuur, P.C., 1998, On a medium-term product recovery and disposal strategy for durable assembly products, *International Journal of Production Research*, 36, 111-140.

Kumar, V., Shirodkar, P.S., Camelio, J.A., and Sutherland, J.W., 2007, Value flow characterization during product life cycle to assist in recovery decisions, *International Journal of Production Research*, 45, 18-19, 4555-4572.

Lambert, A.J.D., 1999, Linear programming in disassembly clustering sequence generation, *Computers and Industrial Engineering*, 36, 4, 723-738.

Lambert, A.J.D., Gupta, S.M., 2002, Demand-driven disassembly optimization for electronic products, *Journal of Electronics Manufacturing*, 11, 2, 121-135.

Lambert, A.J.D., Gupta, S.M., 2005, *Disassembly Modeling for Assembly, Maintenance, Reuse, and Recycling*, Boca Raton, FL: CRC Press.

Langella, I.M., 2007, Heuristics for demand-driven disassembly planning, *Computers and Operations Research*, 34, 552-577.

Lee, D.H., Kang, J.G., Xirouchakis, P., 2001a, Disassembly planning and scheduling: review and further research, *Proceedings of the Institution of Mechanical Engineers—Part B*, 215, 695-709.

Lee, S.G., Lye, S.W., Khoo, M.K., 2001b, A multi-objective methodology for evaluating product end-of-life options and disassembly, *International Journal of Advanced Manufacturing Technology*, 18, 148-156.

McGovern, S.M., Gupta, S.M., 2011, Disassembly Line: *Balancing and Modeling*, New York, NY: McGraw Hill.

Meacham, A., Uzsoy, R., Venkatadri, U., 1999, Optimal disassembly configurations for single and multiple products, *Journal of Manufacturing Systems*, 18, 311-322.

Mulvey, J.M., Vanderbei, R.J., Zenios, S.A., 1995, Robust optimization of large-scale systems, *Operations Research*, 43, 2, 264-281.

O'Shea, B., Grewal, S.S., Kaebernick, H., 1998, State-of-the-art literature survey on disassembly planning, *Concurrent Engineering: Research and Applications*, 6, 345-357.

Östlin, J., Sundin, E., Björkman, M., 2009, Product life-cycle implications for remanufacturing strategies, *Journal of Cleaner Production*, 17, 999-1009.

Özgün, N., Gökçen, H., 2008, A mixed integer programming model for remanufacturing in reverse logistics environment, *International Journal of Advanced Manufacturing Technology*, 39, 1197-1206.

Taleb, K.N., Gupta, S.M., 1997, Disassembly of multiple product structures, *Computers and Industrial Engineering*, 32, 4, 949-961.

Teunter, R.H., 2006, Determining optimal disassembly and recovery strategies, *Omega*, 34, 533-537.

Thierry, M., Salomon, M., Nunen, J.V., Wassenhove, L.V., 1995, Strategic issues in product recovery management, *California Management Review*, 37, 2, 112-135.

Tsai, W.H., Hung, S J., 2009, Treatment and recycling system optimization with activity-based costing in WEEE reverse logistics management: An environmental supply chain perspective, *International Journal of Production Research*, 47, 19, 5391-5420.

Wu, Y., 2006, Robust Optimization Applied to Uncertain Production Loading Problems with Import Quota Limits under the Global Supply Chain Management Environment, *International Journal of Production Research*, 44, 5, 849-882.

Xanthopoulos, A., Iakovou, E., 2009, On the optimal design of the disassembly and recovery processes, *Waste Management*, 29, 1702-1711.

Yu, C.S., and Li, H.L., 2000, A robust optimization model for stochastic logistic problems, *International Journal of Production Economics*, 64, 385-397.

PART 4

Green Information Management Systems

CHAPTER 11
Database for Life Cycle Assessment

CHAPTER 12
Web-Based Information Support Systems

CHAPTER 11

Database for Life Cycle Assessment

Procedure with Database

11.1 Introduction

Life cycle assessment (LCA) is the most commonly used method to identify, quantify, evaluate, and prioritize potential environmental impacts directly attributable to sustainability of products. In ISO 14040 standard (ISO 14040, 1997), LCA is defined as the "compilation and evaluation of the inputs, outputs and the potential environmental impacts of a product system throughout its life cycle."

In recent years, a large amount of fossil fuel has been consumed due to rapid industrial growth, and this has resulted in the emission of a large volume of *greenhouse gases* (GHGs), including carbon dioxide. Thus, global warming started to gain attention in the twentieth century, and methods for reducing GHGs have become the most important environmental issue in this century. The increasing emission of carbon dioxide and other GHGs is the main contributor to global warming, in that the phenomenon can be attributed to human activities.

Figure 11.1 shows a general framework for LCA for which a complete database including BOM of specific product, local life cycle index, and international index are necessary. In addition, the socioeconomic factors and existing directives are the references for audition and overall evaluation of the factors with their weights of importance, which are illustrated in the following sections. The results of assessment can be used for recommendation for improvement in design, as well as in policy making.

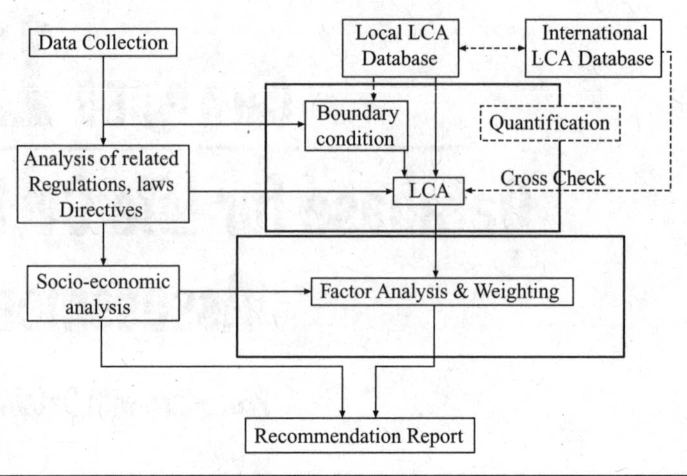

FIGURE 11.1 Framework of LCA.

11.1.1 Applicable International Standards on Product Carbon Footprint

In view of the increasingly serious impact of global warming, the United Nations Environment Programme (UNEP) has set up the Intergovernmental Panel on Climate Change (IPCC) to carry out global monitoring and investigations. Environmental conventions and treaties have been subsequently promulgated to implement specific control on the emission of GHGs. Applicable international standards for a nation's carbon footprint are listed below.

- United Nations Framework Convention on Climate Change and IPCC Directive (IPCC 2007 http://www.ipcc.ch/)
- PAS2050: Commodity and Service Lifecycle Greenhouse Gas Emission Assessment Procedure
- CNS14040: Environment Management—Lifecycle Assessment—Principle and Structure
- CNS14044: Environment Management—Lifecycle Assessment—Requirements and Guidelines
- International Standardization Organization (ISO) and ISO 14064 GHG Management Serial Standard
- ISO 14064-1: Specification with Guidance for Quantification and Reporting of Organization-Level Greenhouse Gas Emission and Removal
- ISO 14064-2: Specification with Guidance for Quantification, Monitoring, and Reporting of Program-Level Greenhouse Gas Emission Reduction or Removal

- ISO 14064-3: Specification with Guidance for the Validation and Verification of Greenhouse Gas Assertions

Since the Kyoto Protocol (2005) became valid in February 2005, major countries in the world have planned and enforced policies addressing GHG emission reduction. To meet the required reduction requirements and to avoid the possibility of international renunciation or trade sanction, energy efficiency had to be improved and renewable energy had to be developed. In contrast with volume control, energy efficiency standard, and other policies that have extensive influence and great impact, case studies have endorsed the GHG inventory and emission reduction to be "more easily accepted by the industry." In fact, the GHG inventory has become the primary policy and technical tool for the preliminary stage of GHG reduction across the world.

11.1.2 Available Software for LCA

There are a number of commercial tools for the above purposes. Table 11.1 lists the commonly used software available, with their functions

	Name	Properties of the Tool	Properties of the Database
www.boustead-consulting.co.uk	Boustead	Interface by words	Energy-related product database
www.pre.nl	SimaPro	Able for impact analysis	Open database
www.gabi-software.com	Gabi3	Interface by figures	European database
www.lcait.com	LCAiT	For assessment	For package and paper products
www.ecosite.co.uk	LiMS	For assessment	For liquid and plastic products
www.ecobalance.com	TEAM	Able to do economic analysis	Steel, package, food products
www.ecosite.co.uk	REPAQ	For assessment	Long-term data
www.edie.net/Library/Features/IEG9736B.html	PEMS	Based on Excel	For package, waste and industrial products
www.aist.go.jp/NIRE/~lca/lcasoft-v2/lca-1.htm	NIRE-LCA	14040 in Japanese	Energy, steel, cement

TABLE 11.1 Web-Available Tools for LCA with Their Properties

and a database for references. Since one of the core applications from the results of LCA is for sustainable design, Richardson et al. (2005) developed a wide range of eco-design tools including the methods and principles of using closed-loop feedback and improving information flows in designs. Although the above-mentioned methods can facilitate the LCA analysis, Yang (2007) pointed out that a full LCA study is still a very complex process involving huge amount of quantitative input data and life cycle inventory analyses. In addition, the regulations and directives announced by international authorities have been the basic guidance for LCA, which will be introduced in the following section.

In order to reduce the global warming effects of the industry, an effective method for the calculation of the carbon dioxide equivalent of industrial products is necessary. Therefore, this chapter aims to introduce the procedure of the carbon dioxide equivalent throughout the entire life cycle of a single product—including stages of raw materials, manufacture, storage, transport, use, recycling, and disassembly with the data required to achieve this task, also to demonstrate the application of the result for the recommendation of improvement.

In addition, ProdTect (KERP Engineering GmbH, 2006) is software used here to demonstrate the content and procedure of LCA because of its ability to estimate the carbon emission of products and analyze the recycling cost based on a database. It is software designed for green product assessment and analysis. It provides a report on how will it conforms to the *reuse, recycling, and recovery* (3R) ratio in WEEE of electronic and electric products.

Furthermore, it can analyze the ease of disassembly of a product and offers 19 evaluation factors to designers as the basis for innovation and improvement of products. An optimum plans option, among others, is also available. ProdTect conforms to WEEE legal requirements, has been accredited by the Recycling System Standards of the European Union, and has been recognized by Hewlett-Packard, an international firm. etc. Based on the framework of LCA illustrated in Fig. 11.1, Fig. 11.2 presents the major applications of the ProdTect software.

Using this software, the recycling ratio, total cost, recycling cost, recycling efficiency, etc., can be easily calculated. In addition, respective functions can be optimized based on the above-mentioned items. Figure 11.3 summarizes some calculation functions and parameters necessary for this study. The software can calculate carbon emissions based on international databases and the operational effectiveness over time. The calculation results where carbon emission is presented are in bar graphs. Before adopting ProdTect for the illustration of LCA, the advantages and disadvantages of this software are listed in Table 11.2 with the suggestion of additional effort for improvement in the end.

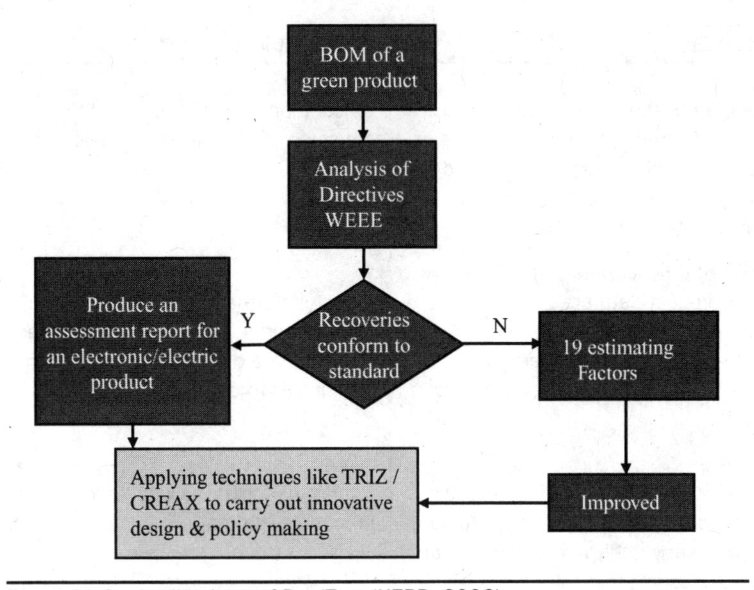

FIGURE 11.2 Applications of ProdTect (KERP, 2006).

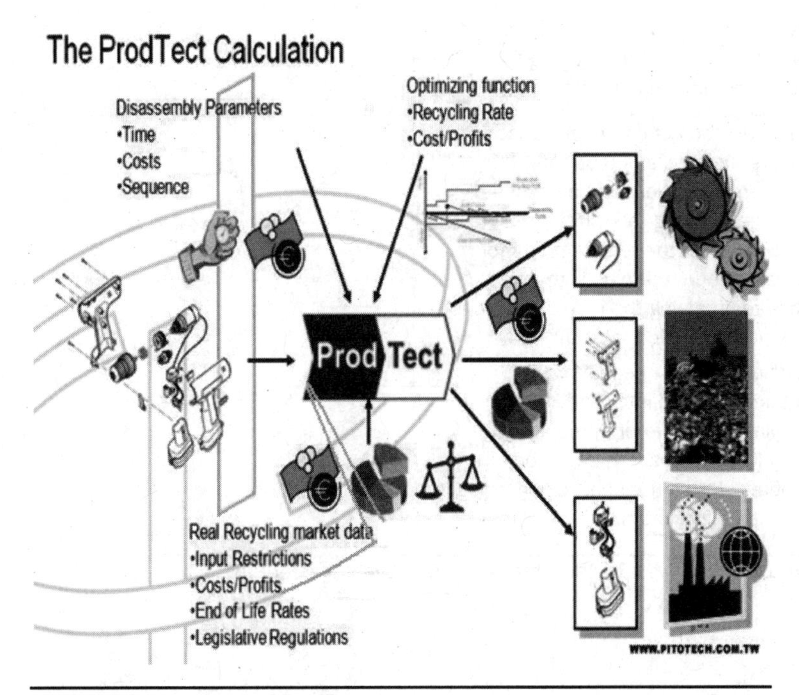

FIGURE 11.3 Functions of ProdTect.

Advantages	Disadvantages
Reliability: Accredited by the recycling system standard of many countries. Analysis of the disassembly report is acknowledged by EU countries, indicating that the report is legal and rational.	Calculation does not include the emission of storage and mobile sources.
Sharing of product data: Similar product data are applied for similar products sold in different countries. Laws and standards of different countries can be selected for analysis.	Certain information about the manufacturing procedure are required, but most are difficult to obtain which are similar to other softwares
Timeliness: Could carry out the assessment and analysis of the WEEE recycling ratio and ease of disassembly during the design stage. Assessment results are obtained, but not necessarily through a common accreditation body. If the verifier reveals that the recycling ratio has not been achieved, the product must be returned to the plant for remanufacturing, thus leading to inefficient time management.	Cannot describe the calculation procedure in detail.
Has the option of optimum plans Adopts mainly plans that meet the requirements of ease of disassembly and the recycling ratio. If the results of analysis meet the requirements for recycling ratio, the plan with lowest cost while maintaining the recycling ratio is selected to reduce recycling cost.	
Additional Effort needed for LCA Obtains detailed information about the circulation of different products. Calculates the carbon emission generated from transport using product logistics information.	

TABLE 11.2 Analysis of Advantages and Disadvantages of ProdTect

11.1.3 Inventory of Product Carbon Footprint

At present, a series of standards on carbon footprints and GHG inventory, as well as reference for management and verification, have been officially promulgated. ISO 14064 Serial Standards (http://csa.carbonperformance.org/index.asp?mode=roadmap&page=roadmap) have been recently formulated to help governments and firms in the inventory, reduction, and management of GHGs.

In this section, the Industrial GHG Inventory Management—Technical Manual by the Environmental Protection Agency headed by Executive Yuan (Taiwan Environment Protection Agency, 2010) is utilized as the reference for procedure of carbon footprint inventory. The manual takes enterprises, businesses, and other organizations as the object of inventory. Figure 11.4 shows a flowchart of the inventory procedure.

The inventory procedure for calculation of the carbon footprint of a single product follows similar principles and methods, as shown in Fig. 11.4. A standard for calculation of the carbon footprint is currently being developed by ISO 14067 (http://www.iso.org/iso/catalogue_detail.htm?csnumber=43278) and is expected to be

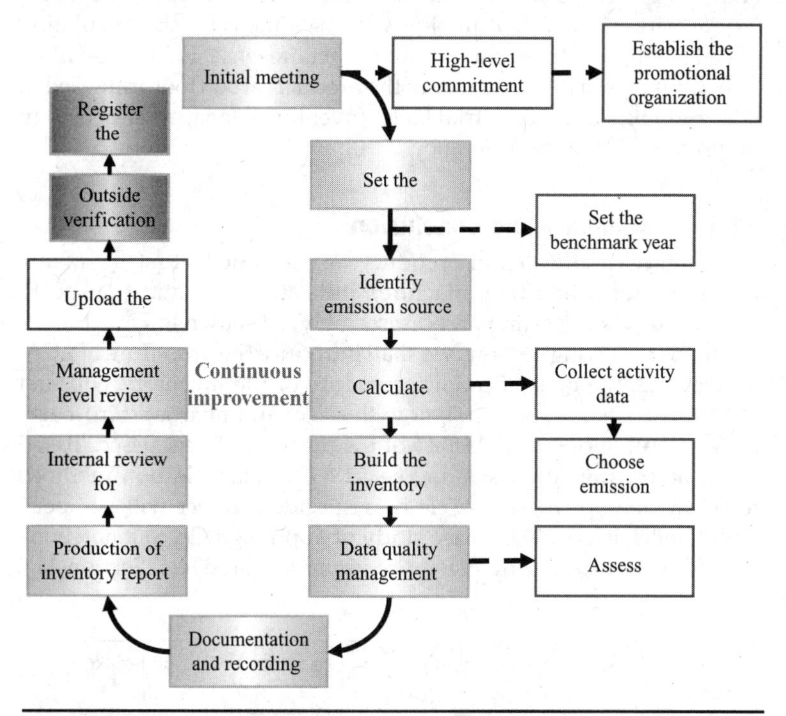

FIGURE 11.4 Flowchart of inventory procedure (Taiwan Environment Protection Agency, 2010).

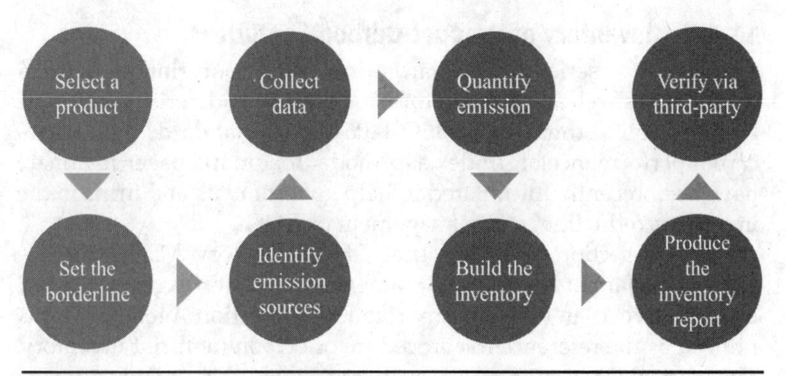

FIGURE 11.5 Procedure for inventory of a product carbon footprint.

completed by 2011. To provide a method for the calculation of the product carbon footprint before the official promulgation of the standard, we take the above-mentioned procedure for GHG inventory as a source reference.

The calculation of a product's carbon footprint is built on life cycle assessment, which can be referred to Yang as quoted as CNS 14040 (Yang, 2010). Assessment of GHG during a product life cycle is specified to ensure consistent methods of assessment for the calculation of GHG emission throughout the product life cycle. Figure 11.5 illustrates the steps in the inventory of the product carbon footprint used in this study based on Industrial GHG Inventory Management (Taiwan Environment Protection Agency, 2010).

11.1.4 Summary and Conclusion

In summary, the life cycle of a product can be divided into stages of raw material, production (manufacture), utilization (use and transport), and end-of-life (recycling and disassembly) as shown in Fig. 11.6.

In the following sections we shall introduce the procedure of LCA. In Sec. 11.2, the quantification procedure of the inventory will first be defined. Then in Sec. 11.3, how to select an impartial third party for verification of the credibility of the inventory is discussed. Based on these preliminary tasks, in Sec. 11.4 the quantification methods for each stage of the life cycle of a specific product will be specified. Finally, in Sec. 11.5 a case study of applying LCA to a notebook will be demonstrated by using the improved ProdTect. Section 11.6

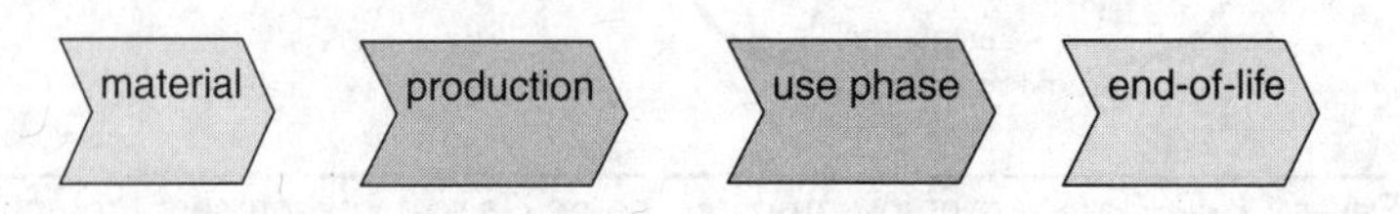

FIGURE 11.6 Stages of a product life cycle.

summarizes the findings of the study and provides recommendations for improvement formulated based on the result.

11.2 Procedure of LCA

Following the framework specified in Fig. 11.1, in this section we shall start to introduce how to proceed with the life cycle assessment of a product. The methods and applications of specific quantification, calculation, and sorting of GHG emission during inventory of a product's carbon footprint will be described with necessary steps and data to maintain the integrity and accuracy of measurement.

11.2.1 Setting the Inventory Target for the Selected Product

Once the product is selected, it has to be classified as either to *business-to-customer* (B2C) or *business-to-business* (B2B) to facilitate the setting of inventory borderline. Their differences can be noted from the following concepts:

Definition 11.1: B2B and B2C Business-to-business (B2B) refers to products "from cradle to grave." GHG emissions are assessed through the following stages: selection of raw materials, product manufacturing, transport to retailers and consumers for use, discarding, recycling, and disassembly. Computers, hand phones, and other electronic products belong to this category.

Business-to-customer (B2C) refers to products "from cradle to portal." GHG emission are assessed through the following stages: selection of raw materials, production, arrival at a new organization—including distribution and transport to the location of consumers without additional production steps, and distribution and transport to consumers for use—until discarding or regeneration and disassembly. Machine work pieces and electronic components and assemblies belong to this category.

While computers, hand phones, and other electronic products often belong to the category of B2B; machine work pieces and electronic components and assemblies belong to B2C category.

11.2.2 Setting the Borderline

After determining whether the product belongs to B2B or B2C, the inventory borderline for inclusion in the inventory can be set. If the product already has a *product category rule* (PCR) built with ISO 14025 as the criteria, the borderline of PCR can be applied to the inventory system. Then, we check and define the type of GHG and product-related emission. Direct and indirect emission of GHG or other sources of indirect emission will be identified based on the borderline set in this stage.

Based on the GHG inventory procedure, we define the source of direct and indirect emission for a firm's operation-related emission (within the organization's borderline) and classify them using the ISO 14064-1/GHG Protocol as follows:

1. **Direct emission (category I):** fixed emission source, manufacture emission source, mobile emission source, fugitive emission source.
2. **Indirect emission (category II):** emission from outsourced energy (i.e., electricity, heat, fossil fuel).
3. **Other indirect emission (category III):** emission from trusted manufacturing, emission not from use of energy, indirect emission from other activities during manufacture.

11.2.3 Identification of Emission Source

Previous sections have mentioned different types of GHG emission sources among the categories of emission. Below are types of emission sources identified in the GHG Proticol.

1. **Fixed burning emission source:** Refers to fuel burning in fixed equipment, including a boiler, smelting furnace, burning furnace, steam turbine, heating furnace, incinerator, engine, and burning tower.
2. **Mobile burning emission source:** Refers to fuel burning by vehicles, including automobiles, trucks, trains, airplanes, and sp on.
3. **Manufacture emission source:** Refers to emission from physical or chemical manufacturing processes, including emission of CO_2 from catalyst cracking during refinery, emission of PFC from dry-etching or from cleaning of the reaction chamber of chemical vapor deposition, burning of exhaust gas in the exhaust burning tower, regeneration of catalyst, removal of coke from pipes of heating furnace, emission from furnace startup or shutdown, and burning of materials by incinerators.
4. **Fugitive emission source:** Refers to intentional or unintentional emission, such as leakage from the joint, sealing, packing, or stuffing materials from the equipment; emissions from coal piles, sewage treatment plants, mining pits, and cooling water towers, and methanol from a gas processing equipment.

Except for category II (emission from outsourced energy, electricity and heat), each emission category can possibly have all four types of emission sources. In addition to emphatic assignments, this step has other purpose, including identification of the type of GHG within the inventory borderline, determination of the possible sources and paths for GHG generation, and confirmation of GHG production.

11.3 Data Collection

Theoretically, all GHG emissions are included in the calculation. To avoid wasting excessive labor and material resources, we will prioritize the emission source producing a large amount of emission and will not consider categories with extremely small or insubstantial emission. This follows so called "Cutoff Principle."

Definition 11.2: Cutoff Principle The volume of emitted substances or energy flow and the environmental significance related to a unit process or product system are beyond the scope of this study

When CO_2 emission during all stages of product life cycle is less than 5%, it could be neglected during inventory analysis.

GHG emission could be measured through several methods, such as mass/energy balance method, and direct monitoring method. In this chapter, the *Emission Coefficient Method* is adopted, which requires two types of data: *activity data* and *emission factor*. Activity data refers to all materials and energy (energy output, input, transport, and so on) involved in a product's life cycle, while emission factor refers to the volume of GHG emission produced by each unit of activity data by converting the activity data to volume of GHG emission. Both types of data can be obtained from primary and secondary sources as illustrated below (Van Oers, 2004):

Definition 11.3: Classification of Data Sources

(1) Primary (Level I) data:

Obtained throughout a product's life cycle or via direct measurement of the supply chain upstream of the product input;

(2) Secondary (Level II) data:

If primary data cannot be acquired, secondary data could be used. Among secondary sources, verified data shall be prioritized. Conversely, secondary data verified by a competent authority, such as a government and environment authority, could also be used.

11.4 Quantification of Emission

After collection, data will proceed to the calculation stage. Emission quantification involves analysis of the activity data collected to obtain the actual emission.

11.4.1 Methods of Quantification

The quantification method used as the principle of calculation must be selected in advance. The method that could minimize the uncertainty of the quantification shall be chosen, as explained in documentation. Common methods include mass balance method, the *direct monitoring method*, and the *emission coefficient method* (Yang, 2007):

1. **Direct monitoring method:** It is mainly used for gaseous emission source (emission of exhaust) and detects the content of exhaust by continuous or regular sampling and measures the volume of GHG emission based on concentration and volume of emission.
2. **Emission coefficient method:** The emission coefficient method is the most extensively adopted method in Taiwan and was selected to be used in this study. It makes use of the consumption of raw materials, fuel, materials, and outsourced electricity multiplied with the corresponding emission coefficient. Below are common formulas:

Volume of emission = proper activity data * emission coefficient (11.1)

Volume of emission from fuel consumption = fuel consumption * caloric value * emission factor (11.2)

Fuel consumption = distance * fuel efficiency factor (11.3)

Emission from outsourced electricity = electricity consumption * electricity emission factor (11.4)

Emission of HCFC = (fugitive amount of HCFC × emission factor) * (1 − elimination factor * use ratio) (11.5)

Fugitive amount of refrigerant = inventory reduction + purchased volume + volume sold − increase in refrigerant inside the equipment) (11.6)

Total volume of fugitive refrigerant = quantity of specific equipment * original filling volume of specific equipment * average fugitive ratio per year of the specific equipment (%) (11.7)

3. **Mass balance method:** Mass balance method could be applied in determining the emission from some manufacturing processes, such as the input, output, consumption, and conversion of material, mass, and energy. This method can be used in lieu of the emission coefficient method if it cannot be used for some industries due to special manufacturing processes or during the preliminary period of technical development. Emission could be calculated with this method for the sintering process of clinker in cement industry (waste tires, waste solvent, and fire retardant), urea used for treatment of De-NOx in cement industry, consumption of electrode for steel production in electric arc furnace, GHG gases used for semiconductor chemicals, and vapor deposition and filling of a small amount of gases (such as fire extinguishers, SF6, air

conditioner's refrigeration, and acetylene for cutting operation) into cylinders.

11.4.2 Selection of the Emission Coefficient

In general, emission coefficients can be classified as the coefficient obtained from measuring/mass energy balance, the coefficient of a manufacturing process/equipment experience, the coefficient provided by the manufacturer, regional emission coefficient, state emission coefficient, and international emission coefficient. The data chosen must be correct (labeled with the source and origin), localized, and timely to ensure the validity of the calculation. Emission coefficients can be further classified into two general types: *transindustrial* and *specific* industry. The emission coefficient of a special industry is chosen based on the industry's features, and is thus prone to wide variations, which makes it difficult to develop a universal criterion. On the other hand, trans-industrial emission coefficients are divided into four major types: fixed burning source, mobile burning source, HFCs fugitive source, and sewage or mud treatment fugitive source.

Appendix lists the emission coefficients of different industries for references.

11.4.3 Summary of Quantification Results

To summarize the results of various emission inventories—findings obtained from a series of GHG inventories were converted into carbon dioxide equivalent (CO_2E) using the *global warming potential* (GWP) with a metric ton as the unit of measure, with reference to ISO 14064-1.5.1. GWP is the value obtained from assessment of environmental pollution caused by a given type of gas compared with CO_2. By multiplying GHG emission with the corresponding GWP, emission can be converted into CO_2 emission, which is then used as the method for summarizing the inventory result.

11.4.4 Construction of Inventory Database

A database of GHG inventory that lists various emission sources and summing up the volume of emission was built. The relative weights of importance for different items can be analyzed by AHP as introduced in Sec. 2.3. Based on these weights, the integrated assessment can be carried out to obtain the total emission.

11.5 Impartial Third-Party Verification

After completion of the inventory of product carbon footprint, the inventory report provided by the inventory unit contains a record of all emission sources, emission categories, data sources, and the necessary details. To ensure the accuracy and impartiality of the inventory results, the report must be validated thoroughly and the accuracy of all information must be verified. Therefore, results must be verified by an impartial third-party verifier.

The validation or verification organization shall complete the task objectively. Qualifications of the certification body used in this study are based on Taiwan Incorporated National Accreditation Foundation (TAF) GHG Validation and Verification Organization Accreditation Procedure (ISO 14065). Section 11.5.1 describes the general requirements and Sec. 11.5.2 introduces the procedures.

11.5.1 General Requirements

To ensure the capability and impartiality of verification, the verifier must satisfy the following basic qualification requirements.

Legal Status

The legal status of the validation or verification organization should be known, including the name of owner and controller (if different from owner) if suitable.

Legal and Contract Considerations The validation or verification organization shall be a legal entity or a specific part of a legal entity, and be able to shoulder legal responsibility of validation or verification. The organization shall sign an enforceable agreement with every client of the validation or verification service. Further, the organization shall maintain its power and responsibilities for all its activities, decisions, and representations of the validation or verification.

Impartiality The validation or verification organization shall be committed to impartiality, and shall operate impartially and avoid any unacceptable conflicts of interests. In addition, the validation or verification organization:

- Has management that is committed to impartially implement validation or verification activities.
- Shall have open and available representation that it understands the importance of impartiality in validation or verification activities and in the management of interests
- Shall have formal rules and/or contractual terms to ensure that every team member performs his or her duties with an impartial attitude
- Shall have documents to specify the management of potential conflicts of interests and risks inside the validation or verification organization or from other channels through the following methods.

a. Identify and analyze potential conflicts of interests from validation or verification activities, including potential conflicts resulting from any relationship
b. Assess the financial and income source to prove that no commercial, financial, or other factors will influence impartiality.
c. Require validation or verification-related persons to disclose any situation of his or the organization that may lead to potential conflict of interests.
d. Avoidance of conflict of interests

The validation or verification organization:

- Shall not employ persons with actual or potential conflicts of interests
- Shall not validate or verify GHG assertions from the same GHG program, unless permitted by the GHG program,
- Shall not validate or verify GHG assertions made by the client to which it provides GHG consultancy services
- Shall not validate or verify GHG assertions when the position as GHG consultancy might cause unacceptable risks (Note 1)
- Shall not employ those involved in providing GHG consultancy services to verify its GHG assertions
- Shall not outsource the review and issuance of validation or verification statements
- Shall not provide products or services for which its impartiality becomes exposed to unacceptable risks
- Shall not state or imply that the verification of GHG assertions will be simpler, easier, faster, or cheaper if a particular type of its GHG consultancy service is used (Note 2).

Note 1. The position described in 4 might be derived from ownership, governance, management, staff, resource sharing, accounting, contract, payment of marketing and sales commission, or referral of new clients.

Note 2. Arrangement or delivery of training services are not regarded as GHG consultancy service, provided that (when the training is related to GHG quantification, GHG data monitoring or recording, GHG information system, or internal verification service) the information provided is only limited to general information readily available from open sources (the trainer shall not provide recommendations or solutions specific to an organization or program).

11.5.2 Monitoring Mechanism of Impartiality

The validation or verification organization shall fulfill the requirement of impartiality through an independent mechanism. The independent mechanism that ensures impartiality and avoids the impact of conflicts of interests and commercial or operation issues on the integrity of validation or verification could be:

- An independent committee
- A GHG program incorporated with the impartiality monitoring function
- A group of directors not at the level of executive officer

The validation or verification organization shall prove that it has assessed the accountability risks related with its activities and has made adequate arrangements (i.e., insurance and deposit) to cover liabilities derived from activities within its business scope. For instance, GHG auditors shall have certain understanding of following topics:

- Introduction to GHG and carbon footprint concepts
- Description of PAS2050 and ISO14067-1 standards
- Procedure for GHG life cycle inventory and carbon footprint calculation
- GHG data quality management
- Carbon footprint validation practice

11.6 An Illustrative Example

In this section, we selected a single product of 3C products and formulated methods for illustration. This coincides with the global trend today, and would be advantageous if we could find methods to reduce carbon emission in this aspect. To demonstrate how to apply the procedure illustrated in the previous sections, only carbon emission was assessed on a notebook-type computer.

11.6.1 Setting the Borderline

First, the selected product is a B2B (i.e., from cradle to grave) product. GHG emission during selection of raw materials, product manufacturing process, transport to retailers and delivery to users for use, and discard, recycling, and disassembly were assessed. Therefore, the entire product life cycle from raw materials to recycling was included in the calculation of carbon emission, including the specific raw materials, transport, manufacturing process, recycling, workshops, vehicles, and even storage. Emission from the manufacturing and use stages were calculated as shown in Fig. 11.6.

Every stage has both direct (i.e., fixed emission source, manufacture process emission process, mobile emission process, and fugitive emission resource) and indirect emission (i.e., emission from outsourced energy, like power, heat, and fossil fuel), as well as other indirect emission (i.e., emission from entrusted manufacture, not because of energy use, or from other activities of the organization). We collected the data on such emission sources, most importantly direct emission source. Before collecting these data, we should identify their emission sources as follows:

11.6.2 Identification of Emission Source

We defined the sources of "direct emission" and "indirect emission" for a firm's operation-related emission (within the organization's borderline) and classified them based on the GHG inventory procedure (ISO 14064-1/GHG Protocol), as shown in Table 11.3.

11.6.3 Application Simulation

As mentioned in the Introduction, there are many databases and software that can be used for analysis. In this example, we adopt ProdTect with its built-in database to roughly calculate the carbon emission of products during the stages of raw material, manufacturing, and recycling. Since this software didn't consider the stage of transportation, we have further included the assessment of this stage. Each step of the simulated calculation of carbon emission by this software was examined.

First, Table 11.4 shows the major components of the notebook.

Note that different products have different transformation coefficients of carbon emission which can be referred to in the Appendix for reference.

In this example based on the major components listed in Table 11.4, the scenario was stated as follows:

Notebook's power is 90 W. Working hours of the notebook is 1388 h/y, sleep-mode hours is 2904 h/y, and shutdown hours is 4468 h/y based on PCR. Assuming that the notebook was used for two years, 90% electricity is saved during sleep-mode hours and shutdown hours are neglected.

First, power must be converted to KW/h. The following calculations were made based on the knowledge that the notebook consumes 249.84 KW/h during working hours and 50.272 KW/h in sleep-mode.

Working status:

$$0.09\ (\text{KW/h}) \times 1388\ (\text{h/y}) = 249.84\ \text{KW/h} \tag{11.8}$$

Sleep-mode:

$$0.09\ (\text{KW/h}) \times 2904\ (\text{h/y}) \times 0.1 = 52.272\ \text{KW/h} \tag{11.9}$$

Lifecycle Stage	Category	Type	Activities/equipment
Raw materials	Category I	Fixed emission source	➢ The raw material is metallic, i.e., boiler, heating furnace, and so on. ➢ The raw material is plastic, i.e., heating furnace, drier, and so on.
		Manufacturing process emission source	➢ Light oil cracking, aromatic hydrocarbon plant, separation of dimethylbenzene, and other nonburning processes
		Mobile emission source	➢ Mobile stackers, cranes, and others in raw material plants ➢ Trucks used for transport of raw materials from the site to the plant, or from the plant to the client, and trucks belonging to the plant
		Fugitive emission source	➢ Cooling air used by the plant ➢ Fire extinguishers or sprayers used inside the plant
	Category II	Outsourced energy	➢ Use of energy from other manufacturers during the manufacturing process
	Category III	Other indirect emission	➢ Outsourced manufacturing process of raw materials
Manufacture	Category I	Fixed emission source	➢ Equipment generator ➢ Steam or heat generator ➢ Fossil fuel-burning facilities, i.e., burning furnace, burning tower, boiler, smelting furnace, exhaust burner, and other equipment or machine that use fossil fuel
		Manufacturing process emission source	➢ Plant catalyst oxidizer and nonburning exhaust treatment facilities

TABLE 11.3 Categories of Emission Sources (*Continued*)

Lifecycle Stage	Category	Type	Activities/equipment
		Mobile emission source	➢ Mobile stackers, cranes, and others in raw material plants ➢ Trucks used for transport of raw materials from the site to the plant, or from the plant to the client, and trucks belonging to the plant
		Fugitive emission source	➢ HFC leakage from air conditioners or refrigerating equipment ➢ Leakage from storage tank, pipeline, pump, or valve of fuel and natural gas ➢ Leakage caused by solvent during cleaning ➢ Fugitive emission from regular overhaul of the equipment (removal of coke from pipeline of smelting furnace, blow valve of aerostatic press, pipeline, or well blowdown, tank cleaning), abnormal emission (emergency shutdown or decompression emission) or pilot burning, and so on ➢ Use of fire extinguishers or sprayers ➢ Fugitive emission from power transmission and distribution lines and devices SF6
	Category II	Outsourced energy	➢ Machinery that consumes outsourced electricity during manufacture ➢ Machinery that provides steam or heat provided by other plants during manufacture
	Category III	Other indirect emission	➢ If the plant has outsourced processes
Transport	Category I	Fixed emission source	➢ Consumption of electricity inside the logistics center

Table 11.3 Categories of Emission Sources (*Continued*)

Lifecycle Stage	Category	Type	Activities/equipment
		Mobile emission source	➢ Trains, vessels, airplanes and etc that belong to the logistics organization ➢ Delivery trucks that belong to the logistics operator
		Fugitive emission source	➢ Use of fire extinguishers or sprayers in the logistics center
	Category III	Other indirect emission	➢ Vehicle fleet outsourced by the logistics center
Use	Category I	Fugitive emission source	➢ Fugitive emission from regular overhaul of the equipment
	Category II	Outsourced energy	➢ Consumption of electricity by households
Recycling	Category I	Fixed emission source	➢ Burning equipment, such as incinerators ➢ Smelting furnace for metal smelting
		Mobile emission source	➢ Vehicles for collection of recycled products from different areas

TABLE 11.3 Categories of Emission Sources

Adopting the software of ProdTect, one may obtain the results of total estimation respective to different stages as shown in Table 11.5. It can be seen that total carbon emission calculated by the software is 440.86 Kg, with majority coming from the use stage. The transport stage shall be added to in later steps to complete the calculation.

11.6.4 Quantification of Emission from Storage and Transport

In this section we assess the storage and transport stages. Results of manual simulation were added to the results using ProdTect, as shown in Table 11.5.

Storage

Since the existing literature has no complete discussion of the calculation of carbon footprint during product storage, we developed a formula under the simple distribution principle to be used in the calculation for illustration (IPCC, 2007):

Component
Battery, Li-ion
Battery, NiMH
Cable, connector for computer
Cable, network
Case, keyboard
CD-ROM/DVD-ROM drive
Generic part for manufacturing
HDD, laptop computer
LCD module
Mainboard
Plugs, inlet and outlet
Plugs, inlet and outlet, computer cable
Power adapter, main parts

Table 11.4 Major Components of the Notebook

Assumptions

- Product storage period $T = 360$ hr
- Current monthly electricity consumption (independent meter) $y = 800$ (KWH)
- Assumed product quantity = 2400 (sets)
- Storage area ratio: 0.05%

Calculation Electricity consumption at the current stage * area ratio * emission coefficient of outsourced electricity/quantity = 800 (KWH) * 0.0005 * 0.623/(2400) = 0.010383 (Kg)

Transport

We calculated the carbon emission of the "transport" part under the Product Carbon Footprint Calculation Principle of PS2050 as follows (see the Appendix).

Assumptions

- Only one type of commodity was carried by the vehicle
- The logistics operator was Retail Support International

Raw Materials	Manufacture Process	Use	Recycling	Discard
189.628 Kg	57.38 Kg	347.521 Kg	47.049 Kg	3.742 Kg

Table 11.5 Calculation of Carbon Emission

- Diesel truck was used
- Diesel fuel was used
- Each truck carried 100 (full load) sets of notebooks
- The retailer required 100 sets
- The truck operated at normal temperature

Calculation

Fuel consumption = distance * fuel efficiency factor = (12.6/100) * 9.8 = 1.2348 (Kg)
CO_2 emission = fuel consumption * calorific value * emission factor = 1.2348 * 2.73 = 3.371004 Kg CO_2)
CH_4 emission = fuel consumption * calorific value * emission factor = 1.2348 * 0.000111 = 0.000137 Kg CH_4)
N_2O emission = fuel consumption * calorific value * emission factor = 3.0968 * 0.000221 = 0.0002729 Kg N_2O)
Emission = fuel consumption * calorific value * emission factor = 3.371004 * 1 + 0.000137 * 25 + 0.0002729 * 298 = 3.45575 Kg

11.6.5 Summary and Discussion

Table 11.6 shows the carbon emission calculated from each stage of which the total emission is 551.627 Kg.

The example revealed that the notebook has the largest carbon emission during its use stage, followed by the stage of raw materials. Therefore, we may make some recommendations for improvement of these two stages as follows:

Use Stage

A large body of evidence indicates that a product's environment-friendliness depends on the design stage. Nonconsumables that work for a long period, such as notebooks and other types of electronic products, cannot avoid GHG emission during the use stage. Energy consumption during the use stage must be reduced to improve GHG emission performance. Specific methods of effectively reducing a product's energy consumption depend on laws and strategies promulgated by state governments and the commitment of the public.

Raw Materials	Manufacture Process	Use	Recycling	Discard	Storage	Transport
189.628	57.38	347.521	47.049	3.742	0.010383	3.45575

TABLE 11.6 Calculation of Carbon Emission (Kg)

Raw Material Stage

The stage of raw materials often causes an increase in a product's carbon emission and is also a stage that a firm must keep confidential. In order to improve carbon emission at this stage, the government must enact relevant laws that firms adhere to by switching to raw materials with low emission instead of pursuing only high profits and neglecting the current crisis of the world.

11.7 Conclusion

This chapter aims to illustrate a procedure of life cycle assessment of a product through an investigation of carbon emission, in order to promote the green label and increase public awareness of the significance of carbon emission reduction. We have found that the environmental-friendliness of most products depends on the design stage. Interventions and careful consideration during the design stage can lead to products with less carbon emission. The decision of manufacturers to move towards environment-friendly products depends on consumer demand to do so.

References

Intergovernmental Panel on Climate Change (IPCC), 2007, http://www.ipcc.ch/

ISO 14040, 1997, *Environmental Management: Life Cycle Assessment: Principles and Framework,* Berlin: Beuth.

ISO 14067-1, Carbon footprint of products-part 1, http://www.iso.org/iso/catalogue_detail.htm?csnumber=43278

IKP University of Stuttgart and PE Europe GmbH, GaBi 4: the software for environmental process and product optimization, http://www.gabisoftware.com/index.html

Kyoto Protocol, 2005, http://unfccc.int/kyoto_protocol/items/2830.php

Van Oers, L, 2004, "CML-IA: Database containing characterization factors for life cycle impact assessment." Centre of Environmental Science (CML), Leiden, available at http://www.leidenuniv.nl/interfac/cml/ssp/index.html

PRé Consultants, SimaPro 7 LCA software, http://www.pre.nl/simapro/default.htm

KERP Engineering GmbH, 2006, ProdTect, Vienna, http://www.KERP-energeering.com

Richardson, J., T. Irwin, and C. Sherwin, 2005, Design and Sustainability, *A Scoping Report for the Sustainable Design Forum.* Taiwan Environment Protection Agency, 2010, http://cfp.epa.gov.tw/carbon/ezCFM/Function/PlatformInfo/FLFootProduct/ProductGuide.aspx

Yang, J. S., 2010, Lecture notes, CNS 14040, www2.kuas.edu.tw/prof/shihtao/ems/powerpoint/.../14040.ppt

Yang, Q. Z, 2007, Life cycle assessment in sustainable product design, *SIMTech Technical*

Appendix: Emission Factors and Coefficient Charts of Different Industries and Nations

The following emission factors and coefficient charts were extracted from literature and are provided here for reference of manufacturers. Due to space limitations, not all of them are presented here.

Comparison of Different Products and Nations

Country	Type of Vehicle	Fuel Efficiency (Km/L)	Target Year (Solar Calendar)
USA	Car	≥11.7	2003 (vehicles)
	Van	≥8.8	
Germany	Car	16.75	2005
Australia	Van	14.7	2010
Japan	All	Improve 15 ~ 20% on the basis of 1995	2010
	Gas-fueled car	12.3 (when at 1995) → 15.1	
	Diesel-fueled car	10.1 (when at 1995) → 11.6	
	Gas-fueled truck (below 2.5 ton)	14.4 (when at 1995) → 16.3	
	Diesel-fueled truck (below 2.5 ton)	13.8 (when at 1995) → 14.7	
South Korea	Domestic car <1500 C.C	≥12.4	2004
	Domestic car >1500 C	≥9.6	
	Imported car	≥8.5	
EU	New car	165 ~ 170 g/km (12.9 ~ 13.3#)	2003/2004
		140 g/km (15.7%)	2008
		120 g/km (18.3#)	2010

NOTE: # assume that gasoline's CO_2 emission coefficient is 2200 gCO_2/L, 165 ~ 170 g/km fuel efficiency is 12.9 ~ 13.3 km/L, 140 g/km fuel efficiency is 15.7 km/L, 120 g/km fuel efficiency is 18.3 km/L, 150 g/km fuel efficiency is 14.7 km/L.

Target Vehicle Fuel Efficiency Across Countries

Fuel	Unit (LHV)	Fixed Burning Source CO_2 Emission Coefficient			
		IPCC	GHG Protocol	API	Canada
Coal	Kg CO_2/Kg	2.53[1]	2.47	2.49[1]	2.21[2]
Diesel	Kg CO_2/L	2.65	2.68	2.73[1]	2.73
Kerosene	Kg CO_2/L	2.54	2.58	2.57[1]	2.55

REMARK:

1. If the calorific value per unit fuel is not available, the data promulgated by Taiwan Energy Bureau was adopted.
2. The emission coefficient of soft coal in Canada is the average value of provinces in Canada.
3. If Kg CO_2/GJ is used as the unit, most parameters in GHG protocol are provided by IPCC.
4. Parameters of API are mainly from IPCC, EIA, and EIIP.

Difference in GHG Emission Between Different Industries

Industry	Major GHGs	Major Path of Generation
Power	CO_2, CH_4, N_2O, SGF_6	➢ Fuel burning of power generation ➢ Fuel burning of transport facilities of raw materials
Petrochemical	CO_2, CH_4, N_2O	➢ Fuel burning of outsourced or self-owned power generation ➢ Direct emission from the manufacture process ➢ Fuel burning of transport facilities of raw materials and products
Cement	CO_2, CH_4, N_2O	➢ Fuel burning of fixed facilities ➢ Fuel burning of outsourced or self-owned power generation ➢ Fuel burning of transport facilities of raw materials and products
Steel and iron	CO_2, CH_4, N_2O	➢ Fuel burning of fixed facilities ➢ Fuel burning of outsourced or self-owned power generators ➢ Burning of carbon-added raw materials ➢ Fuel burning of transport facilities of raw materials and products

Types and Generation Paths of Possible GHG from Industries (*Continued*)

Industry	Major GHGs	Major Path of Generation
Paper-making	CO_2, CH_4, N_2O	➢ Fuel burning of outsourced or self-owned power generators ➢ Fuel burning of outsourced or self-owned manufacture processes ➢ Fuel burning during transport
Pulp	CO_2, CH_4, N_2O	➢ Fuel burning of outsourced or self-owned power generators ➢ Fuel burning of outsourced or self-owned manufacture process ➢ Fuel burning during transport
Photoelectric semi-conductor	CO_2, CH_4, N_2O, SF_6, PFCs	➢ Fuel burning of outsourced power generators ➢ Emission or fugitive emission from manufacture process ➢ Fuel burning during transport
Waste treatment	CO_2, CH_4, N_2O	➢ Fuel burning of fixed facilities ➢ Emission from manufacture equipment
Plastic product manufacture industry	CO_2, CH_4, N_2O	➢ Fuel burning in manufacture process of outsourced or self-owned power or heating facilities ➢ Fuel burning in fixed facilities ➢ Fuel burning during transport
Chemical product manufacture	CO_2, CH_4, N_2O	➢ Fuel burning in manufacture process of outsourced or self-owned power or heating facilities ➢ Fuel burning of fixed facilities ➢ Fuel burning during transport
Automobile	CO_2, CH_4, N_2O, HFCs	➢ Fuel burning of fixed facilities ➢ Fugitive emission of refrigerant additives into vehicles ➢ Fuel burning of outsourced power generators

Types and Generation Paths of Possible GHG from Industries

Industry	Category	Name of Substances	Emission Coefficient	Coefficient Unit	Formula
Cement	Dust	Waste of bypass dust	0.525	$KgCO_2$/Kg waste of bypass dust	CO_2 emission = (EF × d)(1 + EF − EF × d) EF = emission coefficient of the plant's self-owned clinker ($KgCO_2$/Kg clinker) recommended clinker (revised MgO) emission coefficient 0.525 $KgCO_2$/Kg clinker D = lime and cement sintering (percentage of CO_2 in that of carbonate of raw material mixture, generally 100%)
	Dust	Dust of waste and sold cement	(EF × d)/(1 + EF − EF × d)	$KgCO_2$/Kg limestone kiln dust	CO_2 emission = (EF × d)(1 + EF − EF × d) EF = emission coefficient of the plant's self-owned clinker ($KgCO_2$/Kg clinker) recommended clinker (revised MgO) emission coefficient 0.525 $KgCO_2$/Kg clinker D = lime and cement sintering (percentage of CO_2 in that of carbonate of raw material mixture, generally 100%)
	Product	Clinker (revised MgO)	0.525	$KgCO_2$/Kg clinker	CO_2 emission = recommended emission coefficient ($KgCO_2$/Kg clinker) × activity intensity of emission source (Kg clinker)
	Product	Clinker	0.510	$KgCO_2$/KgCaO	CO_2 emission = recommended emission coefficient ($KgCO_2$/Kg clinker × activity intensity of emission source (Kg clinker) MgO content not considered
	Product	CaO	0.784	$KgCO_2$/KgMgO	CO_2 emission = recommended emission coefficient ($KgCO_2$/KgCaO) × activity intensity of emission source (KgCaO) $CaCO_3 \rightarrow CaO + CO_2$
	Product	MgO	1.092	CO_2 ton/year	CO_2 emission = recommended emission coefficient ($KgCO_2$/KgMgO × activity intensity of emission source (KgMgO) $MgCO_3 \rightarrow MgO + CO_2$

Calculation of Carbon Emission Coefficients of Each Industry (*Continued*)

Industry	Category	Name of Substances	Emission Coefficient	Coefficient Unit	Formula
	Product	Cement output	B × C × D × E	CO_2 ton/year	CO_2 emission=recommended emission coefficient (CO_2 ton/clinker) × annual cement output (CO_2 ton/year) Recommended emission coefficient (CO_2 ton/clinker ton) = B × C × D × E B = ratio between clinker and cement (100% port rate cement is 95%, mixture or masonry cement is 75%) C = consumption of raw materials per unit of clinker (recommended value: 1.54) D = ratio of $CaCO_3$ equivalent to raw materials (recommended value is 78%) E = ratio between CaO and $CaCO_3$ (44%)
Steel and iron		Consumption of reducing agent	Coal: 2.5 External coal: 2.8 External coke: 3.1 External petroleum coke: 3.6	CO_2 ton/year	CO_2 emission = recommended emission coefficient (CO_2 ton/ton) × consumption of reducing agent (ton CO_2/year The mass of reducing agent shall be deducted from the burning of natural materials
		Consumption of slagging constituent	Dolomite: 0.477 Limestone: 0.44	CO_2 ton/year	CO_2 emission = recommended emission coefficient (CO_2 ton/ton) × consumption of slagging constituent (ton CO_2/year The slagging constituent includes dolomite, limestone and etc.

		Addition to blast furnace	PVC 1.62 PET: 2.24, PE: 2.85, external grained fuel: 0.44, prebaked anode carbon block and coal electrode 3.6, furnace coal gas: 3.6, converter coal gas: 0.44,	CO_2 ton/year	CO_2 emission = recommended emission coefficient (CO_2 ton/ton) × consumption of reducing agent (ton CO_2/year The mass of reducing agent shall be deducted from the burning of natural materials
		Use waste steel or crude steel as raw materials	mineral ore: 0%, raw iron or cast iron: 4%, steel and iron: 04%	%	CO_2 emission = metal content (%) × 44/12 × outside supply of crude steel or waste steel (ton) Carbon content in crude steel is about 0.5%～2%
		Steel output	1.75	CO_2 ton/steel ton	CO_2 emission = recommended emission coefficient (CO_2 ton/ton) × iron and steel output (CO_2 ton/year) General plants, including coking, iron-making, steel-making, but not including CO_2 generated from consumption of slagging constituent

(Continued)

Industry	Category	Name of Substances	Emission Coefficient	Coefficient Unit	Formula
		Electrode	44/12 × P	$KgCO_2$ Kg electrode carbon content	$C + O_2 \rightarrow CO_2$ Carbon content in electrode is P%
Paper-making	Processed chemical product	Processed chemical product	$CaCO_3$ 0.440; Na_2CO_3: 0.415	CO_2 ton/ton	CO_2 emission = recommended emission coefficient (CO_2 ton/ton) × chemical consumption (ton)
Photo-electronic, semi-conductor	Semi-conductor wafer manufacture or TFT-LCD manufacture	PFCs	(1 – h)(1 – Ci) (1 – Ai) × GWPi + Bi × (1 – ACF4) × GWPCF4]	Kg PFCs/Kg PFCs	PFCi émission = (PFCi) × (1 – h) [(1 – Ci) (1 – Ai) × GWPi + Bi × (1 – ACF4) × GWPCF4] PFCi = purchased amount of PFCi (Kg) h = ratio of residual PFC in steel cylinder (heel), recommended Heel (h) = 0.10 Ci = use ratio of PFCi in manufacture process, (C2F6) Ci = 0.30. (CF4) Ci = 0.20, (CHF3) Ci = 0.70, (SF6) Ci = 0.50, (NF3) Ci = 0.80, (C3F8) Ci = 0.60, (C4F8) Ci = 0.70 Ai = ratio of PFCi being removed by local scrubber = ai × Va, ai = the treatment ratio of local scrubber on PFCi, recommended value 0.9: Va = ratio of PFCi for local scrubber

				GWPi = global warming potentials for one century, with reference to the third assessment report of IPCC (2001) Bi = ratio of CF4 by-products of PFCi, C2F6 → CF4 is 0.10, C3F8 → CF4 s 0.20 ACF4 = the ratio of CF4 (by-product of PFCi by local scrubber = aCF4 * Va) aCF4 = treatment efficiency of local scrubber on CF4
Incinera-tion and treatment	Incineration	Waste (garbage)	CO_2: 0.65 t/t CH_4: 30 $KgCH_4$/TJ N_2O: 4 KgN_2O/TJ (IPCC 2006 coefficient) Emission = activity data × carbon ratio % × 44/12 × 1	Mass energy balance method: if the waste is carried out with carbon ratio analysis, CO_2 GHG is calculated by the carbon ratio formula: CO2 emission equivalent = activity data × content ratio% × 44/12 × 1 If carbon ratio is unknown: CO_2 emission equivalent = activity data × emission coefficient × calorific value × GWP The emission of CH_4 and N_2O could only be calculated with the calorific value If there is a continuous monitor, it could be used as reference for CO_2 emission.

Agriculture

Emission procedure	Emission factor
Synthesis of fertilizer nitrogen	0.0125 Kg N_2O-N/Kg N
Biological nitrogen fixation	0.0125 Kg N_2O-N/Kg N
Use animal waste as fertilizer	0.0125 Kg N_2O-N/Kg N
Decompose residuals of agricultural produce	0.0125 Kg N_2O-N/Kg N
Farming of organic soil	5 Kg N_2O-N/hectare per year
Nitrogen evaporation and redeposit	0.01 Kg N_2O-N/Kg N
Seepage and lost nitrogen	0.025 Kg N_2O-N/Kg N

Emission Factor of Excrement of Animals (Source: IPCC)

Animal Category	Intestinal Fermentation Kg CH_4/Cattle/Year	Fecaluria Treatment Kg CH_4/Cattle/Year
Cattle	—	—
Adult cattle	75[2]	1
Adult dairy cattle	118	36
Adult beef cattle	72[2]	1
Dairy cattle not giving birth yet	56[2]	36
Beef cattle not giving birth yet	56[2]	1
Cow to be slaughtered and not giving birth yet	47	1
Castrated cattle	47	1
Young cattle	47	1
Hog	—	—
Pork hog	1.5	10
Other livestock	—	—
Sheep	8	0.19
Goat	8	0.12
Horse	13	1.4
Poultry	—	—
Chicken	No estimation	0.078
Hen	No estimation	0.078
Turkey	No estimation	0.078

Emission Factor of Fertilizers (Source: IPCC)

Emission Factor of Traffic Fuels in Different Countries

Taiwan

Energy Category	Original Unit	Calorific Value Kcal	CO_2 Emission Index		
			Carbon Volume (Kg) of Calorific Value Per Unit (1 Billion Joule) (see Note)	CO_2 Volume from Each Unit of Energy (Kg)	CO_2 from Each Unit of Calorific Value (Oil Equivalent)
Self-produced coal	Kg	6200	25.8	2.46	3.96
Raw material coal	Kg	6800	25.8	2.69	3.96
Fuel coal	Kg	6400	25.8	2.53	3.96
Coke	Kg	6800	25.8	2.69	3.96
Coal briquette	Kg	3800	26.6	1.55	4.08
blast furnace gas	M^3	777	70.8	0.85	1.09
Crude oil	L	9000	20	2.76	3.07
Refinery gas	M^3	9000	15.7	2.17	2.41
LPG	L	6635	17.2	1.75	2.64
Aviation gasoline	L	7500	19.1	2.2	2.93
Vehicle gasoline	L	7800	18.9	2.26	2.90
Aviation fuel	L	8000	19.5	2.39	2.99
Kerosene	L	8500	19.6	2.56	3.01
Diesel	L	8800	20.2	2.73	3.10
Fuel oil	L	9200	21.1	2.98	3.24
Lubricant	L	9600	20	2.95	3.07
Asphalt	L	10000	22	3.38	3.38
Naphtha	L	7800	20	2.39	3.07
Petroleum coke	L	8200	26.6	3.35	4.08
LNG	L	9900	17.5	2.66	2.69
Natural gas	M^3	8900	15.3	2.09	2.35
Electricity	KWH	860	–	0.66	–

NOTE: Calculated with reference to IPPC (2007) Department Method. IPPC (Intergovernmental Panel on Climate Change).
DATE SOURCE: Energy Industry GHG Information Center, Energy Bureau.

CO_2 Emission Coefficients of Various Categories of Energy

USA

Vehicle Characteristics			CO_2/km
Vehicle Type	**L/100 km**	**Mile/Gallon**	**CO_2 g/km**
New small gas/electricity hybrid vehicle	4.2	56	100.1
Small gas-fueled vehicle, highway	7.3	32	175.1
Small gas-fueled vehicle, city	9.0	26	215.5
Medium gas-fueled vehicle, highway	7.8	30	186.8
Medium gas-fueled vehicle, city	10.7	22	254.7
Large gas-fueled vehicle, highway	9.4	25	224.1
Large gas-fueled vehicle, city	13.1	18	311.3
Medium wagon, highway	8.7	27	207.5
Medium wagon, city	11.8	20	280.1
Small SUV, highway	9.8	24	233.5
Small SUV, city	13.1	28	311.3
Large SUV, highway	13.1	28	311.3
Large SUV, city	16.8	14	400.2
Medium van, highway	10.7	22	254.7
Small van, city	13.8	17	329.6
Large van, highway	13.1	18	311.3
Large van, city	15.7	15	373.5
LPG vehicle	11.2	21	266
Diesel vehicle	9.8	24	233
Light-weight gas-fueled truck	16.8	14	400
Heavy-weight gas-fueled truck	39.2	6	924
Light-weight diesel truck	15.7	15	374
Heavy-weight diesel truck	33.6	7	870
Light-weight locomotives	3.9	60	93
Diesel motorcycle	35.1	6.7	1034.611322

Canada

Types of application	CO_2 g/L fuel	CH_4 g/L fuel	N_2O g/L fuel
Land traffic			
Gasoline			
Light-weight gas-fueled passenger vehicle			
Level 0 Three-way catalyst converter	2360[1]	0.12[2]	0.26[2]
Level 0 New three-way catalyst converter	2360[1]	0.32[2]	0.25[2]
Level 0 Old three-way catalyst converter	2360[1]	0.32[2]	0.58[2]
Oxidized catalyst converter	2360[1]	0.42[2]	0.2[2]
Noncatalyst converter	2360[1]	0.52[2]	0.028[2]
Light-weight gas-fueled truck (LDGT)			
Level 1 Three-way catalyst converter	2360[1]	0.22[2]	0.41[2]
Level 0 New three-way catalyst converter	2360[1]	0.41[2]	0.45[2]
Level 0 Old three-way catalyst converter	2360[1]	0.41[2]	1[2]
Oxidized catalyst converter	2360[1]	0.44[2]	0.2[2]
Noncatalyst converter	2360[1]	0.56[2]	0.028[2]
Heavyweight gas-fueled passenger vehicle (HDGV)			
Three-way catalyst converter	2360[1]	0.17[2]	1[2]
Noncatalyst converter	2360[1]	0.29[2]	0.046[2]
Non-control	2360[1]	0.49[2]	0.08[2]
Motorcycle			
Noncatalyst converter	2360[1]	1.4[2]	0.046[2]
Non-control	2360[1]	2.3[2]	0.046[2]
Diesel			
Lightweight diesel passenger vehicle (LDDA)			
Advanced control	2730[1]	0.05[2]	0.2[2]
General control	2730[1]	0.07[2]	0.2[2]
No control	2730[1]	0.1[2]	0.2[2]

(*Continued*)

Lightweight diesel truck (LDDT)			
Advanced control	2730[1]	0.07[2]	0.2[2]
General control	2730[1]	0.07[2]	0.2[2]
No control	2730[1]	0.08[2]	0.2[2]
Heavyweight diesel passenger vehicle (HDDV)			
Advanced control	2730[1]	0.12[2]	0.08[2]
General control	2730[1]	0.13[2]	0.08[2]
No control	2730[1]	0.15[2]	0.08[2]
Natural gas passenger vehicle	1.89[3]	0.022[2]	6E-05[2]
Propane pasenger vehicle	1500[3]	0.52[2]	0.028[2]
Non–land passenger vehicle			
Other gas-fueled passenger vehicle	2360[1]	2.7[2]	0.05[2]
Other diesel-fueled passenger vehicle	2730[1]	0.14[2]	1.1[2]
Diesel-locomotives	2730[1]	0.15[2]	1.1[2]
Ocean traffic			
Gas-fueled vessel	2360[1]	1.3[2]	0.06[2]
Diesel-fueled vessel	2730[1]	0.15[2]	1.00[2]
Light-fuel vessel	2830[1]	0.3[2]	0.07[2]
Heavy-fuel vessel	3090[1]	0.3[2]	0.08[2]
Aviation			
Traditional airplane	233[1]	2.19[2]	0.23[2]
Jet airplane	2550[1]	0.08[2]	0.25[2]

[1]Jaques, A. (1992), Canada's Greenhouse Gas Emissions: Estimates for 1990, Environmental Protection, Conservation and Protection, Environment Canada, EPS 5/AP/4, December.
[2]SGA (2000), Emission Factors and Uncertainties for CH4 & N2O from Fuel Combustion, SGA Energy Limited, August.
[3]Adapted from McCann, T. J. (2000), 1998 Fossil Fuel and Derivative Factors, prepared for Environment Canada by T. J. McCann and Associates, March.

CHAPTER 12

Web-Based Information Support Systems

12.1 Introduction

Electronic commerce (EC) is widely used by consumers to perform different daily activities on the Internet. Online shopping is one of the most popular applications among these activities. In contrast to conventional shopping, Yang et al. (2004) point out that EC provides alternative ways for users to get information on products such as price, availability, suppliers, substitutes, and even manufacturing process. For competitiveness, both Wang and Archer (2007) and Xu (2008) also emphasize that EC companies need to develop higher business interoperability in their electronic marketplaces by improving the electronic market functions. In particular, Guo (2007) indicates that the enhancement of electronic market functions can lead to an overall reduction of interaction cost for business interoperation on all types of electronic marketplaces.

However, among the numerous EC functions that provide so much available information, it is difficult for online users to make quick and effective decisions. Facing fierce market competition and impatient users, a personalized decision support system is urgent and essential for an EC company. By providing more helpful information to users, faster and more satisfactory decisions can be made; and thus, opportunities for retaining customers and gaining profits are greater.

Many EC suppliers use *recommender systems* (RSs) to find out the preferences of target users so that the right products can be suggested (Schafer et al., 1999). A well-established RS can add value to an EC company in several ways: (1) users can retrieve product information easily; (2) cross-selling for users can be enhanced; and (3) users' loyalty can be sustained by good service. There are numerous studies in the fields of social networks (McPherson et al., 2001) and information-filtering techniques (Sarwar et al., 2000). Hogg (2010) and Lam (2004) both pointed out that in social networks, people with similar characteristics tend to associate with one another. The use of social network structure generally allows the EC to identify the products of likely interest to the target users based on information provided by the members of the network. On the other hand, information-filtering techniques that analyze users' preferences and help EC websites achieve accurate product selection. By filtering the information provided by the users, the techniques aim to track the purchase behavior of users and recommend proper products.

Among information-filtering techniques, researchers such as Kim et al. (2010) and Schafer et al. (2001) all confirm that *collaborative filtering* (CF) is one of the most commonly adopted method. The concept of CF is very highly related to social networks. The CF technique uses collaborative information from "neighbors," which are defined as users with similar behavior to the target user. CF is also regarded as the most effective method for the RS. However, Kohrs et al. (2001) show that CF's drawback is that no recommendation can be made if a user's related data are sparse. On the other hand, excessive emphasis on recommendation performance can lead to the neglect of profit, which is also an essential concern for an EC company. Aside from this, although there are different approaches to retrieve the needed information for recommendation, a systematic and comprehensive decision module is still lacking. Therefore, the time spent on data retrieval can be long, and the recommended products may not match the users' desires. In particular, without a structural module, documenting the recommending procedure becomes difficult, and achieving the goal of "the right goods for the right person" becomes nearly impossible.

With these concerns, Wang and Wu (2010) introduced a strategy-oriented operation module that could be comprehensively applied to EC websites as a decision support mechanism so that the choice of various marketing strategies that consider profit for both suppliers and users could be developed. Therefore, in addition to introducing the framework of the recommender module, a clique-effects collaborative filtering (CECF) technique to predict users' purchase behavior is also illustrated in this chapter.

Furthermore, this chapter presents the modeling perspective to the e-service system, i.e., the recommender system. The introduced RS module aims to fulfill the profits of the customers and suppliers; the final stage of product selection is described as a linear bi-objective model, of which all required arguments are derived from the offline database and the CECF.

12.1.1 Infrastructure of Recommender Systems

Figure 12.1 shows the framework of the recommender system that includes three submodules: (1) **Input sources** of the users' profiles; (2) **output of recommendations,** and (3) **recommendation methods** as the interface between the two. The input submodule deals with the input profiles of a target user; the types of profiles considered in the system would be the demographic information, the binary basket

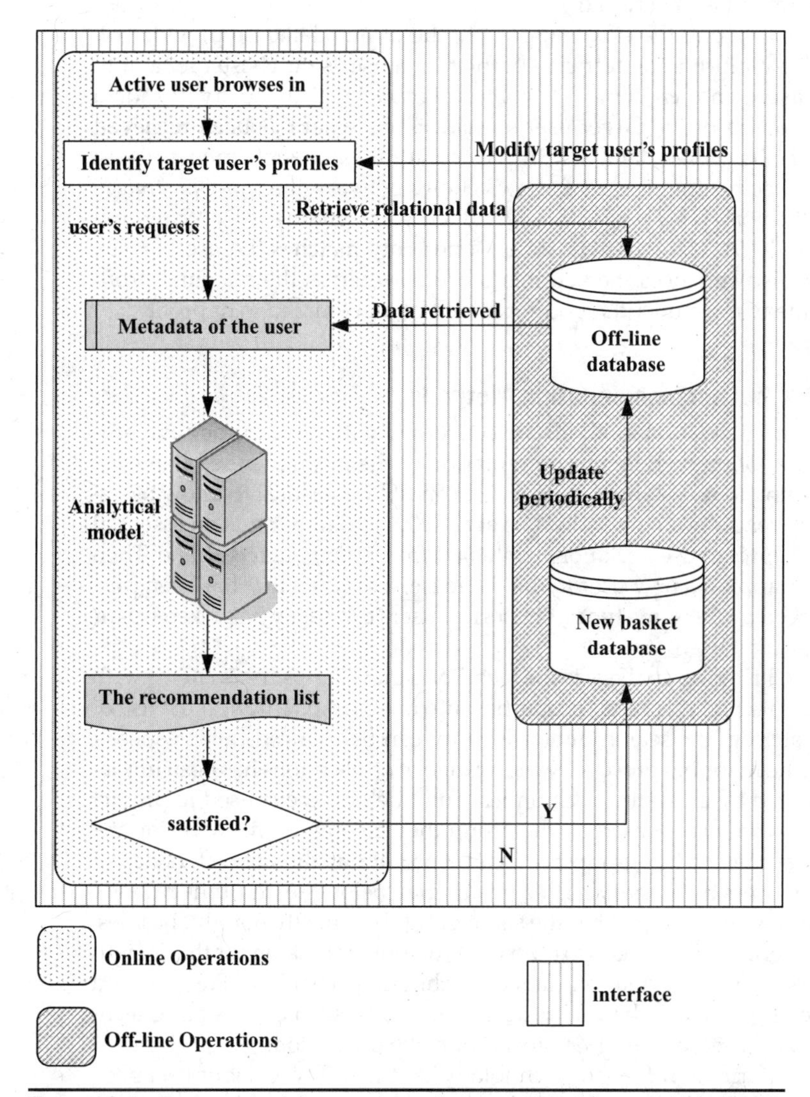

FIGURE 12.1 Framework of the recommendation module (Wang and Wu, 2009).

data, and the target user's requests of the desired satisfaction level and budget limit.

The output submodule would provide the recommended items from the result of the recommendation method. Both input and output submodules are categorized into *online* operations. The recommendation method, which is the core of the recommending module, functions with an *online* analytical model under the *offline* database constructed from three parts—user-group end, item-group end, and the relations in between.

Exploiting the CECF approach, the offline database provides required information retrieval of the target user's purchase probability measure on each item. The analytical model is then run by metadata composed of the target user's request and what has been retrieved from the offline database. In particular, the analytical model uses a bi-objective function that allows choice between the *win-win strategy* and the *maximal profit strategy*, which were proposed by Wang and Wu (2009). The win–win strategy not only matches the user's taste, but also enhances the supplier's profit, whereas the maximal profit strategy recommends products based on maximization of profit.

12.1.2 Recommendation Methods

Recommendation methods are concerned with the accuracy and efficiency of prediction and presentation of the recommended items according to users' input sources. For an RS, it is critical to know users' preferences systematically. Lee et al. (2001) point out that an essential concept is to use a relational database that is constructed offline. Then by mapping a new user to the database, a product that has been purchased by the same type of historical users can easily be picked up for the target use.

Clustering analysis is the technique that groups users/items with similar characteristics/properties into one group. Through the use of clustering, the search dimensionality can be reduced, which speeds up the mapping process. A wide range of applications has been implemented by clustering techniques, and one of these is used to predict unknown users based on the group they belong to. By analyzing the properties of the groups, we can learn about the characteristics of new users by identifying the group they belong to and thus provide them with the items that the same group has mostly bought. Besides, clustering analysis is also a very useful tool for looking for the "neighbors" in the information-filtering technology. That is, the users called the *neighbors* are chosen by certain methods, such as clustering techniques (Breese et al., 1998) to support the prediction.

Information-filtering technology has the ability to define user preferences with little effort. It is divided into two main categories (Kohrs et al., 2001)—*collaborative filtering* (CF) and *content-based filtering* (CBF).

CF is the most popular approach to predict the probability that a user will purchase a specific item based on other users' preferences. A CF method functions by matching people with similar interests and then making recommendations. However, in the initial state of an RS, the main problem would be insufficient users' profiles to sustain the prediction basis while using CF. Consequently, the drawback of CF is the requirement of some relevant rating data given by the target user.

Usually, by clustering users into groups before predicting, group influences can be utilized by recommendation methods on the target user to prevent poor prediction due to rarely relevant information. Furthermore, because the conventional CF approach utilizes preferences of neighbors to make a prediction for a target user, it leaves additional influences of non-neighbors out of consideration. As a result, research such as Kim et al. (2006) tends to discriminate the impacts of neighbors from non-neighbors; by integrating the effects caused by the two sources, better performance can be also expected.

CBF is the technology of analysis based on terms in the content such as texts or documents on the website. It considers term frequency in the content and its relation to the user's preference. However, with other media such as music or movies, its performance is not as good as text content because these objects are not easily indexed. In addition, Guo (2009) indicates that the maintenance of numerous heterogeneous electronic product catalogues on the Internet is still a tough task. Nevertheless, CF is still most commonly used since it is flexible and easily adaptable to an EC's RS. Therefore, in this chapter, we would incorporate the concept of CF into our system as the basic recommendation mechanism.

In addition to CF and CBF, another technique requires the private information of a user. *Demographic filtering* (DF) explains users by their personal demographics. A DF approach uses descriptions of people to learn the probability that an item is most preferred by what type of persons. Therefore, this method would lead to the same recommendation if the users have similar personal data. However, the DF approach requires more information regarding a user's privacy; therefore, DF is confronted with the problem that it is not easy to collect users' demographic descriptions. Therefore, Montaner et al. (2003) suggest that the DF method requires collaborating with other methods such as CF or CBF.

Besides the aforementioned filtering techniques, rules derived from the market basket analysis between items in large databases also account for an RS. Market basket analysis has been a popular system in finding the correlation among baskets (Aggarwal et al., 2009; Russel et al., 2000).

Aggarwal et al. (2009) were the first group who applied the famous *association rules* method to find the pattern of the probability of buying a specific product when another product is purchased. In such a recommending environment, many rules have been developed for how the different purchase behaviors of users can be treated (Hsu et al., 2004). Therefore, Sarwar et al. (2000) also proposed a method of *association rule–based recommendation* (ABR) in 2000. However, for the huge amount of transaction data, there may be many biased rules that would affect the precision of the recommendation. Therefore, the market basket analysis shall be conducted with the aid of filtering techniques such as CF, and the common concept of the CF method adapted to the binary market basket data as proposed by Mild and Reutterer (2003).

12.1.3 Roles and Their Goals in a Recommender System

In the current RS, there are three common roles involved: (1) the supplier, (2) the system developer, and (3) the user. In Table 12.1, we list possible considerations for constructing an RS. In the fields of EC trading, Li and Wang (2007) proposed a multiagent-based model with a win–win negotiation approach in which agents seek to strike a fair deal that also maximizes the payoff for everyone involved. However, such a win–win negotiation mechanism has not been discussed in the RSs with more comprehensive scale. For the existing research, the "performance of recommendation" is an attribute that benefits users. Therefore, when "more is better" is stressed, only the number of sold products is maximized, but not necessarily the profit. In other words, an RS is usually constructed from a user's standpoint. Only a few RSs can be regarded as being built from a supplier's perspective. Therefore, it is also necessary to construct an RS that allows both parties to justify their priorities.

	User	Supplier	System Developer	
			Win-win strategy	*Maximal profit strategy*
Objective	*O(u)*	*O(s)*	*O(u)* & *O(s)*	*O(s)*
Constraints	*C(u)*	*C(s)*	*C(u)* & *C(s)*	*C(u)* & *C(s)*
Problem types	Maximization problem	Maximization problem	Multi-objective problem	Maximization problem

NOTE: (*u*) user; (*s*) supplier.
O(u), objective of the user: Fulfill the demands of oneself.
O(s), Objective of the supplier: Maximize profit or products sold.
C(u), Constraint of the user: Budgets in hand.
C(s), Constraint of the supplier: Fulfill demands of users.

TABLE 12.1 Roles and Resolution in Recommender Systems

12.1.4 Summary and Discussion

After specifying the basic concept of an RS within the infrastructure of a RS, we shall introduce a strategy-oriented operation module for the RS comprising (1) an offline database, (2) CECF, and (3) the analytical model. An offline database that could be mathematically supported for the RS is developed. The database consists of three parts: (1) user-group data, (2) item-group data, and (3) the relations in between. The offline database is designed with the following two characteristics:

1. The users and the items are classified into groups according to their respective features/attributes. We regard any individual in a group as an information provider, which is especially important to a start-up RS with rare data.
2. The group effects are much easier to be retrieved. By bringing out additional effects from the groups of users and items, we aim to dilute the imprecise prediction caused by rare data, and to prevent inconsistent inputted data such as average scores.

However, to avoid the inputted group effects predominating over prediction, the priority of group effects shall be well arranged. Therefore, under the proposed offline database, we use CF to propose a *clique effects approach* (CECF). With the scheme of adjustable weights between the individual's and the group's effects, CECF is likely helpful in solving the situation of sparse data and the so-called "ramp-up" problem. In addition, we also introduce an analytical model proposed by Wang and Wu (2009). The analytical model could allow the system developer to actively adjust the priority between the supplier's profit and the user's satisfaction level.

Therefore, in the next section, we shall first specify the construction of the offline database including the user-group and item-group data. Then the proposed clique effects approach based on CF (CECF) would be presented in Sec. 12.2.2. Finally, we would clarify online and offline operations as well as present the analytical model in Sec. 12.2.3.

12.2 Operations in the Submodules of the System

Two major submodules of off-line and on-line operations are illustrated below:

12.2.1 Offline Operations

In this section, we specify the construction of the offline database including the user-groups data and item-groups data.

Item-Groups with Their Properties

Let D be the items in the market basket, with each item denoted as p_d, where $d = 1, \ldots, D$. Define $\Psi_{p_d} = [\alpha_1, \alpha_2, \ldots, \alpha_k, \ldots, \alpha_K]_{p_d}$

Class	Attribute Labels	Class	Attribute Labels
1	non	5	$\{\alpha_1, \alpha_2\}\backslash\{\alpha_3\}$
2	$\{\alpha_1\}\backslash\{\alpha_2, \alpha_3\}$	6	$\{\alpha_1, \alpha_3\}\backslash\{\alpha_2\}$
3	$\{\alpha_2\}\backslash\{\alpha_1, \alpha_3\}$	7	$\{\alpha_2, \alpha_3\}\backslash\{\alpha_1\}$
4	$\{\alpha_3\}\backslash\{\alpha_1, \alpha_2\}$	8	$\{\alpha_1, \alpha_2, \alpha_3\}$

TABLE 12.2 Classification Rules When $K = 3$

to be an attribute vector of p_d; then the set of items in the database is $P = \{p_d(\alpha_k)|d = 1, 2, \ldots, D\}$. All items in the database are further classified into mutually exclusive item-groups as $P^i = \{p_{d^i}(\alpha_k)|d^i = 1^i, 2^i, \ldots, D^i, i = 1, 2, \ldots, I\}$, each with $|P^i| = D^i$, and thus $\bigcup_{i=1}^{I} P^i = P$ and $\sum_{i=1}^{I} D^i = D$. In particular, we classify the items with respect to the item attributes. A threshold of each attribute value is given; each item with specific attribute values above those thresholds will be assigned to the corresponding group.

The number of attributes (K) is referred with its power set and then 2^K item-groups are generated. For instance, in Table 12.2, the number of item-groups generated is 8 when K is 3; an item is distributed into Class 5 only if its attribute values in $\alpha_1, \alpha_2$are higher than the thresholds of α_1, α_2 as well as its α_3 value lower than the thresholds of α_3 as denoted by $\{\alpha_1, \alpha_2\}\backslash\{\alpha_3\}$. Thus, the classification rules will provide exclusive groups so that one item belongs to only one group. Properties of each item-group can be also easily and clearly be identified by observing attribute labels. The *selling prices* of items in the market basket are defined as $\mathbf{s} = [S_{d^i}|d^i = 1^i, 2^i, \ldots, D^i, i = 1, 2, \ldots, I]$; possible *profits* are defined as $\mathbf{c} = [c_{d^i}|d^i = 1^i, 2^i, \ldots, D^i, i = 1, 2, \ldots, I]$, where s_{d^i} and c_{d^i} represent the corresponding price and profit of p_{d^i}. Therefore, for the items database, it will be stored by each item-group with its items and specified properties.

User-Groups and Their Profiles

Denote a user as u_f with $f \in N$. Let $U = \{u_f(\omega_g)|f \in N\}$ be a set of users labeled by the demographic features $\omega_g \in \{\omega_1, \omega_2, \ldots, \omega_g, \ldots, \omega_G\}$. To facilitate analysis—providing solutions for the "new user" problem and exploiting clique effects, the users are classified into mutually exclusive user-groups and assumed to behave similarly as the DF method suggests. The user-groups are formed by the following rules: assume each demographic feature ω_g can be divided/categorized into v_g intervals/categories denoted by $\omega_g^{v_g}$, and then we define $U^j : U \rightarrow \omega_1^{v_1} \times \omega_2^{v_2} \times \ldots \times \omega_G^{v_G}$, we have $U^j = \{u_f(\omega_g)|\omega_g \in \omega_g^{v_g}, g = 1, 2, \ldots, G, j = 1, 2, \ldots, J\}$. Then each user-group can be represented as $U^j = \{u_{f^j}(\omega_g)|f^j = 1^j, 2^j, \ldots, F^j, j = 1, 2, \ldots, J\}$, $|U^j| = F^j$ and thus $\bigcup_{i=1}^{J} U^j = U$. For instance, we define the demographic

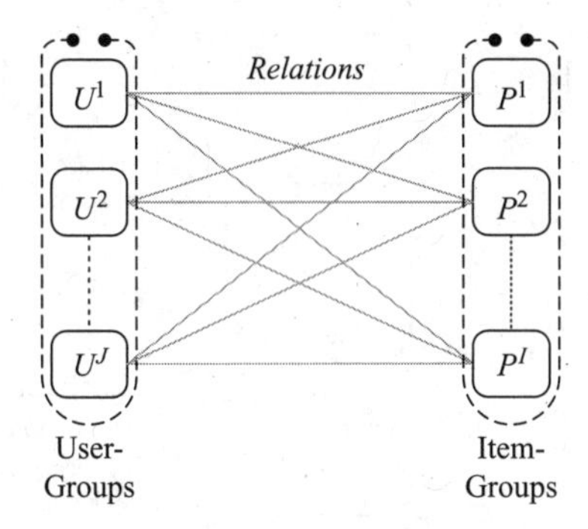

FIGURE 12.2 Relations among user-groups and item-groups.

features to be Gender (ω_1) and Age (ω_2); ω_1 is categorized into $\nu_1 = 2$ categories as male and female; ω_2 is divided into $\nu_2 = 4$ intervals as (0, 20], [20, 30], [30, 40], [40, ∞). Then we define the user-groups as $U^j : U \rightarrow \omega_1^2 \times \omega_2^4$ and eight user-groups yield as U^j, $j = 1, 2, \ldots, 8$.

Derivation of Relations Among Users and Items: CECF

To relate item-groups with user-groups, Figure 12.2 shows a bipartite grouping that connects users and items.

The relations embedded in the framework are regarded as clique effects of the purchase probability measured for a target user. The clique effects result mainly from the grouping of users. Users in the same clique with the target user (the so-called *neighbors* in CF) can provide collaborative information to measure purchase probability. However, users in different cliques may also provide collaborative information to the target user to a certain degree. In this respect, we propose the following concept to measure the purchase probability of the target user with respect to a predicted item as Eq. (12.1):

$$P_{r(\text{user, item})} = \theta \cdot P_{r(\text{user, item})}^{\text{in-clique}} + (1 - \theta) \cdot P_{r(\text{user, item})}^{\text{out-of-clique}} \tag{12.1}$$

where the probability ($P_{r(\text{user, item})}$) is a convex combination of two distinct probabilities: one is the purchase probability predicted by collaboration of users in the same clique (the *neighbors*) with the target user, and the other is predicted by collaboration of users in the different cliques. The composition of the proposed probability measure is illustrated in Fig. 12.3.

Let us refer to Fig. 12.3. First, note that arrows 3 and 4 jointly represent the "in-clique" purchase probability measure used by conventional CF. The common concept of the CF method with the binary market basket data adapted to Mild et al. (2001) is presented as Eq. (12.2):

$$P_{r(u_{fj}, p_{di})}^{\text{in-clique}} = \kappa_1 \sum_{u_{f\tau} \in U^j} sim(u_{fj}, u_{f\tau}) \times C_{u_{f\tau}, p_{di}} \tag{12.2}$$

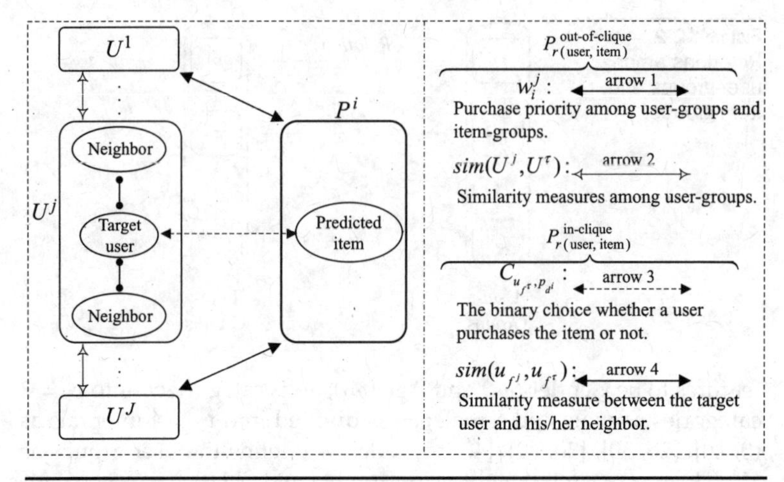

FIGURE 12.3 Various probability measurements of the target user on the predicted item.

where $P_{r(u_{fj}, p_{di})}^{\text{in-clique}}$ is the probability that target user u_{fj} purchases item p_{d^i} by using a collaboration of neighbors' preferences; κ_1 is a normalizing factor to ensure the absolute values of probability sum to unity; $sim(u_{fj}, u_{f^\tau})$, which refers to arrow 4, is the similarity between the target user u_{fj} and the neighbors u_{f^τ}; and $C_{u_{f\tau}, p_{di}}$, which refers to arrow 3, is the binary choice whether a user u_{f^τ} purchases p_{d^i} or not. It is noteworthy that for the similarity measure between the target user u_{fj} and the neighbors u_{f^τ}, as specified in Eq. (12.2), the neighbors are chosen from the user-group to which the target user belongs; this is in compliance with the structure of our proposed RS, which assumes that users in the same demographic group will tend to behave similarly.

Second, for the probability measure of "out-of-clique" based on the concept of CF, two factors should be considered: (1) the similarity between the target user-group and other user-groups as well as (2) other user-groups' purchase priorities on the predicted item-group. For the former, the similarity measures would refer to arrow 2 in Fig. 12.3. For the latter, that refers to arrow 1 in Fig. 12.3; the relative purchase frequency in the binary basket analysis has been adopted from Chen et al. (2008) as the prediction of purchase priority in Eq. (12.3):

$$w_i^j = C(U^j, P^i)/S(U^j) \tag{12.3}$$

where $C(P^i, U^j)$ is the relative frequency that users in U^j purchase items in P^i; $S(U^j)$ is the total number of market baskets for U^j. Therefore, the probability measure of "out-of-clique" purchase can be presented as Eq. (12.4):

$$P_{r(u_{fj}, p_{di})}^{\text{out-of-clique}} = \kappa_2 \sum_{\tau \neq j} sim(U^j, U^\tau) \times w_i^\tau \tag{12.4}$$

where $sim(U^j, U^\tau)$, which refers to arrow 2, is the similarity measure between the target user-group U^j and other user-group U^τ; κ_2 is a normalizing factor to ensure the absolute values of probability sum to unity. Therefore, the probability measure of a target user u_{fj} purchasing item p_{d^i} would be represented as Eq. (12.5):

$$P_{r(u_{fj},p_{di})} = \partial_{p_{di}}^{u_{fj}} = \theta \cdot \overbrace{\left(\kappa_1 \sum_{u_{f\tau} \in U^j} sim(u_{fj}, u_{f^\tau}) \times C_{u_{f\tau},p_{di}}\right)}^{\text{in-clique}} + (1-\theta) \cdot \overbrace{\left(\kappa_2 \sum_{\tau \neq j} sim(U^j, U^\tau) \times w_i^\tau\right)}^{\text{out-of-clique}}, \quad (12.5)$$

where the probability measure $P_{r(u_{fj},p_{di})}$ is replaced by $\partial_{p_{di}}^{u_{fj}}$ for simplicity; and θ is an adjustable weight on the in-clique probability measure. The way of the probability measure in Eq. (12.5) would lead us into the consideration on how to select similarity functions. Note that the CF performance depends on the choice of similarity measures. By taking into the effect of both common and noncommon item set, Wang and Wu (2010) define the similarity function as in Eq. (12.6):

$$sim(U^J, U^\tau) = \left| \frac{\bigcup_{j=1}^{J} S(U^j) - (S(U^j) \cup S(U^\tau))}{\bigcup_{j=1}^{J} S(U^j) - (S(U^j) \cap S(U^\tau))} \right| \quad (12.6)$$

where $sim(U^j, U^\tau)$ is the similarity measure between the target user-group U^j and the other user-group U^τ.

Summary of Offline Module

In summary, offline operations are mainly based on CECF which contains users' purchase probability measures as defined in Eq. (12.5), which is a convex combination of two distinct probability measures from the in-clique effects of Eq. (12.2) and out-of-clique effects of Eq. (12.4). The classification of the target user into in-clique users as well as out-of-clique users, the proposed probability measure function provides different insight from that of conventional CF method.

As for the probability measure of in-clique users, we adopt the traditional CF method, whereas for the measure of out-of-clique users, we propose an alternative similarity function by incorporating the items not purchased simultaneously by each pair of compared users to

	Function of User Similarity	Function of User-Group Similarity
Schemes	In-clique effects	Out-of-clique effects
CF	Common item set	—
CECF-C	Common item set	Common item set
CECF-NC	Common item set	Non-common item set

TABLE 12.3 Recommendation Schemes

find the similarity among user-groups. Then the proposed probability measure is predicted by the purchase and nonpurchase behaviors of the users, which is expected to provide more information in expounding the users. Therefore, to facilitate flexible applications, under the proposed CECF, we have two schemes in the recommendation method, namely, CECF-C and CECF-NC. Both C and NC represent the choice of similarity functions applied in computing the similarities among user-groups. *C* is based on the *c*ommon item set, whereas *NC* is based on *n*on-*c*ommon item set. It is worth discussing the hybrid of *C* and *NC* in measuring similarities among user-groups. We would not focus on a hybrid approach currently since the adjustment of weights would make the module more complex for analysis. Note that measuring similarities between in-clique users still apply the concept of common item set since their basket sizes are much smaller. In Table 12.3, we list all recommendation schemes that would be compared in Sec. 12.3.

12.2.2 Online Operations

In order to describe the online operation, we shall introduce the analytical model proposed by Wang and Wu (2009) before the operation procedures is demonstrated.

Analytical Model for Optimal Recommendation

After the offline operations, three databases were constructed, namely, the item-group database defined by $P^i = \{p_{d^i}(\alpha_k) | d^i = 1^i, 2^i, \ldots, D^i, i = 1, 2, \ldots, I\}$; the user-group database defined by $U^j = \{u_{f^j}(\omega_g) | f^j = 1^j, 2^j, \ldots, F^j, j = 1, 2, \ldots, J\}$; their relations are constructed by CECF. When a user is online, we can identify a user's preferences through the corresponding information retrieved from the databases. The retrieved data as well as the user's requests (desired satisfaction level and budget limit) are regarded as the user's metadata input into the analytical model below to obtain the optimal recommendation which satisfies customer's preference under the budget constraint:

Recommendation Model

$$\text{Maximize} \quad \sum_{j=1}^{J} \sum_{f^j=1}^{F^j} \mathbf{c}\mathbf{x}^{f^j} \tag{12.7}$$

s.t.

$$\mathbf{a}^{f^j}\mathbf{x}^{f^j} \geq b^{f^j}, \quad f^j = 1^j, 2^j, \ldots, F^j, j = 1, 2, \ldots, J \tag{12.8}$$

$$\mathbf{s}\mathbf{x}^{f^j} \leq B^{f^j}, \quad f^j = 1^j, 2^j, \ldots, F^j, j = 1, 2, \ldots, J \tag{12.9}$$

$$\sum_{i=1}^{I} \sum_{d^i=1}^{D^i} x_{id^i}^{f^j} \geq 1, f^j = 1^j, 2^j, \ldots, F^j, j = 1, 2, \ldots, J \tag{12.10}$$

$$x_{id^i}^{f^j} \in \{0, 1\} \tag{12.11}$$

where $\mathbf{x}^{f^j} = [x_{d^i}^{f^j}]_{\sum_i D^i \times 1}$, $i = 1, 2, \ldots, I$, $d^i \in \{1, 2, \ldots, D^i\}$, $j = 1, 2, \ldots, J$, $f^j = 1^j, 2^j, \ldots, F^j$, $x_{d^i}^{f^j} = 1$ if item p_{d^i} is recommended to u^{f^j}; otherwise, $x_{d^i}^{f^j} = 0$. **c** and **s** are the corresponding profit and price of p_{d^i}. b^{f^j} is the satisfactory level requested by u^{f^j}; B^{f^j} is the budget limit given by u^{f^j}. $\mathbf{a}^j = [\partial_{p_{d^i}}^{u_{fj}}]_{1 \times \sum_i D^i}$, $\partial_{p_{d^i}}^{u_{fj}}$ to be the purchase probability measure of user u_{fj} on p_{d^i}. This model maximizes the profits of an EC company (12.7) when the items recommended to users satisfy their satisfactory level as shown in constraint (12.8); the total prices spent on the items should not exceed the budget of the user as shown in constraint (12.9). Constraint (12.10) provides a tool for strategic uses by recommending different number of items of which at least one item should be recommended to a user at each time.

Strategies of Recommendation

Under the basic model, two strategies can be provided for different marketing strategies—the *maximal profit strategy* and the *win–win strategy*. When the recommending processes use only the supplier viewpoint, the goal will be to *maximize the profits* of the goods under a set of items that satisfy the users' preferences and budgets. When this is intended, denote the reduced decision-variable vector and the corresponding coefficients by "$'$" to mean that all items left for consideration are at least above the requested satisfactory level, namely b^{f^j}. The maximal profit model below will immediately reflect such strategy.

Maximal Profit Model

$$\text{Maximize} \quad \sum_{j=1}^{J} \sum_{f^j=1}^{F^j} \mathbf{c}'\mathbf{x}^{f^j\prime}$$

s.t.

$$\mathbf{s}'\mathbf{x}^{f^j\prime} \leq B^{f^j}, \quad f^j = 1^j, 2^j, \ldots, F^j, j = 1, 2, \ldots, J \tag{12.12}$$

$$\sum_{i=1}^{I} \sum_{d^{i\prime}=1}^{D^i} x_{id^{i\prime}}^{f^j} \geq 1, f^j = 1^j, 2^j, \ldots, F^j, j = 1, 2, \ldots, J, \quad x_{id^{i\prime}}^{f^j} \in \{0, 1\}$$

Although the *maximal profit strategy* will bring about the highest income to suppliers, from the viewpoint of management, it only passively satisfies the users' desire for minimal levels and thus is not a strategy to provide good services. Alternatively, the *win–win strategy*, which actively takes both suppliers' profit and users' preferences into account, is proposed. The win–win model (12.13) realizes such a strategy in which the first objective function maximizes the supplier's profit as previously done; meanwhile, the second objective function represents the maximization of the user's satisfaction.

Model (12.13) is a bi-objective programming model. Since there are a lot of prominent literatures discussing and solving this kind of bi-criterion problems (e.g., Adulbhan et al., 1977; Wang, 2004), we do not focus on how to solve the proposed models. In the manner of convex combination of the two objectives: introducing a weighting parameter $\beta, \beta \in [0, 1]$, model (12.13) can be transformed into a single objective programming model as model (12.14). While model (12.13) with $\beta = 1$ yields model (12.12) for implementing *maximal profit strategy*, that with $\beta = 0$ will emphasize the users' benefit as *best service strategy*, and depending on the marketing preference the suppliers adopted, β can be given by any values between 0 and 1 as *win–win strategy*. Note that in model (12.14), $\mathbf{c}''$ is further normalized from $\mathbf{c}'$ into [0, 1] to match the same scale with $\mathbf{a}^{j'}$.

$$\begin{aligned} &\textit{Maximize} \quad \sum_{j=1}^{J} \sum_{f^j=1}^{F^j} \mathbf{c}'\mathbf{x}^{f^{j'}} \\ &\textit{Maximize} \sum_{j=1}^{J} \sum_{f^j=1}^{F^j} \mathbf{a}^{j'}\mathbf{x}^{f^{j'}} \end{aligned}$$

s.t.

$$\mathbf{s}'\mathbf{x}^{f^{j'}} \leq B^{f^j}, \quad f^j = 1^j, 2^j, \ldots, F^j, \ j = 1, 2, \ldots, J \tag{12.13}$$

$$\sum_{i=1}^{I} \sum_{d^{i'}=1}^{D^i} x_{id^{i'}}^{f^j} \geq 1, \ f^j = 1^j, 2^j, \ldots, F^j, \ j = 1, 2, \ldots, J, \quad x_{id^{i'}}^{f^j} \in \{0, 1\}$$

$$\textit{Maximize} \quad \beta \left(\sum_{j=1}^{J} \sum_{f^j=1}^{F^j} \mathbf{c}''\mathbf{x}^{f^{j'}} \right) + (1 - \beta) \sum_{j=1}^{J} \sum_{f^j=1}^{F^j} \mathbf{a}^{j'}\mathbf{x}^{f^{j'}}$$

s.t.

$$\mathbf{s}'\mathbf{x}^{f^{j'}} \leq B^{f^j}, \quad f^j = 1^j, 2^j, \ldots, F^j, \ j = 1, 2, \ldots, J \tag{12.14}$$

$$\sum_{i=1}^{I} \sum_{d^{i'}=1}^{D^i} x_{id^{i'}}^{f^j} \geq 1, \ f^j = 1^j, 2^j, \ldots, F^j, \ j = 1, 2, \ldots, J, \ x_{id^{i'}}^{f^j} \in \{0, 1\}$$

12.2.3 Measures of Recommendation Performance

To evaluate the performance of information retrieval, three measures of recall, precision, and F1 are usually employed (Cho et al., 2004; Schein et al., 2002). They are defined as follows and will be used to evaluate our recommendation system as well.

$$\textbf{Recall} = |S(\text{user}) \cap \text{Rec}(\text{user})|/|\text{Rec}(\text{user})| \quad (12.15)$$

$$\textbf{Precision} = |S(\text{user}) \cap \text{Rec}(\text{user})|/|S(\text{user})| \quad (12.16)$$

$$\textbf{F1} = 2 \times \text{Recall} \times \text{Precision}/(\text{Recall} + \text{Precision}) \quad (12.17)$$

where S(user) is the actual basket for the compared user; Rec (user) is the recommendation item set. Recall is the ratio of items successfully recommended, whereas precision measures the user's satisfactory degree. F1 is a leverage measure when recall and precision conflict with each other.

12.2.4 Summary of Offline and Online Operation Procedures

After introducing the individual submodules of the proposed RS, we would summarize the operation procedures for the proposed RS. The procedures are categorized into offline and online operations.

Offline Operation Procedures

Step 1. Construct user-groups through users' demographic features and item-groups by item attributes to obtain $U^j = \{u_{fj}(\omega_g) | f^j = 1^j, 2^j, \ldots, F^j, j = 1, 2, \ldots, J\}$ and $P^i = \{p_{d^i}(\alpha_k) | d^i = 1^i, 2^i, \ldots, D^i, i = 1, 2, \ldots, I\}$.

Step 2. Compute relative purchase priorities (w_i^j) between user-groups and item-groups by Eq. (12.3).

Step 3. Compute similarity measures between user-groups [Eq. (12.6)]. The similarity function is used from the common item set [Eq. (12.2)] or noncommon item set [Eq. (12.4)].

Step 4. Derive out-of-clique probability measures by Eq. (12.5).

Online Operation Procedures

Step 1. Set up parameters on in-clique effects (θ) and profit consideration (β) if adopting one of the following:
1.1. Maximal profit strategy, set $\beta = 1$;
1.2. Win–win strategy, set a $\beta \in (0, 1)$;
1.3. Best service strategy, set $\beta = 0$.

Step 2. Online inquiry of target users' profiles of demographic features ($u_{fj}(\omega_g)$), binary basket data ($C_{u_{f\tau}, p_{di}}$); the desired satisfaction level (b^{f^j}), and the budget limit (B^{f^j}).

Step 3. Classify target user into proper user-group by $U^j = \{u_{f^j}(\omega_g) | f^j = 1^j, 2^j, \ldots, F^j, j = 1, 2, \ldots, J\}$.

3.1 A *historical user* with basket data ($0 < \theta \leq 1$)—compute purchase probabilities on each item with CECF-C [Eq. (12.2)] or CECF-NC [Eq. (12.4)].

3.2 A *new user* without basket data ($\theta = 0$)—retrieve out-of-clique probability measures as purchase probability on each item.

Step 4. Derive metadata from purchase probabilities [Eq. (12.5)] and user's request as input to Step 5.

Step 5. Run the analytical model and yield a recommendation list.

12.3 An Illustrative Case: Laptops RS of a 3C Retailer

The 3C industries of Taiwan have the most advanced technologies in the world. Among various electronic products, experiments of our proposed RS are conducted specifically with laptops because of three reasons: (1) Laptop transactions are usually fewer than those of other electronic products, so introducing an RS would be meaningful to attract new users; (2) fewer transactions are difficult to exploit when introducing the RS, so our proposed RS aims to solve this situation by incorporating clique effects; and (3) laptops are all highly priced so that the profit consideration would be more applicable.

Following the provided data of a 3C retailer, the prototype of the system was established and evaluated in this section by first describing the given database; the laptop data set contains 915 market baskets including 227 customers and 192 items. The types of items in the basket are ranged from two to eight for each user. The user's information is revealed by user types (defined with users' demographic features by the 3C retailer) and five user-groups are yielded (U^1, U^2, U^3, U^4, U^5). The item attributes (k) are denoted as: (1) central processing unit (CPU); (2) random-access memory (RAM); (3) brand; (4) storage capacity; and (5) weight. By our classification rules with $K = 5$, the item-groups consist of 32 exclusive groups. Due to incomplete data, there are only 16 nonempty item-groups.

12.3.1 Experiments of Strategy Implementation

In the experiments, 227 customers are divided randomly into 20%/80% as testing and training data in an echo. We conduct three experiments with different goals. In the first experiment, we compare the recommendation performance of conventional CF with our proposed recommendation approach CECF in two cases of CECF-C and CECF-NC, and the three schemes are all with a fixed neighborhood size of 20. In the second experiment, we compare the recommendation performance as well as the profit gained with respect to the supplier's market strategies as: (1) $\beta = 1$ yields *maximal profit strategy*, and (2) $\beta \in (0, 1)$ yields the *win–win strategy*, (3) $\beta = 0$ emphasizes the customer's benefit of the *best service strategy*. In the third experiment, we compare the sensitive F1 values with respect to the neighborhood sizes (3, 5, 7, 10, and 20) under three schemes of CF, CECF-NC with profit consideration ($\beta = 0.2$) and CECF-NC with nonprofit consideration.

Three measures of recall, precision, and F1 will be used for evaluation. Different values of parameters are chosen to demonstrate their impacts as sensitivity analysis. We pick one of the echoes for illustration in the following section. All experimental procedures would be shown in compliance with the procedures proposed in Sec. 3.3.3.

Offline Operation Procedures (Training Data)

Step 1. (i) Construct user-groups, $U^j, j = 1, 2, \ldots, 5, |\cup_j U^j| = 182.(227*0.8 \cong 182)$
(ii) Construct item-groups, $P^i, i = 1, 2, \ldots, 16, |\cup_i P^i| = 192$

Step 2. Compute relative purchase priority w_i^j.

Step 3. Compute similarity measures between user-groups by Common item set function, i.e., Eq. (12.2) and *non*-common item set function, i.e., Eq. (12.4).

Step 4. Derive out-of-clique probability measures by Eq. (12.5) as shown in Table 12.4. Note that the probability measures in each row are normalized and ensure that they sum up to 1.

Online Operation Procedures (Testing Data)

Step 1. Set up parameters on in-clique effects (θ) and profit consideration (β), respectively. For implementation, the system could set up θ and β as arbitrary values. In the experiments, we set up θ to be 0, 0.1, 0.2, ..., 1 and β to be 0, 0.2, 0.4, ..., 1 for testing.

Step 2. The users are tested as *new users* or *historical users* by setting $\theta = 0$ or $0 < \theta \leq 1$, respectively. Satisfaction levels (b^{f^j}) are also defined to be 0.7, 0.8, 0.9, for experiments. Budget limits (B^{f^j}) are set as arbitrary values that are lower than the summation of all items' prices.

Step 3. Classify target user into one user-group by $U^J = \{u_{f^j}(\omega_g) | f^j = 1^j, 2^j, \ldots, F^j, j = 1, 2, \ldots, 5\}$.

3.1 The situation is simulated in a manner where some historical users are recommended when we set $0 < \theta \leq 1$.

3.2 The situation is simulated wherein some new users (u_1^1, $u_2^1 \ldots$) are recommended by CECF-NC or CECF-C, respectively, when we set $\theta = 0$, which is shown in Tables 12.5 and 12.6. Note that when a target user is regarded as a

NC	P^1	P^2	P^3	P^4	P^5	P^6	P^7	P^8	P^9	P^{11}	P^{12}	P^{13}	P^{14}	P^{15}	P^{16}	Total
U^1	0.003	0.082	0.005	0.222	0.005	0.060	0.365	0.001	0.054	0.029	0.029	0.054	0.026	0.007	0.034	1
U^2	0.003	0.085	0.004	0.244	0.019	0.053	0.335	0.000	0.060	0.024	0.020	0.072	0.023	0.006	0.030	1
U^3	0.003	0.087	0.005	0.278	0.021	0.062	0.326	0.001	0.057	0.022	0.022	0.072	0.012	0.000	0.021	1
U^4	0.004	0.095	0.001	0.270	0.018	0.062	0.323	0.001	0.050	0.024	0.024	0.063	0.016	0.006	0.022	1
U^5	0.000	0.088	0.006	0.264	0.016	0.058	0.331	0.001	0.055	0.017	0.020	0.070	0.022	0.006	0.029	1
C	P^1	P^2	P^3	P^4	P^5	P^6	P^7	P^8	P^9	P^{11}	P^{12}	P^{13}	P^{14}	P^{15}	P^{16}	Total
U^1	0.003	0.081	0.005	0.219	0.005	0.060	0.366	0.001	0.055	0.029	0.029	0.055	0.025	0.006	0.035	1
U^2	0.004	0.077	0.005	0.216	0.013	0.054	0.359	0.000	0.063	0.031	0.025	0.065	0.025	0.005	0.034	1
U^3	0.004	0.081	0.006	0.254	0.015	0.062	0.347	0.001	0.058	0.027	0.026	0.066	0.015	0.000	0.025	1
U^4	0.007	0.088	0.001	0.239	0.012	0.063	0.349	0.001	0.053	0.034	0.031	0.054	0.016	0.004	0.024	1
U^5	0.000	0.077	0.009	0.233	0.008	0.057	0.360	0.001	0.059	0.022	0.023	0.067	0.026	0.004	0.036	1

Table 12.4 Out-of-Clique Probability Measures

	p_1^2	p_2^2	p_{13}^2	p_2^4	p_3^4	p_5^{16}	p_6^{16}
u_1^1	0.0026	0.0026	0.0026	0.0070	0.0070	0.0008	0.0008
u_2^1	0.0027	0.0027	0.0027	0.0078	0.0078	0.0007	0.0007
u_3^1	0.0028	0.0028	0.0028	0.0088	0.0088	0.0004	0.0004
u_4^1	0.0031	0.0031	0.0031	0.0087	0.0087	0.0007	0.0007
u_5^1	0.0028	0.0028	0.0028	0.0084	0.0084	0.0006	0.0006

TABLE 12.5 Purchase Probabilities of New Users by CECF-NC ($\theta = 0$)

new user, the probability measures for him/her could be only derived from out-of-clique measures. For instance, in Table 12.4, the probability of U^1 to $P^{16}$is 0.025, which shall be the same with that of u_1^1 to p_5^{16} and p_6^{16} in Table 12.5. The value is 0.0008 instead of 0.025 due to normalization.

Step 4 and Step 5. In the two steps, the target user's metadata is obtained and fed to the analytical model, and the output of recommendations is then yielded. We skip the list of the recommendation results and directly compare the performance of the proposed operation module by the following experiments.

Experiment 1

The performance of the recommendation results on CECF-C, CECF-NC, and CF is shown in Table 12.7, with evaluation of recall, precision, and F1 under sample values θ, with a neighborhood size of 20. Note that when $\theta = 1$, CECF-C and CECF-NC both become the CF since out-of-clique effects no longer exist. In Table 12.7, the results of an average performance show that CECF-C and CECF-NC perform better than CF except $\theta = 0$. In addition, it could be also observed that CECF-NC performs slightly better than CECF-C. In Fig. 12.4, CECF-NC has been compared with CF; in the figure, the CECF-NC performs much better than CF in Recall and F1 (p-value < 0.001, 95% confidence level), and slightly better in Precision.

	p_1^2	p_2^2	p_{13}^2	p_2^4	p_3^4	p_5^{16}	p_6^{16}
u_1^1	0.0025	0.0025	0.0025	0.0069	0.0069	0.0008	0.0008
u_2^1	0.0024	0.0024	0.0024	0.0068	0.0068	0.0007	0.0007
u_3^1	0.0025	0.0025	0.0025	0.0079	0.0079	0.0004	0.0004
u_4^1	0.0028	0.0028	0.0028	0.0076	0.0076	0.0008	0.0008
u_5^1	0.0024	0.0024	0.0024	0.0073	0.0073	0.0005	0.0005

TABLE 12.6 Purchase Probabilities of New Users by CECF-C ($\theta = 0$)

	CECF-NC			CECF-C		
θ	Recall	Precision	F1	Recall	Precision	F1
0	0.297	0.458	0.325	0.297	0.458	0.325
0.1	0.877	0.928	0.939	0.867	0.925	0.932
0.2	0.900	0.942	0.962	0.893	0.938	0.955
0.4	0.908	0.943	0.963	0.903	0.942	0.959
0.5	0.910	0.945	0.968	0.910	0.945	0.968
0.6	0.911	0.945	0.968	0.907	0.945	0.967
0.7	**0.910**	0.945	0.968	0.910	0.945	0.968
0.8	**0.910**	0.945	0.968	0.910	0.945	0.968
0.9	0.910	0.945	0.968	0.910	0.945	0.968
1(*CF*)	0.457	0.930	0.569	0.457	0.930	0.569

TABLE 12.7 Average Performance of CECF-C, CECF-NC and CF

Experiment 2

In Experiment 1, the average performance is better and more stable when CECF-NC and $\theta = 0.6$ are used. Therefore, we set up θ to be 0.6 and continue experimenting on the analytical model by introducing β to be 0, 0.2, 0.4, . . . , 1 and satisfaction level (bf^j) to be 0.7, 0.8, 0.9 under users' budget limits. We compare the CECF-NC with profit consideration as well as nonprofit consideration in terms of recall, precision and F1 as shown in Fig. 12.5; and the difference of profit gained in the two cases are presented in Fig. 12.6. In Figure 12.5, the results show that even when profit consideration is introduced, the recommendation performance would not be poorer (p-value <0.05, 95% confidence level). In Figure 12.6, the results show that profit increases along β increases from 0 to 1.

Experiment 3

In this experiment, we compare three recommendation schemes of CF, CECF-NC with profit consideration ($\beta = 0.2$) and CECF-NC with nonprofit consideration in terms of their F1 measures. Figure 12.7 shows that the F1 values increase as the neighborhood size increases from 3, 5, 7, 10, to 20. In addition, Fig. 12.7 showed consistent results we obtained from the previous two experiments, that is, the CECF-NC with profit/nonprofit consideration outperforms conventional CF.

12.3.2 Summary and Remarks of Experiments

We have conducted three experiments in the case study for the proposed strategy-oriented operation module of a 3C retailer in Taiwan.

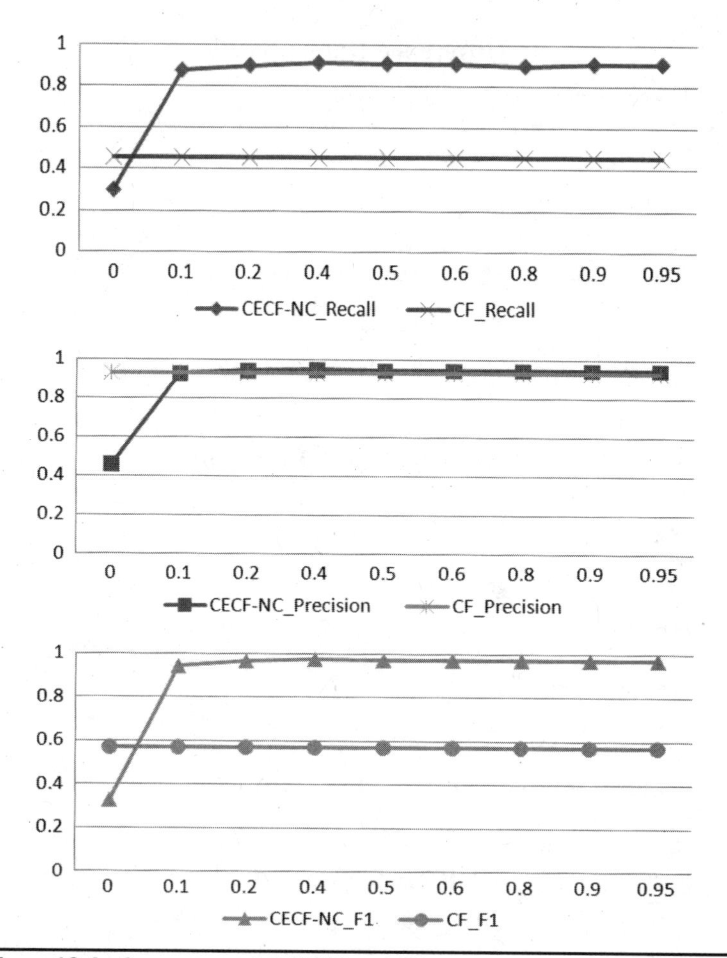

FIGURE 12.4 Comparison of CECF-NC and CF.

In the first experiment, we compared the performances of CECF-C, CECF-NC, and CF by three measures of recall, precision, and F1. In Table 12.7 and Figure 12.4, it showed that the proposed CECF-NC and CECF-C perform better than conventional CF except for $\theta = 0$, which was the situation in which new user recommendations was simulated. It has been mentioned that *CF* could not recommend while the target user is without basket data. Besides, the results in Experiment 1 also showed slightly better performance of CECF-NC as compared to CECF-C. The reason is that in the relatively sparse user's basket data, the noncommon item set would show additional information for recommendation. In Table 12.7, we could observe that when $\theta = \{0.7, 0.8, 0.9\}$, the performances are almost the same; when $\theta = 0.6$, the recommendation performance reaches the highest level.

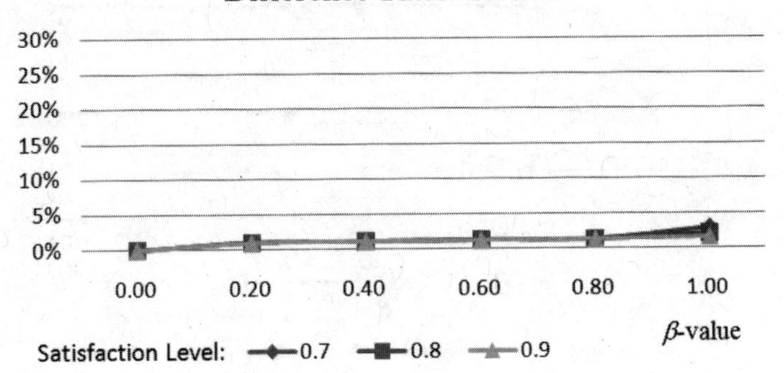

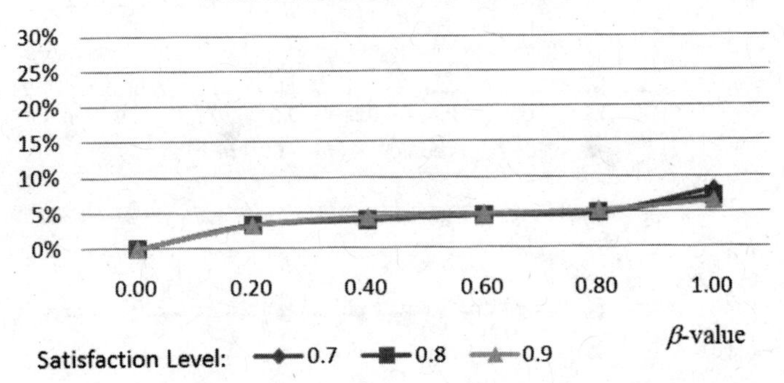

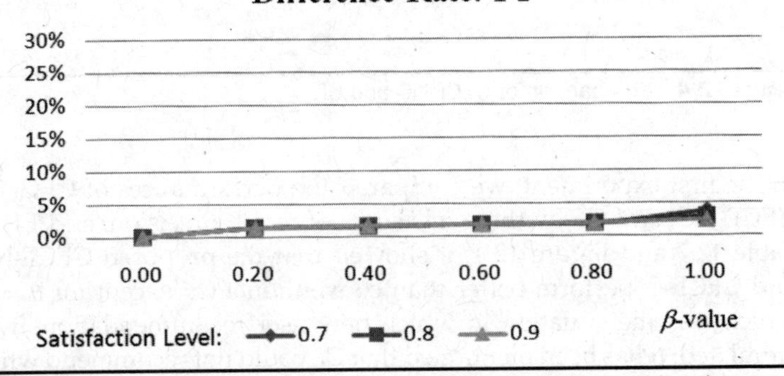

FIGURE 12.5 The difference rates while introducing profit consideration.

This phenomenon, which would be data-specific, tells that the effects from non-neighbor groups would not enhance but maintain the performance while $\theta \geq 0.6$.

In the second experiment, while introducing the profit parameter β and the user's satisfaction level (b^{f^j}), we set up the recommendation

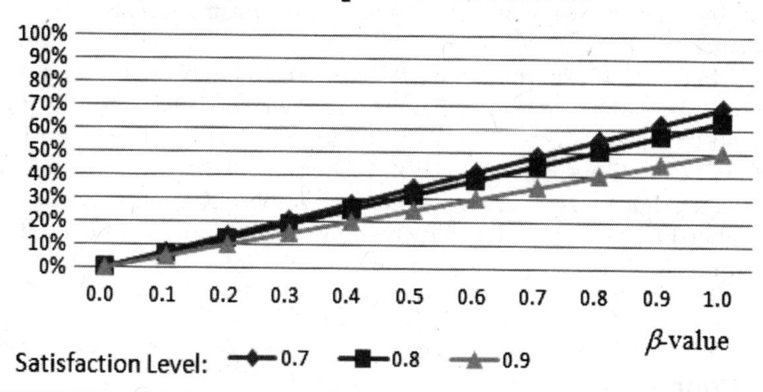

FIGURE 12.6 Profit gained under different β-values.

environment by $\theta = 0.6$ for better and more stable performance. It is very important to note that while we introduce the profit parameter β in the recommendation process, the recommendation performance with $\beta \in (0, 1)$ would probably decrease since the goal of recommendation was no longer to emphasize the user's benefits with $\beta = 0$ only. Then the emphasis should be on whether the **service level decreases** while **the profit gain increases**.

In Figs. 12.5 and 12.6, the results show that when we increased β, the profit gained increases without losing recommendation performance. This phenomenon can be attributed to the analytical model since the recommended items are aligned with the user's satisfaction

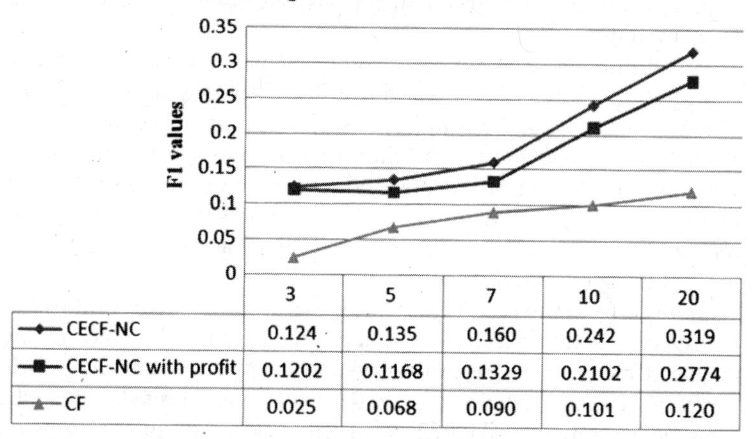

	3	5	7	10	20
CECF-NC	0.124	0.135	0.160	0.242	0.319
CECF-NC with profit	0.1202	0.1168	0.1329	0.2102	0.2774
CF	0.025	0.068	0.090	0.101	0.120

FIGURE 12.7 Comparison of F1 values with respect to the neighborhood sizes.

level. Therefore, even if we look forward to increasing the profit gains of the supplier, we would still maintain the recommendation performance for services.

In the third experiment, we tested the effect of neighborhood size on F1 measure. The three schemes all showed consistent results that the F1 values were positively related to the neighborhood sizes; and the CECF-NC outperformed CF in F1 measure. In particular, while the neighborhood sizes were small, i.e., 3, 5, 7, the CECF-NC still reached higher F1 values than CF, which shows the advantage of using the clique effects to compensate for rare information.

12.4 Conclusion

In addition to the insufficient knowledge of the green products for general consumers, the demand-driven property of green products requires an effective website to provide necessary information for both producers/sellers and consumers. Therefore, developing a website that actively recommends suitable products for consumers according to their needs and preferences is important to promote a green market.

In the field of RSs, there have been numerous studies proposed in order to find the best recommendation for users. Among these studies, CF has been regarded as the most effective method for its recommendation accuracy and flexibility. However, in practice, it is confronted with the problem that target users with rare information could not get recommendations from the system. Although many approaches based on CF have been proposed to pursue better performance by increasing service levels and solving the problem of sparse data, the excessive emphasis on recommendation performance would lead to overlooking the profit consideration, which is also an essential concern for an EC company.

In addition, a systematic and comprehensive module for an RS is still lacking. In this regard, we introduce a strategy-oriented operation module that can be comprehensively applied to EC websites as a decision support module so that the choice of various marketing strategies combining profit consideration for suppliers and users can be developed. Consequently, under the framework of the proposed recommender module, we also have developed a method called CECF to predict users' purchase behavior.

The performance of the introduced system is assured by the experiments conducted in the case of a laptop, where possible recommendation strategies that are profit-oriented for producers and win–win for both sides are specified. From the experimental results, it can be seen that the introduced CECF is not only able to provide a promising solution to the "new user" problem, but it also facilitates marketing strategies to be performed.

References

Adulbhan, P., Tabucanon, M.T., 1977, Bicriterion linear programming. *Computers & Operations Research*, 4, 2, 147-153.

Aggarwal, C.C., Wolf, J.L., Yu, P.S., 1999, A new method for similarity indexing of market basket data. *SIGMOD Conference*, 407-418.

Breese, J.S., Heckerman, D., Kadie, C., 1998, Empirical analysis of predictive algorithms for collaborative filtering. In *Proc. of the 14th Conf. Uncertainly in Artificial Intelligence*, Jul. 24-26 1998, Madison, WI, 43-52.

Chen, L.S., Hsu, F.H., Chen, M.C., Hsu, Y.C., 2008, Developing recommender systems with the consideration of product profitability for sellers. *Information Sciences*, 178, 1032-1048.

Cho, Y.H., Kim, J.K., 2004, Application of web usage mining and product taxonomy to collaborative recommendations in e-commerce. *Expert Systems with Applications*, 26, 233-246.

Guo, J., 2007, Business-to-business electronic market place selection. *Enterprise Information Systems*, 1, 4, 383-419.

Guo, J., 2009, Collaborative conceptualization: Towards a conceptual foundation of interoperable electronic product catalogue system design. *Enterprise Information Systems*, 3, 1, 59-94.

Hogg, T., 2010, Inferring preference correlations from social networks, *Electronic Commerce Research and Applications*, 9, 1, 29-37.

Hsu, P.Y., Chen, Y.L., Ling, C.C., 2004, Algorithms for mining association rules in bag databases. *Information Sciences*, 166, 31-47.

Kaufman, L., Rousseeuw, P.J., 1990, *Finding Groups in Data: An Introduction to Cluster Analysis*. New York: John Wiley.

Kim, B.M., Li, Q., Park, C.S., Kim, S.G., Kim, J.Y., 2006, A new approach for combining content-based and collaborative filters. *Journal of Intelligent Information System*, 27, 79-91.

Kim, H.T., Ji, A.T., Kim, Ha. I., Jo, G.S., 2010, Collaborative filtering based on collaborative tagging for enhancing the quality of recommendation, *Electronic Commerce Research and Applications*, 9, 1, 73-83.

Kohrs A., Meriadlo B., 2001, Creating user-adapted websites by the use of collaborative filtering. *Interacting with Computers*, 13, 695-716.

Lam, C., 2004, SNACK: Incorporating social network information in automated collaborative filtering. In *Proceedings of the Fifth ACM Conference on Electronic Commerce (EC'04)*. ACM Press, New York, NY, USA, 254-255.

Lee, C.H., Kim, Y.H., Rhee, P.K., 2001, Web personalization expert with combining collaborative filtering and association rule mining technique. *Expert Systems with Applications*, 21, 131-137.

Li, H., Wang, H., 2007, A multiagent-based model for a negotiation support system in electronic commerce. *Enterprise Information Systems*, 1, 4, 457-472.

McPherson, M., Smith-Lovin, L., Cook, J.M., 2001, Birds of a feather: Homophily in social networks. *Annual Review of Sociology*, 27, 415-444.

Mild, A., Reutterer, T., 2001, Collaborative filtering methods for binary market basket data analysis. *Lecture Notes in Computer Science*, 2252, 302-313.

Mild, A., Reutterer, T., 2003, An improved collaborative filtering approach for predicting cross-category purchases based on binary market basket data. *Journal of Retailing and Consumer Services*, 10, 123-133.

Montaner, M., López, B., Rosa, J.L.D.L., 2003, A taxonomy of recommender agents on the Internet. *Artificial Intelligence Review*, 19, 285-330.

Russell, G.J., Petersen, A., 2000, Analysis of cross category dependence in market basket selection. *Journal of Retailing*, 76, 367-392.

Sarwar, B.M., Karypis G., Konstan, J.A., Riedl J., 2000, Analysis of recommendation algorithms for E-commerce. In *Proc. of 2nd ACM EC*, Minneapolis, MN, USA, 158-167.

Schafer, J.B., Konstan, J.A., Riedl, J., 1999, Recommender systems in E-commerce. In *Proc. of 1st ACM EC*, Denver, CO, USA, 158-166.

Schafer, J.B., Konstan, J.A., Riedl, J., 2001, E-commerce recommendation applications. *Data Mining and Knowledge Discovery*, 5, 115-153.

Schein, A.I., Popescul, A., Ungar, L.H., Pennock, D.M., 2002. Methods and metrics for cold-start recommendations. In *Proc. of the 25th ACM SIGIR on Research and Development in Information Retrieval*, Tampere, Finland.

Wang, H.F., 2004. *Multicriteria Decision Analysis: From Certainty to Uncertainty*. ISBN 986-7777-55-7, Ting Lung Book Co., Taiwan, ROC.

Wang, H.F., Wu, C.T., 2009, A mathematical model for product selection strategies in a recommender system, *Expert Systems with Applications*, 36, 3, 7299-7308.

Wang, H.F., Wu, C.T., 2010, A strategy-oriented operation module for recommender systems in e-commerce, to appear at *Computers and Operations Research*. (http://dx.doi.org/10.1016/j.cor.2010.03.011)

Wang, S., Archer, N.P., 2007, Electronic marketplace definition and classifications: Literature review and clarifications. *Enterprise Information Systems*, 1, 1, 89-112.

Xu, X., 2008, Effects of two national environmental factors on e-commerce functionality adoption: A cross-country case study of a global bank. *Enterprise Information Systems*, 2, 3, 325-339.

Yang, H.W., Pan, Z.G., Wang, X.Z., Xu, B., 2004, A personalized products selection assistance based on e-commerce machine learning. In *Proc. of 3rd Conf. on Machine Learning and Cybernetics*, 26-29.

Index

Note: Page numbers followed by *f* and *t* denote figures and tables, respectively.

A

Activity data, 239
Aggregate planning, 106–107, 106*f*
Alliance development, 91–92
Analytic hierarchy process (AHP), 17–30, 45
basic concepts and pairwise comparison in, 18–23, 18*t*
improvement of consistency in, 22–23
measure of inconsistency in, 21–22, 22*t*
multiplicative comparison matrix in, 18
multiplicative preference matrix in, 19
pairwise comparison matrix in, 18
pairwise comparison scale in, 18–19, 19*t*
weight of importance derivation in, 20–21, 21*t*
basic properties of, 17
example of: determining consumer preference, 24–30
attribute hierarchical structure in, 24, 24*f*
most relevant attributes in, 24
step 1 in, matrix comparing attributes, 24, 25*t*
step 2 in, comparison matrices of intensity level, 26, 26*t*–27*t*
step 3 in, first-level priority intensities, 27, 27*t*, 28*t*
step 4 in, vector of desired attribute intensities, 27–28
step 5 in, perceived product standards, 28, 28*t*–29*t*
step 6 in, second-level priority intensities, 28–30, 30*t*
step 7 in, overall priorities, 30
green procurement in:
numerical example of, 93–98
conclusion of, 101
Analytic hierarchy process (AHP), green procurement in, numerical example of (*Cont.*):
pairwise matrices for, 102*t*–103*t*
risks and ranking of two suppliers in, estimation of, 93–98, 94*t*–96*t*, 98*f*, 98*t*
sensitivity analysis in, 99–100, 99*t*, 100*t*
pairwise matrices for AHP of, 102*t*–103*t*
green vendor selection in, 80, 82
risk analysis in, 80–89
criteria of selection with hierarchical relations in, 81, 82*f*
measures of attributes in, 83–89. *See also under* Vendor evaluation and selection, green
weighting of criteria in, objective (analytical hierarchical process), 82
weighting of criteria in, subjective (risk utility function), 81–82
summary, 92
importance of, 17
origins and use of, 17
procedure of, 23–24
sensitivity analysis on, 92, 99–100, 99*t*, 100*t*
steps of, 17
vendor selection in, 80
AND module, 207, 208*t*, 211, 211*t*
AND/OR graph, 204, 205*f*
Animal excrement emissions, 260*t*
ANOVA test, 181–182
among clusters, of target green customers, 192, 192*t*
significance in, 181–182

Association rule–based recommendation (ABR), 270
Association rules, 269–270
Auditing, suppliers' green performance, 10
Auto industry, GSCM on, 7
Automation, in disassembly, 204
Automobile emissions, 252*t*–253*t*
Auxiliary crisp model, equivalent, 119, 120*t*

B

Best service strategy, 278
Bill of manufacturing (BOM), 9, 45
Bill of materials (BOM), 205
Bill of materials (BOM), reverse, 206
Blue ocean strategy, 10
Business-to-business (B2B), 237
Business-to-customer (B2C), 237

C

C-mean method:
  dissimilarity in, 186
  fuzzy, 185–186
  for targeting green customers, 185–187, 188*f*, 190–192, 192*t*
Capacitated dynamic lot-sizing problem:
  vs. existing models, 109, 110*t*
  history of, 109
  issues covered by, 109
Capacity, 107–108
Carbon dioxide, 229
Carbon dioxide equivalent (CO_2E), 241
Carbon emission coefficients, by industry, 255*t*–256*t*
Carbon footprint, product:
  international standards on, 230–231
  inventory of, 235–236, 235*f*, 236*f*
  summary and conclusion on, 236–237, 236*f*
Case-based reasoning (CBR), 45
Ceramics, 52–53
Certification, 88
Chebyshev inequality model, 181
Client–supplier interaction, increased, 9
Clique-effects collaborative filtering (CECF), 266, 271, 273–276, 274*f*
Closed-loop logistics, 125. *See also* Logistics, green
  conditions in, 127–128
  deterministic modeling of, 127–133
    conditions in, 127–128
    Deterministic Closed-Loop Logistics Model in, 128–131. *See also* Deterministic Closed-Loop Logistics Model (DCLL)
    transformed integer linear programming model in, 131–133
  recycling rate in, 128
Closed-loop modeling for uncertain logistic, 137–164
  fuzzy programming model in, 137–140
Closed-loop modeling for uncertain logistic (*Cont.*):
  numerical illustration of, 155–164
    post-analysis and discussion of, 162–164, 163*t*
    step 1 in, 155, 155*t*
    step 2 in, 155–157, 156*f*, 157*t*, 158*f*
    step 3 in, 157–158, 158*f*, 159*t*, 168*t*
    step 4 in, 158–160, 159*t*
    step 5 in, 159*t*, 160–161, 161*f*
    step 6 in, 159*t*, 161–162
  resolution of uncertainty in, 140–154
    general interval programming model in, 145
    general fuzzy programming model in, 141–142
    information reserved in defuzzified number in, 142–145, 143*f*
    interval programming model solution in, 145–147, 146*f*
    numerical illustration of risk attitude in, 153–154, 153*f*, 154*t*
    overview and diagram of, 140, 141*f*
    preference model in, 147–148
    risk level of interval programming with analysis in, 148–150, 148*f*
    risk property in, 150–151, 150*f*
    summary and conclusion in, 151–152
    trade-off analysis between risk and cost in, 152–153
Closed-loop supply chain, 198
  components of, 208–209
  issues with, 198–199, 199*f*
  product returns in, 198–199, 199*f*
  purpose of, 199*f*, 208
  remanufactured product market development in, 199, 199*f*
  remanufacturing operational issues in, 199, 199*f*
  system boundary for, 209, 210*f*
Clumping, 40
Cluster analysis, 169, 182, 268
Cluster analysis, of green customers, 182–187
  c-mean method in, 185–187, 188*f*, 190–192, 192*t*
  hierarchical clustering method, 183–185, 185*f*
  problem definition in, 182–183
  purpose of, 182
Cluster centers, 186
Collaborative filtering (CF), 266, 268–269
Collaborative filtering, clique-effects (CECF), 266, 271, 273–275, 274*f*
Company–supplier collaboration:
  green product design, 9
  innovative product design, 9
Competition intensification, 10
Composites, 54
Consistency, improvement of, 22–23
Consistency index (CI), 22
Consistency ratio (CR), 22

Contaminating materials minimization, harmful, 41
Content-based filtering (CBF), 268–269
Convergence properties, 205
Cost:
  increased prime, 10
  total disposal, 59–60
  total processing, 59
Cost–benefit function, in design for disassembly index, 56, 58–60
  total disposal cost, 59–60
  total processing cost, 59
  total recycling revenue, 59
  total resale revenue, 58–59
Crisp mixed-integer programming model (CMIP), 113–114, 120*t*
Customers, green, 169–194
  analytical methods in, 172, 180–187
    cluster analysis in, 182–187
      c-mean method in, 185–187, 188*f*
      hierarchical clustering method in, 183–185, 185*f*
      problem definition in, 182–183
      purpose of, 182
    data analysis in, 181–182
    sample size determination in, 172, 180–181
  cluster analysis and, 169
  definition of, 170
  features of, 169–172
    demographic characteristics in, 170–171
    psychogenic characteristics in, 171–172
  questionnaire design for, 172, 173*t*–180*t*
  target consumer identification in, numerical illustration of, 187–194
    ANOVA test among clusters in, 192, 192*t*
    cluster analysis (c-mean method) in, 190–192, 192*t*
    conclusion of, 192–194
    factor analysis in, 189, 191*t*
    sample size determination in, 187–188
    sociodemographic variable distribution in, 189, 190*f*
    survey data quantification in, 189, 189*t*
Cutoff principle, 239

D

Data. *See also specific topics*
  dynamic, 72
  static, 72
Data mining, sensor, 74
Decarbonization, 5
Decision process, routes in, 15–16, 16*f*
Decisions, types of, 15
Demand-driven disassembly planning, 198, 208
Demand-driven lot-sizing problems for disassembly planning, 207
Demand phases, in product life cycle, 200, 200*f*
Dematerialization, 4
Demographic characteristics, green customer, 170–171
Demographic filtering (DF), 269
Design:
  functional unit, 40
  product, on retirement strategy, 56
Design for disassembly (DfD), 39
Design for disassembly index (DfDI), 39, 55, 56–71
  application (example) of, 60–67, 61*f*, 61*t*–66*t*
  calculation procedure in, 60, 67
  cost–benefit function in, 56, 58–60
    total disposal cost, 59–60
    total processing cost, 59
    total recycling revenue, 59
    total resale revenue, 58–59
  criteria for judging design in, 56
  development of, 55
  fundamentals of, 55
  nomenclature in, 57–58
  optimization model in, 67–69
    constraints of, 68–69
    mathematical formulation of, 69
    objective function in, 67–68
  optimization procedure application (example) for, 69–71
Design for disassembly representation, 204–207, 205*f*–207*f*, 208*t*
Design efficiency, 55
Design efficiency measurement, 56
Design, environmental, 55–74. *See also specific topics*
  design for disassembly index of, 55, 56–71
  product monitoring framework benefits in, 74
  sensor embedded products in, 56,71–74
Design for environment (CfE), 38–39
Design, green, 38–40
  design for X in, 38–39
  life cycle assessment in, 39
  materials selection in, 39–40
Design guidelines, green, 40–43
  ease of separation in, 41
  functional units design in, 40
  materials selection in, 41
  modular product structure in, 40
  steps of, 41–43
  waste and harmful contaminant minimization in, 41
Design for recycling (DfR), 39
Design for X (DfX), 38–39

Deterministic closed-loop logistics, 128–133
Deterministic Closed-Loop Logistics Model in, 128–131. *See also* Deterministic Closed-Loop Logistics Model (DCLL)
transformed integer linear programming model in, 131–133
Deterministic Closed-Loop Logistics Model (DCLL), 128–131
Gauss symbol in, 129–131
model description in, 128–131
objective and constraints in, 131
parameters and notations in, 128–129
Deterministic modeling of closed-loop logistics, 127–133
conditions in, 127–128
Deterministic Closed-Loop Logistics Model in, 128–131. *See also* Deterministic Closed-Loop Logistics Model (DCLL)
transformed integer linear programming model in, 131–133
Detoxification, 4–5
Direct emission, 238
Direct monitoring method, 239–240
Direct reuse, 43
Disassembly, 197–198, 202–208
categories of, 203–204
degree of, 203
design for, 39
design for disassembly representation in, 204–207, 205*f*–207*f*, 208*t*
importance of, 203
remanufacturing in, 202–203
Disassembly center, 73–74
Disassembly configurations, for modules, 210–211, 211*t*
Disassembly effort index (DEI), 81, 82*f*, 85
Disassembly effort index (DEI) example, green vendor selection, 95–96, 96*t*
Disassembly for demand (DfD):
constraints in, 219–221
maximize in, 218–219
Disassembly line balancing, 203
Disassembly planning, demand-driven lot-sizing problems for, 207
Disassembly product structure, 206
Disassembly representation, design for, 204–207, 205–207*f*, 208*t*
Disassembly to demand (D2D) modeling and analysis, 208–221
closed-loop supply chain system in, 199*f*, 208–209, 210*f*
disassembly configurations for modules in, 210–211, 211*t*
mathematical model of, 211–221
constraints in, 219–221
mixed-integer programming optimization model in, 218–219
notations in, 212–217
Disassembly to demand (D2D) modeling and analysis (*Cont.*):
recovery options restrictions in, 199*f*, 211, 212*t*
representation product structure in, 209–210, 211*f*
Disassembly-to-order (DtO) system, 203, 208
Disassembly-to-order (DtO) system problems, 207
Disassembly tree (DT), 55
Dismantlers, 125, 126*f*
Disposal:
end-of-life, 37
product materials, 43
Disposal center, 74
Disposal cost, total, 59–60
Dissimilarity, in c-mean method, 186
Divergence properties, 205
DTO problems, 207
DTO system, 203, 208
Dynamic data/information, 72
Dynamic lot-sizing model, 107
Dynamic lot-sizing problem, single-item capacitated, 106
Dynamic programming (DP), 43–44

E

Ease of separation, 41
Eco-architecture analysis, 39
Ecologically conscious consumer behavior (ECCB), 170
Economic order quantity (EOQ), 107
Efficient set, 31–32
Eigensystem tableau, 21, 21*t*
Elastomers, 53
Electric commerce, 265
Electrical and electronics industry, GSCM on, 8
Electronic Industry Code of Conduct (EICC), 8
Emission:
carbon, by industry, 255*t*–256*t*
direct, 238
indirect, 238
other indirect, 238
Emission coefficient method, 239–240
Emission factors and coefficient charts, by industry and nation, 239, 252*t*–264*t*
animal excrement emission factor, 260*t*
automobiles, 252*t*–253*t*
carbon emission coefficients by industry, 255*t*–256*t*
fertilizer emission factor, 260*t*
GHG emission by industry, 253*t*–259*t*
traffic fuels emission factor:
Canada, 263*t*–264*t*
Taiwan, 261*t*
USA, 262*t*

Emission quantification, 239–241
construction of database on, 241
emission coefficient selection in, 241
illustrative example of, 248–250, 249*t*
methods of, 239–241
summary of results of, 241
Emission source:
fixed burning, 238
fugitive, 238
identification of:
illustrative example of, 245, 246*t*–248*t*
method, 238
manufacture, 238
mobile burning, 238
End-of-life (EOL) design for disassembly index (DfDI). *See* Design for disassembly index (DfDI)
End-of-life (EOL) disposal, 37
End-of-life (EOL) management, 197–223
current developments in, 198–202
closed loop supply chain issues in, 198–199, 199*f*
demand-drive disassembly planning in, 208
EOL and end-of-use recovery selection in, 201–202, 202*t*
life cycle effects on returned product quantity and quality in, 199–200, 200*f*
product life cycle in, 199–200, 200*f*
demand-driven disassembly planning in, 198
disassembly in, 197–198, 202–208
categories of, 203–204
design for disassembly representation in, 204–207, 205*f*–207*f*, 208*t*
remanufacturing in, 202–203, 203*f*
disassembly to demand (DtD) modeling and analysis in, 208–221
closed-loop supply chain system in, 199*f*, 208–209, 210*f*
disassembly configurations for modules in, 210–211, 211*t*
mathematical model of, 211–221
constraints in, 219–221
mixed-integer programming optimization model in, 218–221
notations in, 212–217
recovery options restrictions in, 199*f*, 211, 212*t*
representation product structure in, 209–210, 211*f*
future directions in, 223
illustrative example of, 221, 221*f*–223*f*, 222*t*
product recovery in, 197
waste minimization in, 197
End-of-life product recovery, 43–45
ENDLESS, 39
Energy consumption minimization, 51
Engineering, green, 37–46. *See also specific topics*
definition of, 37
green design guidelines for, 40–43. *See also* Design guidelines, green
green design in, 38–40
history of, 37
interactions among activities in product life cycle in, 37, 38*f*
product recovery at end-of-life in, 43–45
scope of, 37
12 principles of, 45–46
Environment, design for, 38–39
Environment-protection behavior, 171–172
Environmental concern (EC), 171
Environmental design, 55–74. *See also specific topics*
design for disassembly index in, 55, 56–71
product monitoring framework benefits in, 74
sensor embedded products in, 56, 71–74
Environmental information, supplier, 9
Environmentally conscious manufacturing (ECM), 37
Environmentally conscious product recovery (ECMPRO), 37
Equivalent auxiliary crisp model (ACM), 119, 120*t*
Equivalent matrix, 184
Ergonomics, in disassembly, 204
Extended production responsibility (EPR), 4

F

Factor analysis, of target green customers, 189, 191*t*
Fertilizer emissions, 260*t*
Fixed burning emission source, 238
FL (fuzzy-logic) based MCDM methodology, 44
Forward logistics, 125
Fossil fuel consumption, 229
Four Rs, 4, 80
Fugitive emission source, 238
Functional units design, 40
Fuzzy addition, 15
Fuzzy arithmetic, 15
Fuzzy c-mean method, 185–186
Fuzzy division, 15
Fuzzy mixed integer programming model (FMIP), 115–118, 117*f*
Fuzzy multiplication, 15
Fuzzy numbers, 13–14, 14*f*
in green logistics, 126
trapezoidal, 115, 117*f*

Fuzzy programming:
in closed-loop modeling for uncertain logistic, 137–140
in green logistics, 126
Fuzzy sets, 13
Fuzzy subtraction, 15

G

General fuzzy programming model (GFPM), 141–142
General interval programming model (GIPM), 145
Genetic algorithm (GA), multi-objective, 44
Global recycling index, 39
Global warming potential (GWP), 241
Goal programming (GP), 40
Grade of acceptability, 145–146
Gray relational analysis (GRA), 44
Green engineering. *See* Engineering, green
Green logistics. *See* Logistics, green
Green lot-sizing production model. *See* Lot-sizing production model, green
Green performance, auditing suppliers', 10
Green procurement. *See* Procurement, green
Green production. *See* Production, green
Green productivity (GP), 5
Green supply chain (GSC), 125
Green supply chain (GSC) logistics. *See* Logistics, green
Green supply chain management (GSCM):
adoption of, 7
development of, 3–4, 7
evolution of:
from supply chain management, 4–6
technologies in, 7, 7*f*
extension of, implementation procedure for, 11, 11*f*
framework of, 6, 6*f*
impact on industry of, 6–10
auto industry, 7
electrical and electronics industry, 8
green procurement system, 7
life cycle assessment, 6–7
industrial administration of, 8–10
benefits of, 9–10
competition intensification in, 10
IT for problem solving in, 8
Greenhouse gases (GHGs), 229
emission by industry of, 253*t*–259*t*
Greenhouse Gas Protocol for, 105–106

H

Harmful contaminating materials minimization, 41
Hewlett-Packard Social and Environmental Responsibility (SEM) plan, 8
Hierarchic representation and decomposition, 17
Hierarchic structuring, 17
Hierarchical clustering method, for targeting green customers, 183–185, 185*f*

I

Incineration, 201, 202*t*
Income quality control (IQC), 81, 82*f*, 83–85, 83*f*, 84*t*
Inconsistency, measure of, 21–22, 22*t*
Index of disassembly effort (DEI), 85
Indirect emission, 238
Industrial administration, GSCM on, 8–10
Information:
dynamic, 72
static, 72
Information-filtering technology, 268–269
Information support systems, web-based, 265–288. *See also* Web-based information support systems
Information technology (IT), GSCM problem solving, 8
Inspection reports, 87
Integer linear program (ILP), 132
Integer linear programming model, transformed, 131–133
Intergovernmental Panel on Climate Change (IPCC), 230
Interval programming model (GIPM), 145–147, 146*f*
Inventory, 108
Inventory shortage, 108
Inventory target, in LCA:
B2B *vs.* B2C, 237
setting borderline for:
illustrative example of, 236*f*, 244–245
procedure of, 237–238

L

L-R fuzzy number, 14, 14*f*
Landfilling, 201, 202*t*
Life cycle, product, 199–200, 200*f*
Life cycle assessment (LCA), 6–7
definition of, 39, 229
development and use of, 6–7
framework for, 229, 230*f*
green design in, 39
origins of, 4
standards of, 7
Life cycle assessment (LCA) database, 229–264, 252*t*
data collection in, 239
emission factors and coefficient charts, by industry and nation, 252*t*–264*t*
animal excrement, 260*t*
automobiles, 252*t*–253*t*
carbon emission coefficients by industry, 255*t*–256*t*

Life cycle assessment (LCA) database (*Cont.*):
fertilizer, 260*t*
greenhouse gases, by industry, 253*t*–259*t*
traffic fuels, Canada, 263*t*–264*t*
traffic fuels, Taiwan, 261*t*
traffic fuels, USA, 262*t*
emission quantification in, 239–241
construction of database on, 241
emission coefficient selection in, 241
methods of, 239–241
summary of results of, 241
foundations of, 229
framework for, 229, 230*f*
illustrative example of, 244–252
application simulation in, 239*t*, 245, 248
emission quantification in, from storage, 248–249, 249*t*
emission quantification in, from transport, 249–250, 249*t*
emission source identification in, 245, 246*t*–248*t*
setting borderline in, 236*f*, 244–245
summary and discussion of, overall, 250, 250*t*
summary and discussion of, raw material stage, 252
summary and discussion of, use stage, 251
procedure of, 237–238
borderline in, 237–238
emission source identification in, 238
inventory target for selected product in, 237
product carbon footprint in:
international standards on, 230–231
inventory of, 235–236, 235*f*, 236*f*
summary and conclusion on, 236–237, 236*f*
software for, available, 231–232, 231*t*, 233*f*, 234*t*
third-party verification of, impartial, 241–244
general requirements for, 242–243
impartiality monitoring mechanisms in, 244
qualifications for, 242
requirement for, 241–242
Line balancing, disassembly, 203
Logical consistency, 17
Logistic level, 81, 82*f*
Logistics, 88–89
closed-loop. *See* Closed-loop logistics
forward, 125, 126*t*
reverse, 125, 126*t*
Logistics, green, 125–165
closed-loop, 125
conditions in, 127–128
recycling rate in, 128
Logistics, green (*Cont.*):
closed-loop, deterministic modeling of, 127–133
conditions in, 127–128
Deterministic Closed-Loop Logistics Model in, 128–131. *See also* Deterministic Closed-Loop Logistics Model (DCLL)
transformed integer linear programming model in, 131–133
closed-loop modeling for uncertain logistic in, 137–164. *See also* Closed-loop modeling for uncertain logistic
conclusions on, 164–165
dismantlers in, 125, 126*f*
fuzzy numbers in, 126
fuzzy programming in, 126
illustrative example of, 134–137, 134*t*–136*t*, 136*f*
stages in, 127
uncertain factors in, 125–126
Lot-sizing model:
current, 108–109, 110*t*
elements of, 107–108
static *vs.* dynamic, 107
Lot-sizing problem:
capacitated dynamic
vs. existing models, 109, 110*t*
history of, 109
issues covered by, 109
single-item capacitated dynamic, 106
Lot-sizing production model, green, 111–120
auxiliary crisp model *vs.* crisp mixed-integer programming model in, 120*t*
capacitated dynamic lot-sizing problem in:
vs. existing models, 109, 110*t*
history of, 109
issues covered by, 109
modeling in uncertain environment in, 115–120
equivalent auxiliary crisp model in, 119, 120*t*
fuzzy mixed integer programming model in, 115–118, 117*f*
returned rate as trapezoidal fuzzy number in, 115, 117*f*
numerical illustration of, 120–123, 121*t*–123*t*, 122*f*
periodic closed-loop production system in:
crisp mixed-integer programming model in, 113–114
framework of, 111, 112*f*
modeling in certain environment in, 111–114, 112*f*
product recovery in, 106

Lot-sizing production model, green (*Cont.*):
  production planning in, 106–107, 106*f*
    aggregate, 106–107, 106*f*
    operational, 106*f*, 107
    strategic, 106, 106*f*
  United Nations Framework Convention on Climate Change and, 105

M

Maintenance center, 73
Manufacture emission source, 238
Market expectations, adding value in, 6
Mass balance method, 240–241
Material requirements planning (MRP), 205–206
Material requirements planning (MRP), reverse, 206–207, 207*f*
Materials:
  disposal of, 43
  green, 51–54. *See also* Materials, green
    ceramics, 52–53
    composites, 54
    elastomers, 53
    general principles, 51
    metals, 52
    natural organic, 53
    polymer thermoplastics, 53
    polymer thermosets, 53
    RoHS directive, 51–52
    selection, 39–40, 41, 43, 52–54
    WEEE directive, 51
  minimization of:
    general, 51
    harmful contaminating, 41
  selection of:
    green, 43, 52–54
    green design guidelines for, 41
    green design in, 39–40
    recyclability in, 40
Mathematical background, 13–33
  analytic hierarchy process (AHP) in, 17–30. *See also* Analytic hierarchy process (AHP)
    basic concepts and pairwise comparison in, 18–23, 18*t*, 21*t*, 22*t*
    example of: determining consumer preference, 24–30, 24*f*, 25*t*–30*t*
    importance of, 17
    procedure of, 23–24
  fuzzy arithmetic in, 15
  fuzzy numbers in, 13–14, 14*f*
  optimization programming in, 30–33
    illustrative example of, 32–33, 33*f*, 33*t*
    methods for, 30–31
    multi-object linear program for, 31
  utility theory in, 15–17, 16*f*, 17*f*
Max-dominance, 31
Maximal profit strategy (model), 268, 277–278
Mean consistency index of randomly generated matrices, 22*t*
Measure of inconsistency, 21–22, 22*t*
Measures of risks, green vendor selection in, 94–97, 94*t*–96*t*, 98*t*
Metals, 52
Mixed-integer programming model:
  crisp, 113–114, 120*t*
  in D2D, 218–219
  fuzzy, 115–118, 117*f*
Mobile burning emission source, 238
Modular product structure, 40
Multi-object linear program (MOLP), 31–32
Multi-objective genetic algorithm, 44
Multicriteria decision making (MCDM), 44, 201–202, 202*t*
Multicriteria decision making (MCDM), FL (fuzzy-logic) based, 44
Multicriteria matrix, 44
Multiplicative comparison matrix (MCM), 18
Multiplicative preference matrix, 19

N

Natural organic materials, 53
Neighbors, 268, 273

O

Offline operations, 271–276
  derivation of relations among users and items (CECF) in, 273–275, 274*f*
  item-groups with properties in, 271–272, 272*t*
  summary of, 275–276, 276*t*, 279
  user-groups and profiles in, 272–273
Online operations, 276–278
  analytical model for optimal recommendation in, 276
  maximal profit model for, 277–278
  recommendation model for, 277
  strategies of recommendation for, 277
  summary of, 279–280
Online shopping, 265
Operational planning, 106*f*, 107
Optimization model, in design for assembly index, 67–69
  application of (example), 69–71
  constraints of, 68–69
  mathematical formulation of, 69
  objective function in, 67–68
Optimization programming, 30–33
  illustrative example of, 32–33, 33*f*, 33*t*
  methods of, 30–31
  multi-object linear program in, 31–32
OR module, 207, 208*t*, 211, 211*t*
Organic materials, natural, 53
Original design manufacturing (ODM), green requirement for, 8

Original equipment manufacturer (OEM), 72
Other indirect emission, 238

P

Pairwise comparison matrix, 18
Pairwise comparison scale, 18–19, 19*t*
Perceived consumer effectiveness (PCE), 171
Performance:
green, auditing suppliers', 10
of recommendation, 270
Periodic closed-loop production system:
crisp mixed-integer programming model in, 113–114
framework of, 111, 112*f*
modeling in certain environment in, 111–114, 112*f*
Periodic demand, 108
Planning, production, 106, 106*f*
Planning horizon, 107
Pollution, 105
Polymer thermoplastics, 53
Polymer thermosets, 53
Potential remanufacturing volumes, 200, 200*f*
Preference model, 147–148
Prime costs, increased, 10
Priority discrimination and synthesis, 17
Priority setting, 17
Processing cost, total, 59
Procurement, 80
Procurement, green, 7, 79–103
alliance development in, 91–92
numerical example of, 93–98
conclusion of example in, 101
pairwise matrices for AHP of, 102*t*–103*t*
risks and ranking of two suppliers in, estimation, 93–98, 94*t*–96*t*, 98*f*, 98*t*
sensitivity analysis in, 99–100, 99*t*, 100*t*
purchasing in, 80
sensitivity analysis of, 91–92, 99–101
of analytic hierarchy process, 92, 99–100, 99*t*, 100*t*
issues in, 91–92
ProdTect, 230*f*, 232, 233*f*, 234*t*
Product carbon footprint:
international standards on, 230–231
inventory of, 235–236, 235*f*, 236*f*
summary and conclusion on, 236–237, 236*f*
Product category rule (PCR), 237–238
Product design:
green, 38–40
design for X in, 38–39
life cycle assessment in, 39
materials selection in, 39–40
on retirement strategy, 56
Product life cycle, 199–200, 200*f*
definitions of, 199–200
demand phases in, 200, 200*f*
interactions among activities in, 37, 38*f*
potential remanufacturing volumes in, 200, 200*f*
on returned product quantity and quality, 199–200, 200*f*
stages of, 200
Product monitoring framework, 56, 71–74. *See also* Environmental design; Sensor embedded products (SEPs)
Product recovery, 37, 106, 197
end-of-life, 43–45
environmentally conscious, 37
Product remanufacturing systems, 203, 203*f*
Product returns management, 198–199, 199*f*
Product structure representation, 207, 207*f*
Production, green, 105–123
green lot-sizing production model in, 111–120. *See also* Lot-sizing production model, green
Greenhouse Gas Protocol and, 105–106
lot-sizing model for
current, 108–109, 110*t*
elements of, 107–108
static *vs.* dynamic, 107
numerical illustration of, 120–123, 121*t*–123*t*, 122*f*
Production planning, 106–107, 106*f*
aggregate, 106–107, 106*f*
operational, 106*f*, 107
strategic, 106, 106*f*
Psychogenic characteristics, green customer, 171–172
Purchaser-in-charge, 93
Purchasing, 80. *See also* Procurement, green

Q

Qualitative method, of periodic demand, 108
Quality control (IQC), income, 81, 82*f*, 83–85, 83*f*, 84*t*
Quantitative method, of periodic demand, 108
Questionnaire design, for targeting green customers, 172, 173*t*–180*t*

R

Random index (RI), 22, 22*t*
Ranking method, of vendor evaluation and selection, 90–91
Recognition, 88
Recommendation methods, 268–270
Recommendation performance, measures of, 279

Recommender systems (RSs), 265–266
infrastructure of, 267–268, 267f
roles and goals of, 270, 270t
Recovery, product, 37, 106, 197
end-of-life, 43–45
environmentally conscious, 37
restrictions of options for, 199f, 211, 212t
Recyclability, in materials selection, 40
Recycle, 201, 202t
Recycling:
design for, 39
of product materials, 43
Recycling center, 74
Recycling rate, in closed-loop logistics, 128
Recycling revenue, total, 59
Red ocean strategy, 10
Redesign, 4, 80
Reduction, 4, 80
Reflective properties, 183
Remanufacture, 4, 43, 80
Remanufacturing, 201, 202–203, 202t, 203f
center for, 74
potential volumes of, 200, 200f
Remote monitoring center (RMC), 72–73, 72f
Repair, 43, 201, 202t
Report evaluation, rule of, 86–87
Representation product structure, 209–210, 211f
Resale revenue, total, 58–59
Resource minimization, 51
Returned products:
life cycle effects on quantity and quality of, 199–200, 200f
options for, 108–109
Reuse, 4, 80, 201, 202t
Reuse, direct, 43
Revenue:
total recycling, 59
total resale, 58–59
Reverse bill of materials, 206
Reverse logistics, 125
Reverse material requirements planning, 206–207, 207f
Reverse supply chain, 125, 126t
Reverse Wagner-Whitin's dynamic production planning and inventory control model, 108
Risk aggregation, green vendor, 94, 94t
Risk aggregation method, for vendor evaluation and selection, 89–90, 90f
Risk measures, green vendor, 94–97, 94t–96t, 98t
Risk utility function, green vendor:
example of, 93–94, 94t
theory of, 81–82
RoHS (Restriction of Hazardous Substances) directive, 51–52
4R's, 4, 80
Rule of report evaluation, 86–87

S

Sample size determination, for target green customers, 187–188
Scheduling, disassembly, 203
Sensitivity analysis, green procurement, 91–92, 99–100, 99t, 100t
on analytic hierarchy process, 92, 99–100, 99t, 100t
issues in, 91–92
Sensor data mining, 74
Sensor embedded products (SEPs), 56, 71–74
benefits of, 74
components of, 72
definition and use of, 71–72
in disassembly center, 73–74
in disposal center, 74
for life cycle management of products, 71, 72f
in maintenance center, 73
in recycling center, 74
in remanufacturing center, 74
in remote monitoring center, 72–73, 72f
in sensor data mining center, 74
Separation, ease of, 41
Sequencing, disassembly, 203
Shopping, online, 265
Significance, ANOVA test of, 181–182
Single-item capacitated dynamic lot-sizing problem, 106
Social and Environmental Responsibility (SEM) plan, Hewlett-Packard, 8
Social networks, 266
Sociodemographic variable distribution, for target green customers, 189, 190f
Software, life cycle assessment of, 231–232, 231t, 233f, 234t
Specific industry emission coefficient, 241
Static data/information, 72
Static lot-sizing model, 107
Strategic planning, 106, 106f
Subassemblies, 206
Submodule operations, system, 271–280
measure of recommendation performance, 279
offline operations, 271–276
online operations, 276–278
summary of offline and online operations, 279–280
Supplier risk map, 98, 98f
Suppliers:
auditing green performance of, 10
environmental information from, 9
risks and ranking of, estimation of, 93–98, 94t–96t, 98f, 98t
Supply chain management (SCM):
definition of, 4
green supply chain management evolution from, 4–6
scope of, 3–4

Sustainability:
  Asian Productivity Organization, 5
  green supply chain management, 4
Sustainable development adoption, new methods for, 5
Symmetric properties, 183
System submodule operations, 271–280
  measure of recommendation performance in, 279
  offline operations in, 271–276
  online operations in, 276–278
  summary of offline and online operations in, 279–280

T

Target consumer identification, green customer illustration of, 187–194
  ANOVA test among clusters in, 192, 192*t*
  cluster analysis (c-mean method) in, 185–187, 188*f*, 190–192, 192*t*
  conclusion of, 192–194
  factor analysis in, 189, 191*t*
  sample size determination in, 187–188
  sociodemographic variable distribution in, 189, 190*f*
  survey data quantification in, 189, 189*t*
Total disposal cost, 59–60
Total processing cost, 59
Total recycling revenue, 59
Total resale revenue, 58–59
Traffic fuels emission factor:
  Canada, 263*t*–264*t*
  Taiwan, 261*t*
  USA, 262*t*
Trans-industrial emission coefficient, 241
Transformed integer linear programming model, 131–133
Transitivity properties, 184
Trapezoidal fuzzy number, 14, 14*f*, 115, 117*f*
Tree network model, 206–207, 206*f*
Triangular fuzzy number, 14, 14*f*
12 principles of green design, 45–46

U

United Nationals Environment Programme (UNEP), global greenhouse gas monitoring by, 230
United Nations Framework Convention on Climate Change (UNFCCC), 105
Utility function:
  computation of, 16, 17*f*
  definition of, 15
Utility theory, 15–17, 16*f*, 17*f*

V

Value chain, 4
Vendor evaluation and selection, green, 80–91. *See also* Procurement, green; Vendor evaluation and selection, green
  evaluation of, 89–91
    ranking method in, 90–91, 90*f*
    risk aggregation method in, 89–90, 90*f*
  risk analysis in, 80–89
    criteria of selection with hierarchical relations in, 81, 82*f*
    measures of attributes in, 83–89. *See also under* Vendor evaluation and selection, green
      income quality control in, 83–85, 83*f*, 84*t*
      index of disassembly effort in, 85
      inspection reports for, 87
      logistics of, 88–89
      recognition and certification of, 88
      rule of report evaluation for, 86–87
      vendor management in, 85–86, 86*t*
    selection procedure in, summary of, 92–93
    weighting of criteria in:
      objective (analytical hierarchical process), 82
      subjective (risk utility function), 81–82
  selection procedure in, summary of, 92–93
Vendor management (VM), 81, 82*f*, 85–86, 86*t*

W

Waste Electrical and Electronic Equipment (WEEE) Directive, 197
Waste minimization, 41, 197. *See also* End-of-life management
Web-based information support systems, 265–288
  clique-effects collaborative filtering in, 266, 271, 273–275, 274*f*
  collaborative filtering in, 266
  electric commerce in, 265
  illustrative case of, 280–288
    3C industries in, 280
    strategy implementation experiments in, 280–284, 282*t*–284*t*
    summary and remarks on experiments in, 284–288, 285*f*–287*f*
  recommendation methods in, 268–270
  recommender systems in, 265–266
    infrastructure of, 267–268, 267*f*
    roles and goals of, 270, 270*t*

Web-based information support systems (*Cont.*):
social networks and, 266
summary and discussion on, 271
system submodules in, operations of, 271–280
measures of recommendation performance in, 279
offline operations in, 271–276. *See also* Offline operations
online operations in, 276–278
summary of offline and online operations in, 279–280
WEEE (Waste Electrical and Electronic Equipment) directive, 51
Weight of importance, derivation, 20–21, 21*t*
Weighting method, 32
Win-win strategy, 268, 270, 277–278

Y

Yield matrix, 210
Yield parameters, 209–210